MOBILE INFORMATION SYSTEMS

For a complete list of *The Artech House Telecommunications Library,* turn to the back of this book. . .

MOBILE INFORMATION SYSTEMS

JOHN WALKER

Editor

Artech House

Boston · London

Library of Congress Cataloging-in-Publication Data

Mobile information systems / edited by John Walker.
p. cm.
Includes bibliographical references.
ISBN 0-89006-340-0
1. Mobile communication systems. 2. Information technology.
I. Title.
TK6570.M6W35 1990 90-30131
621.3845'6--dc20 CIP

British Library Cataloguing in Publication Data

Mobile information systems.
1. Communication sytems
I. Walker, John
621.38

ISBN 0-89006-340-0

685 Canton Street
Norwood, MA 02062

International Standard Book Number: 0-89006-340-0
Library of Congress Catalog Card Number: 90-30131

10 9 8 7 6 5 4 3 2 1

Contents

Preface

Mobile information systems have been my major professional preoccupation since the early 1980s—because of Racal's involvement in cellular radio, *private mobile radio* (PMR), and radiopaging, and particularly because of my own involvement in the Mobile Information Systems Demonstrator project (which is partly funded by the Information Engineering Directorate of the UK Department of Trade and Industry).

The subject is diverse, and it is advancing and expanding at a remarkable rate, driven by developments in information technology, by increases in personal mobility (i.e., more people have cars), and by the willingness of governments and regulatory authorities to make more of the radio spectrum available for mobile applications. However, the radio spectrum is a very scarce and valuable resource, and we know that, unlike oil or minerals, there are no large deposits lying on or under the seabed waiting to be discovered (although perhaps an analogous process is occurring in that higher and higher frequencies are being brought into service, or proposed for new services, in order to relieve the congestion lower down in the spectrum).

There is a correspondence between spectrum congestion and road transport congestion in that the more traffic you permit—by opening roads or opening frequency bands—the more traffic you stimulate, so the congestion, in practice, is not alleviated. This is because there is a great deal of suppressed demand. We also do not generally realize how important the road transport sector is to the economy. The UK road transport industry is a $50 billion per year business, and 10 to 15% of the cost of any article you buy can be attributed to the cost of its transportation. We expect that similar figures apply throughout the developed world.

Congestion is bad news! So is the increased pollution from exhaust fumes and the increased costs and wasted time that result. So, too, are the increases in road accidents that occur because there is more traffic. The good news is that information technology and mobile communication can help (as they are doing) to alleviate some of these problems.

All these issues are considered in more detail throughout the book, which is aimed primarily at technical management as well as the professional engineer who is new to the field and wants an up-to-date overview before plunging into the details. (To aid in this process, most chapters have a list of references and further reading.) We hope that the book will also be useful to the nontechnical executive who needs to know some of the technical background in the important and expanding field of *road transport informatics* (RTI), and to the layperson who is just interested.

Acknowledgements

The Racal authors are grateful to Racal Electronics PLC, and in particular to Mr. G.J. Lomer for permission to publish the book. However, the opinions expressed in Chapters 1, 3, 4, 8, and 11 are the personal views of the authors, and do not necessarily represent positions of Racal. We want to thank our colleagues within Racal who have helped in the book's compilation in various ways, by making diagrams and papers available, and by commenting on manuscripts: Terry Barwick, Ted Beddoes, Peter Blair, Tim Burnett, Alan Cox, Chris Gent, Nigel Morgan, Peter Munday, Jon Noble, Dave Targett, and Keith Thrower. In addition, we are grateful to Steve Hannigan, Val Herring, and Leela Damodaran of the HUSAT Research Centre at Loughborough University for the material on which is based the table on cellular data applications in Chapter 3.

I also want to thank the Society of Automotive Engineers and the Philips organization for permission to base Chapter 8 on papers published in their journals, and Mr. Aylot, Derek Wilsdon, and M.F. Zuurveen of Philips, who helped with other aspects of that chapter. Ron Tridgell (Chapter 2) wrote Appendix C and most of Appendix B.

I am grateful to all my colleagues within the Mobile Information Systems Alvey Demonstrator project, and to the UK Department of Trade and Industry's Alvey Directorate and its successor, the Information Engineering Directorate, for their indirect contributions to the book.

Finally, the book would have been impossible without the support, encouragement, and forbearance of my family; to Liz, Christopher, Nicola, and Stephanie, thank you.

JOHN WALKER
READING, BERKSHIRE, UK
AUGUST 1989

The Authors

P.D. Britten graduated in 1974 with an honors degree in electronic engineering from Surrey University. From 1974 to 1977 he worked for Decca Navigator Ltd. on the development of electronic navigation equipment. He earned an M.Sc. in cybernetics from University of London in 1978. From 1977 to 1984 he worked for Sira Ltd. on satellite payload development. He joined Racal Decca Advanced Development (now combined with Racal Research Ltd.) in 1984, and is now responsible for the aeronautical satellite communication and satellite navigation developmental activities in that company.

Ian Catling formed the Ian Catling Consultancy (ICC) in 1983, having previously worked in government and for a systems house, particularly on real-time and traffic-related systems. Under his direction, ICC has become established as a specialized consultancy in the application of information technology to road transport. He was responsible for the system design and software development of the Hong Kong Electronic Road Pricing pilot project, and ICC subsequently became the UK Department of Transport's consultant on Autoguide, coordinating the implementation of the London Autoguide Demonstration Scheme. He played an active role in the definition of the DRIVE workplan, and ICC became the prime contractor for two DRIVE projects, concerned with the development of the Integrated Road Transport Environment. Mr. Catling has also been active throughout the PROMETHEUS project, and is the deputy international coordinator of the PRO-GEN subprogram. He has acted as specialist adviser to the Dutch and Swedish governments on the implementation of road-based advanced technology systems, and is active in similar developments in the US.

S.R. Ely graduated with first class honors from the University of Liverpool in 1972 and remained there to work for his doctorate on error statistics and error control in high-speed data communication. He joined BBC Research Department at Kingswood Warren in 1975, and worked in the Carrier Systems Section on the development of the radio-data system now known as RDS. He was responsible for the development and execution of the mobile measuring techniques and statistical analysis used by the EBU in the field tests of 1980 and 1982, which led to the design and optimization of RDS. He was one of the principal authors of the EBU's RDS specification document. He is currently Head of the Carrier Systems Section at BBC Research Department and Chairman of EBU Specialist Group (R/RDS), which has responsibility for further technical development of the RDS system. He is UK representative on CENELEC TC107, which is preparing a European standard for RDS. Dr. Ely is a chartered engineer, a member of the Institution of Electrical Engineers (IEE), and a member of Professional Group E14 (Television and Sound) of the IEE.

Brian Gardner earned a second class honors degree in physics from Oxford University in 1969. Since then he has worked for several companies in the Racal Electronics Group, concentrating mainly on the field of communication. His early experience was in the design of an HF radio, which won him a Queen's award for technology. Subsequent work centered mainly on military topics involving the design and development of surveillance and direction-finding equipment; systems for electronic countermeasures and counter-countermeasures; and secure, survivable communication networks. Since 1986 he has been a Technical Manager at Racal Research Ltd., working on many aspects of civil communication, such as cordless telephones and the forthcoming pan-European cellular radio system. Mr. Gardner holds numerous patents, ranging over all the fields in which he has worked.

K.W. Huddart graduated in physics and was trained as an electrical engineer with Reyrolle, a switchgear firm, and subsequently worked on power transmission for the Central Electricity Generating Board. He joined the Greater London Council in 1966 specifically to install traffic control systems, including area traffic control. By 1986, as Chief Traffic Engineer, he was responsible for design, maintenance, and operation of most of London's 2200 traffic signals and their supporting computer systems. He was also responsible for all specialist traffic engineering, including bus priority, cycling schemes, road safety, and traffic order-making. During this time, he also worked as a consultant for the World Bank, specifying and reviewing traffic management projects, mainly in the Far East, including the traffic signal systems of Manila, Bangkok, and Calcutta. He reviewed Australian traffic signaling for the Australian Research Board. More recently, Mr. Huddart has specified traffic signaling for a new light rail system in Hong Kong, reviewed and introduced novel schemes for improving traffic flow of two major tunnels in Hong Kong, and devised an alternative system of direction signing for London. He was involved in managing the DRIVE project for the Commission of the European Community.

D.J. Jeffery joined the UK government's Transport and Road Research Laboratory (TRRL) in 1962, and was sponsored by them to work for his B.Sc. in physics and technology of electronics at the Polytechnic of North London. After graduating, he returned to join the Drivers Aids and Abilities Division, where he worked on a range of research topics, including fog detection, driving simulators, and the development of systems and instrumentation for driver's aids. In 1976 he became head of the

Driver Information Systems Section in the Highway Traffic Division of TRRL, where he led a team investigating the potential of the Information Technologies for developing improved traffic control systems, and in particular for providing dynamic information and guidance advice for drivers. His work culminated in 1988 with proposals for AUTOGUIDE, an electronic system of in-vehicle route guidance in London, and RDS-TMC, a pan-European system for broadcasting traffic information. He then spent six months in Brussels helping to launch the EEC's DRIVE program before returning to TRRL at the beginning of 1989 to become head of the Vehicle Safety Division. He is a member of the IEE and serves on Professional Group C12 for Transport Electronics and Control. He has been a member or chairman of several international committees set up by the OECD and the EEC to coordinate international work on driver information and guidance systems. Mr. Jeffery is the author of numerous papers on the subject.

J.G. Schoenenberger graduated in 1977 with an honors degree in electronic engineering from University College London. From 1977 to 1982 he carried out bistatic radar research at University College London, having gained his Ph.D. in 1981. He joined Racal-Decca in 1982, working on side-looking airborne radar, HF radar, and military satellite communication. Dr. Schoenenberger joined Racal Avionics in 1986, with responsibility for the commercial aeronautical satellite communication products development in that company, including data, voice, and voice plus data equipment and associated antennas. He joined PA Consulting Group in 1989, with responsibility for business development in the aerospace sector.

M.L.G. Thoone graduated in 1978 with a degree in electrical engineering from the Eindhoven University of Technology, the Netherlands. After military service in the Dutch Royal Navy he joined Philips Research Laboratories in Eindhoven where, in 1984, he was named leader of the CARIN project, a research and development project coordinated jointly by Philips Research Labs and their Product Division Consumer Electronics. Since September 1989, Mr. Thoone has been the manager of the Philips pre-development activities on Car Information Systems in Wetzlar, West Germany.

R.H. Tridgell joined the British Post Office Research Station in 1942. Projects included protection of cables against induced voltages, telegraphs, and automatic character recognition. In 1963, he moved to BPO headquarters to lead engineering development of telegraphic ARQ and data transmission, in which position he led the UK delegation to CCITT on data transmission during the period when many of the early CCITT V-Series Recommendations were drafted. From 1973, he led the Post Office (later British Telecom) engineering development of mobile services. This included implementation of the BPO national radiopaging system, and a radiophone system. While in that position, he instituted the Post Office Code Standardization Advisory Group, and attended CCIR as a UK expert in radiopaging. He retired from BT in 1984, and currently chairs a committee drafting the UK standards for private mobile radio trunked systems. Mr. Tridgell gained a degree in electrical engineering in 1949, and is a member of IEE.

John Walker has a first degree in physics from Oxford University, an M.Sc. in applied solid state physics from Brighton Polytechnic, and a Ph.D. in solid state physics from Reading University. He spent 18 months as a Royal Society European Programme Research Fellow with the Solid State Physics Group of the Ecole Normale Superieure, at the University of Paris VII. His industrial experience includes three years at GEC-Marconi in the 1960s, as well as a period in technology consultancy and microprocessor training. Since 1980, he has been a Technical Manager with Racal Research Ltd, the research company of the Racal Electronics Group, working on various projects, including the early phases of cellular radio in the UK. In 1983, he was the joint proposer, and subsequently the Project Manager, of a project titled "Mobile Information Systems," which is partly funded by the UK Department of Trade and Industry under its Alvey Programme. He is also the manager of a new project to apply artificial intelligence and natural language understanding techniques to the field of traffic information broadcasting. Dr. Walker has published numerous technical papers. This is his second book.

M.H. Westbrook is Manager of the Technological Research Department of Ford of Europe. He is a graduate in electrical and electronic engineering from the University of Southampton, where he is now a Visiting Professor in the Department of Mechanical Engineering. Mr. Westbrook is a Fellow of the IEE, IMechE and Institute of Physics, a Past Chairman of the IEE Computing and Control Division, and was Chairman of the Organizing Committee of the IEE/IMechE International Automotive Electronics Conference in 1976, 1979 and 1987.

PART I
INTRODUCTION

Chapter 1
Introduction To Mobile Information Systems

John Walker
Racal Research Ltd.

1.1 BACKGROUND AND HISTORY

1.1.1 Mobility and its Problems

Mobility seems to be increasingly important in our society as we approach the 21st century. The social mobility characteristic of the 1960s and 1970s has been complemented by the physical and transportational mobility of the 1980s. More people travel, by car and by airplane, for business and pleasure.

This mobility, although desirable, creates problems: more road traffic means more traffic jams, more money needed to build roads, and greater deterioration of the environment through pollution. While we are in transit we are out of reach of the communication links we have come to expect and on which we depend—whether it is the businessperson unable to contact his or her client to close that vital deal, or the private motorist having difficulty finding his or her favorite VHF radio station as the vehicle moves out of range of the local transmitter.

Transportation is also "big business." The UK road transport industry is a $50 billion per year business, and 10 to 15% of the cost of any item that we purchase is due to transportation costs. Any delays in delivering goods, whether due to traffic jams or the driver not taking the optimum route to his or her destination, increase that cost. According to the UK Institution of Civil Engineers (1989), congestion costs the UK $16 billion per annum. In the US, the amount of "excess travel" due to inefficient route-finding amounts to 84 billion miles and 914,000 person-years per annum, costing (i.e., wasting) more than $45 billion every year (King and Mast, 1987). Finally, for reasons of cost as well as conservation, we are reluctant to build new roads, and any new road schemes can be guaranteed to raise a great deal of opposition from people directly affected in addition to conservation groups and "green parties."

1.1.2 The Information Technology Revolution

A revolution paralleling that of mobility, which was predicted in the 1960s but only became a reality at the end of the 1980s, is the *information technology revolution*. As yet, that revolution has had less effect on mobility and transportation than on other sectors of business and commerce, and what effects there are have been invisible, or largely so, to the general public. Motorists have been unaware that a computer-based *urban traffic control* (UTC)[1] system has been expediting their progress along the route, and probably do not realize that their antilock braking system uses a microprocessor.

There is, however, a great deal more to come. There will be a solution to (or at least some alleviation of) the traffic and travel problems. The restoration of communication to the individual in transit, the removal of the need to travel at all for some people (by computer conferencing, live video links, "teleworking" [the ability to work from home without visiting the office, using a computer terminal, electronic mail, facsimile, and communication links]), and the means to improve the efficiency of travel by way of broadcast traffic information, route guidance, and urban traffic control will all be parts of this solution. The broadcasting of timely information on traffic tie-ups and congestion has been estimated (Chapter 6) to save $50 million per annum in the UK, and a dynamic system of route guidance may save a staggering $1.7 billion per annum in the UK alone, by finding improved routes for drivers. Urban traffic control schemes, which maximize traffic flow on roads by dynamically controlling traffic lights to match the flow of traffic, can return the investment in less than one year in favorable cases (see Chapter 7). Apart from financial savings, there is also a concomitant reduction in noise and environmental pollution, and fewer new roads need to be built.

Some of the benefits of applying information technology to the mobile environment are elaborated in subsequent chapters and are outlined in Table 11.5.

1.2 INTERNATIONAL CONTEXT

1.2.1 Europe

Such potential benefits have not gone unnoticed in the "corridors of power." In 1987, the European Economic Community (EEC) announced its *DRIVE* program, a research program aimed at employing information technology to improve the efficiency

[1]A glossary of abbreviations and technical terms is found in Appendix A at the end of the book.

of road transport and road safety, and to reduce its environmental effect. The program has a budget of 60 million ECU (about £40 million or $60 million) to promote road transport information technologies and establish standards. A related project, *PROMETHEUS* (Programme for a European Traffic with Highest Efficiency and Unprecedented Safety), has been set up by European motor manufacturers under the *Eureka* program. Thus, Europe has a leading position in research and development into *road transport informatics* (RTI).

Similarly, in cellular radio, the 16 CEPT nations (see Table 3.4) are specifying a second-generation pan-European Digital system (also known as GSM because it comes under CEPT's Group Special Mobile subcommittee). There will also be a pan-European paging system. Covered in detail are DRIVE and PROMETHEUS in Chapter 9, the GSM cellular radio system in Chapter 3, and paging in Chapter 2.

1.2.2 United States

The United States, and in particular Bell Laboratories, can claim the distinction of inventing cellular radio. The US also has more cellular and paging subscribers than any other nation, although Norway has the highest penetration of cellular radio per capita (see Tables 11.1 to 11.3). There are route guidance and traffic information systems in operation in the US (some on a commercial basis), but the technical lead seems to be elsewhere.

1.2.3 Japan

Japanese companies have leading positions in the supply of cellular radios (i.e., the subscriber units) to the US and European (particularly UK) markets. Japanese motor manufacturers have developed route guidance systems for their vehicles. The National Police Agency, in collaboration with the Ministry of Posts and Telecommunications and a private group, has been developing an *Advanced Mobile Traffic Information and Communications System* (AMTICS), which is described in more detail in Chapters 6 and 11. Unlike Europe, there does not seem to be a concerted and coordinated effort in RTI in Japan.

1.3 AIMS OF THE BOOK

Mobile Information Systems, as we are beginning to see, is a book covering a very diverse subject. The aims of this book are:

to bring together in one volume all these diverse strands, and indicate how they relate to each other;

to give an up-to-date review of the current technology by experts in each field;

to identify synergies between the topics; and

to predict future developments.

Despite the diversity, there are links, and indeed overlaps in some cases, between the various subjects addressed. For example, pagers (see Chapter 2) fulfill some, but obviously not all, of the functions of cellular radiotelephones (see Chapter 3); some pagers can receive messages, and pagers that can send messages are now under development. Traffic information can be broadcast by the Radio Data System (see Chapter 5) or by paging (see Chapter 2); traffic reports can also be dialed (and, in the case of one system, be sent automatically) over cellular radio; pagers can feed into route guidance systems (see Chapter 6) and urban traffic control systems (see Chapter 7). However, for these different systems to communicate effectively, standards are necessary (see Chapter 9), both inside (see Chapter 10) and outside (see Chapters 2 to 8) the car.

1.4 COVERAGE OF THE BOOK—WHICH TOPICS AND WHY

Because the field of mobile information systems is large and diverse, some selection of topics is necessary:

- the book is primarily about personal communication (i.e., person-to-person, broadcast-to-person, or computer-to-person in the case of route planning and guidance);
- the communication systems are primarily land-based, although one chapter covers airborne systems;
- the book covers only civil communication. Military communication, as well as communication and navigation systems used by ships and aircraft, are excluded;
- *automatic vehicle location* (AVL) *per se* is excluded, although aspects of the topic are covered under route guidance, where appropriate, because efficient route guidance may necessitate a knowledge of the vehicle's position;
- "communication" means primarily "radio communication," at least so far as the mobile aspects are concerned (although infrared data links are used in the Autoguide system, described in Chapters 6 and 9).

Thus, *civil land mobile communication* applies to most of the topics in the book, although here the term is used in a much wider sense than normal (traditionally meaning person-to-person radio communication as used by police forces, public utilities, and certain sectors of industry).

1.4.1 Essential Characteristics of Each Type of Communication System

The communication systems described in the book are primarily narrowband rather than broadband; they use relatively low-bandwidth channels of speed or low-bit-rate data communication. Some systems are broadcast (in the sense that the message can be received simultaneously by many users), others are point-to-point or point-to-multipoint, and some show aspects of all three (see Table 1.1).

Table 1.1
Characteristics of Mobile Communications Systems

	Broadcast	*Point-to-Point*	*Point-to-Multipoint*
Paging	N	Y	Y
Cellular Radio	N	Y	1
Satellite Mobile Systems	N	Y	
Traffic Information	Y	Y	Y
Radio Data System	Y	N	N
Route Guidance	N	Y	
Urban Traffic Control	N	N	Y

NOTE

1. Most cellular radio systems are solely point-to-point, in that each caller is connected to one other caller. In the emerging pan-European digital system, however, there will be a "broadcast short message service" so that messages can be sent to all subscribers or to closed user groups.

1.5 RELATION OF TOPICS TO EACH OTHER

The topics covered fall into three groups, albeit with some overlap:

paging and mobile telephony;

traffic information, guidance, and control; and

information technology in vehicles.

We will look briefly at each of these groups in turn.

1.5.1 Paging and Mobile Telephony

This group of topics covers point-to-point communication, and comprises the following:

Chapter 2—paging and messages;

Chapter 3—cellular radiotelephony; and

Chapter 4—satellite mobile communication (i.e., airborne [passenger] telephony).

Most people are familiar with paging (which incidentally seems to have been given a boost by the advent of cellular telephony, when we might have expected the reverse). A pager is compact and easy to carry around; its chief drawbacks are two-fold: the one-way nature of the "conversation," and the need to find a telephone in order to respond. (However, message pagers alleviate the former and the author of Chapter 2 argues that the latter can be an advantage rather than a drawback.)

Cellular radiotelephones completely remove both drawbacks, putting their owners in immediate contact with the worldwide telephone system; there has been explosive growth in cellular phone usage since its introduction in the early 1980s, and growth consistently has exceeded expectations. Unfortunately, as explained in more detail in Chapter 3, mobility is only possible within a country (or small group of countries), because various nations have chosen to use different frequency bands and cellular systems. So, the day when we pack our mobile phone along with our toothbrush is still some years away for most of us (although, provided you confine your travels to the CEPT countries of Western Europe, this facility will be available in 1991 when the pan-European system becomes operational).

Travelers, however, particularly business travelers, want to telephone not only from the office and the car, but from airplanes as well. Various satellite-based systems have been proposed and are becoming operational. Siting the antenna is one of the technical features that is more problematical on an aircraft than it is on an automobile, although it has been solved by Racal, as the photograph in Chapter 4 indicates.

All three of these communication systems are aimed at business rather than consumer applications. That will change, however, as costs decline—particularly in the case of cellular radio and its more recent and less expensive cousin, the CT2 cordless telephone and the *telepoint* system—(see Chapter 3), which may ultimately become comparable in cost to the *public switched telephone network* (PSTN).

1.5.2 Traffic Information, Guidance and Control

The next group of chapters are much more consumer-oriented; they also relate particularly to road traffic. There are four chapters in this group, namely:

Chapter 5—traffic information broadcasting (including the radio data system (RDS));

Chapter 6—route guidance;

Chapter 7—urban traffic control; and

Chapter 8—the CARIN project.

Traffic information broadcasting (Chapter 5) is the sending of timely and relevant messages to drivers in their vehicles about the current state of the road network—to warn them about traffic jams ahead, for example—thereby allowing them to take an alternative route. The RDS is one particular means of broadcasting information on road traffic conditions that is being implemented in Europe. (RDS also does other things such as automatic tuning of radios to a particular program.) Historically, RDS and paging share a common technology in that the techniques used to superimpose the RDS information onto a normal radio channel were first used by the Swedish PTT to broadcast paging information.

Traffic information broadcasting exists now, of course, in that information on road and traffic conditions is already broadcast on national and local radio, and in some cases on special channels devoted to traffic information, as with the ARI system (see Chapter 5) in West Germany. However, these systems have their drawbacks; information may not be very timely because broadcasters will not normally interrupt their program to send a traffic message, and the intended recipients may not be listening—they may be tuned to another channel, listening to a tape, or the radio may be switched off. Hence, the systems described in Chapter 5 (such as RDS) have some automatic means of overriding whatever other channel to which the driver is listening, as well as ensuring that the information is relevant so that information on traffic conditions in London is not given to drivers in Liverpool.

Route guidance (see Chapter 6) is *not* the automatic piloting of unmanned vehicles, but the calculation, on behalf of a driver, of the "best" route to take to go from point A to point B, with instructions to help navigate the route. "Best" may mean fastest, shortest, or most fuel-efficient, which are not necessarily the same. Route guidance is still in its infancy. Current systems, such as the Etak system sold on the US West Coast, display an electronic map and indicate the vehicle's position, but leave the driver to find his or her own route to the destination. Although this may be an improvement over the current situation of paper maps, the driver is unlikely to find the optimum route because maps help to find only the shortest route; to find the quickest route, you need information on average traffic speeds on roads, which maps do not normally tell you. Furthermore, actual speeds are variable, depending on congestion.

Route guidance usually works on an individual basis; you specify your start and end points and the system calculates the best route for you. However, one system, *Autoguide,* which is now being tested in London and Berlin, can recognize individual vehicles. If it finds that vehicles are taking a long time to traverse a particular route, Autoguide can take that into account and send the next vehicle a different way. Hence, Autoguide (described in Chapters 6 and 9) has some characteristics of an *urban traffic control system.*

Urban traffic control (Chapter 7) means optimizing the use of the existing road network by means of information technology (especially computer-controlled traffic signals), maintaining traffic flows, and obtaining the best throughput.

CARIN (see Chapter 8) is a *car information and navigation* system developed by Philips, based on optical disc technology and vehicle location; CARIN involves primarily route guidance, but also, to some degree, traffic information. Although the CARIN project has been in existence for several years, it has recently become part of the PROMETHEUS program (see below).

All four chapters and topics are closely related. Route guidance will be more effective if it can utilize up-to-date information about traffic conditions on the road network. Efficient traffic information broadcasting and route guidance would facilitate, and could be weapons in the arsenal of, urban traffic control. CARIN is concerned primarily with vehicular matters, while Chapters 5 and 6 are concerned with "the system" and the infrastructural aspects of the topics.

1.5.3 PROMETHEUS, DRIVE, and the Intelligent Car

The third group of chapters addresses some of the topics covered earlier, but more from the "point of view" of the vehicle:

Chapter 9—PROMETHEUS and DRIVE—European initiatives; and

Chapter 10—The intelligent car.

PROMETHEUS is a project set up under the Eureka program by European motor manufacturers to develop and standardize various kinds of automotive systems; a central concept is the intelligent automatic vehicle "co-pilot." DRIVE is an initiative of the Commission of the European Community with the aims of improving road safety, improving road transport efficiency and reducing pollution. The two programs are closely linked, both formally and informally.

"The intelligent car" concentrates on the description of how electronics and the development of better sensors, displays, and control systems can improve the performance of the mechanical and engine-related aspects of motor vehicles, making them more efficient, safer, and more "user-friendly," as well as reducing pollution.

1.6 COMMON THEMES

Figure 1.1 indicates schematically how the topics covered in each chapter overlap, and Table 1.1 highlights some common themes. Thus, route guidance and traffic information necessitate improved electronic displays and speech synthesizers, ideally within the vehicle's console. For example, Autoguide uses LCDs and speech synthesis.

Reduced pollution, reduced fuel consumption, and increased road safety can all be realized by the use of information technology applied to both the vehicle (Chapter 10) and the "road + vehicle system" via route guidance, traffic information broadcasting, and urban traffic control (Chapters 5 to 8). These are also major themes of the PROMETHEUS and DRIVE programs (Chapter 9). Better communication, a

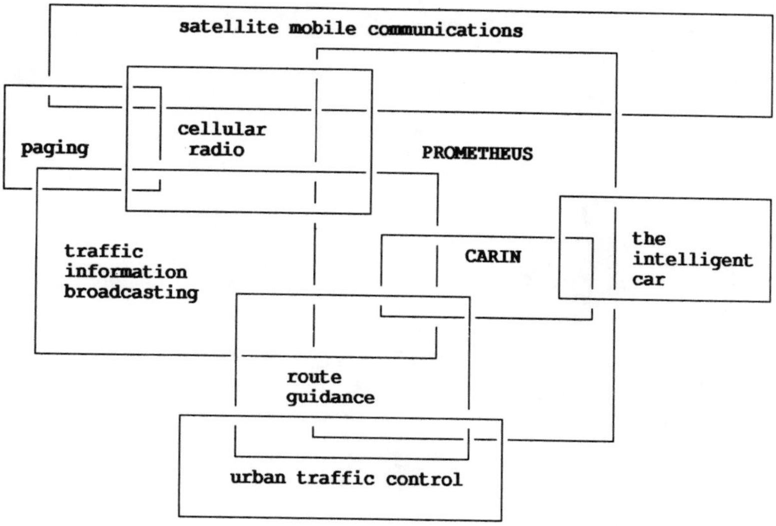

Figure 1.1 *Mobile Information Systems*—How the book's topics overlap.

concern of Chapter 9, is covered in detail in Chapters 2 to 4. We will return to some of these common themes, and attempt some crystal-ball gazing, in Chapter 11.

REFERENCES

Institution of Civil Engineers (1989), "Congestion," Thomas Telford, London.

King, G.F., and Mast, T.M., "Excess travel: causes, extent and consequences," *Proc. 66th Annual Meeting of the Transportation Research Board,* Washington, DC, pp. 126–134.

PART II
PAGING AND MOBILE TELEPHONY

Chapter 2
Radiopaging and Messaging

R.H. TRIDGELL
MOBILE RADIO CONSULTANT

2.1 INTRODUCTION

2.1.1 Basic Radiopaging

Radiopaging is probably the simplest of all the major organized mobile radio services. As such, it forms an ideal platform from which the reader can be introduced to a number of the technicalities and facets of mobile radio, which accounts for its position as an early chapter in this book. Radiopaging performs the same function as the hotel pageboy who goes around calling the name of a person because they have a telephone call or a message waiting. Of course, this personal call can cover a much greater geographic area and is far faster when radio is used.

Radiopaging is defined in CCIR* Recommendation 584 (CCIR, 1982) as: "a non-speech, one-way, personal selective calling system with alert, without message or with defined message such as numeric or alphanumeric." This definition is restricted only for political reasons, and radiopaging systems often include speech message transmission.

Figure 2.1 shows the basis of any radiopaging system. Each user carries a paging receiver (pager), light and small enough to be carried in a handbag, slipped into a jacket or shirt pocket, or clipped to a dress, tie, or trousers. The pager is powered by one or two small batteries, which usually last several months before needing to be changed.

---

*CCIR—Consultative Committee for International Radio. This very important body is part of the International Telecommunication Union (ITU) and authorized by the United Nations to make *regulations* and *recommendations* on international radio matters, which are vital for the harmonious employment of radio throughout the world. See Appendix B for a fuller description.

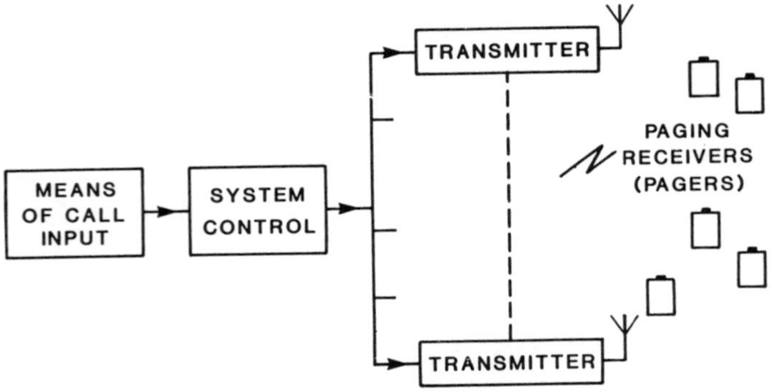

Figure 2.1 Basic layout of paging system.

Each pager has at least one *address* (i.e., a specific digital value), which the pager can recognize when broadcast on its radio channel. The caller must know the wanted address and specify it when making a call. The system control then transmits that address as soon as possible (not necessarily immediately). Any pager that receives its address then alerts its user by beeping (or by vibrating silently). The alerted user takes whatever action seems appropriate (e.g., he or she might call the office from a public telephone, or hurry to join the crew of a fire engine). Note that an address may be unique to one pager or common to several pagers so that a group of users can be contacted by a single transmitted call. Most pagers can have several addresses, each with its distinctive alert (e.g., differing beep cadences).

As the definition above suggests, some systems and pagers can deal with messages. In this case, the message is transmitted immediately after the address, and the pager delivers that message (either audibly or by display) and alerts the user.

Occasionally, a paging call is not received or its message is received incorrectly. Because no transmission return path is provided, the caller only knows that the call has been received when he or she detects some action by the called user. Callers must take this into account and act accordingly (e.g., by making a repeat call if no response is observed within, 20 minutes). Despite this limitation, users find radiopaging to be very attractive.

2.1.2 Public (Wide-Area) and Private (On-Site) Systems

Radiopaging systems can be roughly classified into two categories: public wide-area and private systems (which are often restricted to on-site coverage). *Public wide-area systems* use several or many medium-power (or even high-power) transmitters to cover a city or large area, or even a whole country. In contrast, *private on-site systems* (e.g., covering a hospital), are often specially engineered and tend to use one (or very few) low-power transmitter.

Another difference frequently found among these types of paging system is that of paging data throughput and the coding adopted to convey that data load. Because of the large number of users on a public system, wide-area systems tend to carry a high data load, and so use sophisticated, high-rate codes. Frequently, the pagers are supplied by several manufacturers, and so a standard, nonproprietary code is used. For private systems, the data load tends to be low, so a simple, low-rate, proprietary transmission code may be employed, and a single manufacturer provides and maintains the entire system. A final difference may be the method of call origination. In public systems, the paging calls usually originate via the PSTN or a data network. The calls are queued and then transmitted in batches by a complex controller. In private systems, the call input often is via a manual operator or a PBX, with immediate transmission.

Of course, there are private systems with wide-area coverage. There is also a growing tendency for private systems to require the facilities of a sophisticated code. Bearing in mind the blurred technical boundary between private and public paging systems, henceforth only techniques for public systems will be discussed because those for private systems can be assumed to be a subset.

2.1.3 Pager Facilities

Pagers vary greatly in their facilities. Facilities available now include:

- one address and simple beep (typically of about eight seconds duration.)
- beep stop button
- multiple addresses with corresponding different beep patterns
- escalating loudness of beep
- silent alert by vibrator
- call memory which suppresses the immediate beep, but which can later be interrogated
- multiple memory for multiple addresses
- numbered display of memorized calls (helpful with silent alert)
- selective cancellation of multiple call memory
- extended intermittent beep or flash (or both) for uncancelled memorized calls
- beep plus an immediate voice message
- beep plus linkage with a shared voice message storage system
- beep plus storable numeric messages with display (typically used for conveying the caller's telephone number)
- beep plus storable alphanumeric messages with display
- alphanumeric message printer (rarely used)
- intrinsically safe pagers for inflammable atmospheres

- included digital clock

Other facilities which could be available in the future are mentioned in Section 2.7.

2.1.4 Importance of Radiopaging

The importance of radiopaging stems from its advantages. In particular, because of the lightness and smallness of a pager, carrying it does not generate any user resistance. Moreover, in contrast to the immediate action needed to answer a telephone call, after reception of a paging call, the user can choose an appropriate time to take action. This is particularly important to drivers, who thus are not distracted. The vibrating alert or call memory can make call reception completely discreet. Also, if a displayed message is involved, the pager stores it to be read again as often as necessary, thus relieving the user of the need to remember the message or write it down. All of these features combine to make the pager so attractive that user resistance is extremely rare.

A paging system can be installed, operated, and expanded at relatively low cost. Indeed, a pager is cheaper than any other organized mobile personal communication system. By taking this into account, pagers on public systems have been estimated eventually to be worn by between 2% and 8% of the population in the covered areas (see Section 2.5 for more statistics on usage). Apart from analog voice paging systems, paging is a highly efficient method of utilizing the radio spectrum. In view of the rapidly growing demands for radio spectrum for mobile services of all types, this aspect of paging is particularly important. For example, simulations (see Appendix 2.C, Section 3.2) indicate that a single 1200 b/s radio channel could carry (without frequency reuse) the entire paging needs of a population of about 16 million persons, assuming a 2% use within the covered population and employing normal paging traffic figures. (A beep plus analog voice paging system would cover a population of only about 120,000 persons. Hence, we have the political restrictions in the CCIR definition.)

As demonstrated above, radiopaging can make a very large contribution to the business efficiency and convenience of any institution or nation if personal contact is required. Indeed, any developing nation considering implementing some sort of public mobile radio service might well start with a paging system.

2.2 HISTORY OF RADIOPAGING

The first well documented paging systems came into being in the mid-1950s. The first one in the UK was installed in St. Thomas's hospital in London in 1956, and it was an immediate success. Initially, an audio frequency loop was placed around the building. The pagers were very simple tuned frequency receivers, as was appropriate for the technology then available. However, the audio frequency loop caused interference to other equipment and eventually the system was changed to use a 35 kHz carrier modulated by audio tones (i.e., the carrier was *tone coded*).

As the demand arose to page over wider areas such as oil refineries, there was a change from 35 kHz to a radio carrier radiated from a central transmitter. Frequencies in the 80–1000 MHz region were found to be particularly suitable in that they permitted the use of integral antennas in the pagers and also gave relatively good penetration into buildings. Typically, tone coding was used, with 2 of 30 possible tones being transmitted sequentially to yield one address. (This permits only 870 addresses because the two tones in any call must differ.)

The first public wide-area paging systems developed in the US and Canada in the early 1960s. At first, callers phoned an operator who keyed in the wanted address to the system. However, as was quickly realized, the caller could directly dial a number that represented the wanted address, and the dial pulses could be received on an automatic terminal, which could check the validity of the call and acknowledge this to the caller, memorize and queue the calls, switch on the transmitters, and transmit the waiting calls in a batch. Accounting could also be included if desired. One such example was the Bell Canada *System Wide Area Paging* system (SWAP). In Europe, wide-area paging systems were opened in Holland and Belgium (1964) and Switzerland (1965), both using vehicle-mounted pagers.

As demand rose, there was a change from two-tone coding (i.e., any two from about 70 tones, to give nearly 5000 addresses) to five-tone coding with 100,000 address capacity (the five tones are transmitted sequentially, and each tone represents a decimal digit). However, tone-coded systems are prone to false calling because there is little redundancy in the signal, and a filter and timer system is needed in each pager to detect the wanted tones. Thus, the circuitry needed is not very amenable to design that uses low-cost digital integrated circuits. Moreover, the code is limited and unsuitable for complex messages such as alphanumeric text. Thus, binary digital signaling systems began to appear in the late 1960s. Production of pagers utilizing early examples of binary codes has long since ceased, but the Motorola *Golay Sequential Code* [CCIR Report 900, 1982] is still produced.

In Japan many district systems were opened, each on a different channel if within interference distance of one another. Therefore, national paging was not generally possible. A digital, alert-only code emerged about 1978 with a capacity of 65,000 addresses (two per pager) [CCIR Report 900, 1982]. Later the code was extended to include numeric display paging. In the UK, the British Post Office (BPO; now British Telecom) opened its first paging service in 1973 and initiated the start of a national system in 1976 in London. The system used two different proprietary codes at first, transmitting, in turn, a batch of calls for one code and another batch for the other code. However, it quickly became apparent that this multicode arrangement was wasteful. The BPO generated an industry standard code by the simple expedient of calling likely manufacturers (from various parts of the world) together into the *Post Office Code Standardisation Advisory Group* (POCSAG) to reach agreement on a standard. The resultant POCSAG code, which included full numeric and alphanumeric message capabilities, was so promising that it was offered to the CCIR as a prospective world standard. CCIR adopted it in February 1982 as the CCIR Radiopaging Code No. 1 (RPC1), which is now its correct name. RPC1 is manufactured in many parts of the world and is more popular than any other pager code now in production. The benefits of using a standard paging code are dealt with in Appendix 2.C. Numeric and alphanumeric message pagers using the RPC1 standard became available in 1983 and 1984, respectively [Tridgell, 1987].

Another form of radiopaging, with numeric message capability, was developed in Sweden in 1978. Here the paging code is carried (together with other types of data, see Chapter 5) on a 57 kHz subcarrier on broadcast FM radio [CCIR Report 900, 1982]. The advantage of this is that separate transmitters are not needed, but the penetration into buildings given by the broadcast service does not necessarily match that required by the paging service. The pagers automatically tune over the broadcast channels until clear paging reception is obtained. Thus, the system is inherently suitable for nationwide paging, but at added cost and complication in the pagers. The system is used in the US and Canada as well as Sweden.

National systems now exist in many countries, especially those with high population densities. The British Telecom system (in the UK) is probably the largest, with well over 400,000 pagers in service, and said to be growing at over 3000 per month. In the US and similar large-area countries, to provide this type of service is not practical, but nationwide service can be organized by linking various localized paging systems. There now are estimated to be over nine million pagers in the world, of which some five million are on public systems.

2.3 AN OUTLINE OF MOBILE RADIO AND PAGING ENGINEERING

This book is intended mainly for business readers, and thus this outline of radio and paging techniques is sufficient only to enable the busy reader to follow the subject. Avoidable (but interesting) technical detail is included in appendices (for deep study, see standard textbooks e.g., Jakes, 1974). The first radio aspects described are common to all mobile radio services (e.g., cellular radio, see Chapter 3) and recur throughout this book. Later sections (from 2.3.4) refer particularly to radiopaging.

2.3.1 Transmission Properties of Radiopaging Frequency Bands

At low frequencies, radio waves can bend (diffract) around large objects quite well. However, for wide-area paging (and many other mobile services), the most usual frequencies are in the 80–960 MHz band. Emissions in this and higher frequency bands do not diffract well, and a deep radio shadow occurs on the unilluminated side of any obstruction (e.g., hill, building, truck, or human being; see Figure 2.2). The most pervasive obstruction is the radio horizon (due to the earth's curvature), which is about 30% farther from the transmitting antenna than the equivalent visible horizon due to the reduction of refraction in the upper atmosphere as compared to that at ground level. Beyond a transmitter's horizon, the signal strength falls very rapidly so that in areas well beyond the horizon the same frequency can be reused without causing interference (see Chapter 3 and Appendix C at the end of this book). This ability to reuse frequency is helpful because radio spectrum is scarce compared to the potential demand for it.

Of course, the higher the transmitter is above ground level, the farther away is its radio horizon. The area (not radius) covered is approximately proportional to the individual antenna heights of both transmitter and receiver, with consequent reduction of system costs. However, the far flung horizon of a high transmitter prevents close reuse of that frequency and a balance of these opposing considerations often must be found. Between the transmitter and horizon, in open, flat country, the received power reduces approximately as the inverse fourth power of distance from the transmitter.

Shadowing by buildings obviously causes very patchy reception in built-up areas. Fortunately, this frequency band reflects well off of most hard-surfaced objects, and most local shadows are filled in by reflections from other nearby objects. Indeed, in urban situations, to be able to receive a direct ray from the transmitter is comparatively rare, and reception nearly always consists of several reflected rays (i.e., it is *multipath*).

Multipath propagation has the curious property that, while the rays generally add to strengthen the received signal, at some points, they subtract from one another so that the signal is severely weakened. These weak points in the radio field are termed *nulls* or *fades,* and occur at approximately half-wavelength intervals. The field power levels are approximately Rayleigh distributed [Jakes, 1974, Chapter 1]. Compared to the local average field strength, fades tend to be up to 20 dB weaker (see Figure 2.3), but fades that are 40 dB weaker than the local mean level are not uncommon. The combination of shadowing and multipath fading produces a radio field that varies wildly over quite short distances (up to 60 or 70 dB difference between the maximum and minimum street level value within a 100 m square).

Figure 2.2 Typical radio shadows (not to scale).

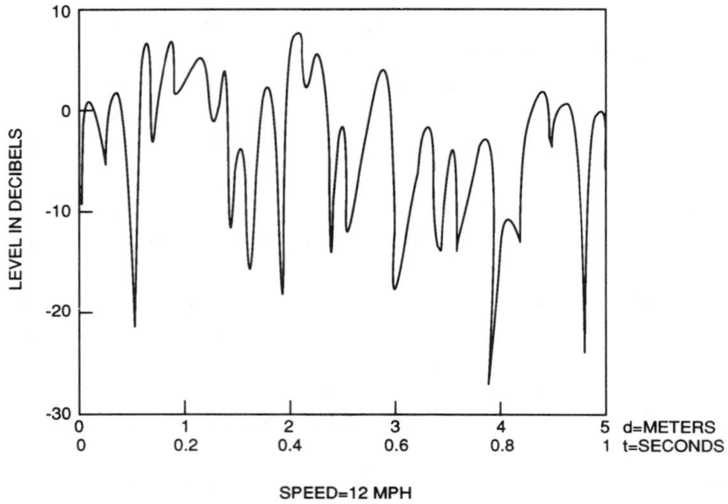

Figure 2.3 Typical multipath (Rayleigh) fading (at 900 MHz).

Mobile services, especially paging, require penetration into buildings. All these frequencies penetrate buildings fairly well, mostly through doors, windows, and thin nonmetallic roofs. Typical mean building penetration losses are 10 to 20 dB, but penetration losses as high as 40 dB have been encountered. Radio waves also may need to penetrate through the wearer's body to reach the pager. Pager antennas tend to be somewhat directional, and differences between the best and worst body orientation of a pager (as worn) are typically 15 dB (see Figure 2.4) and can approach 30 dB. As we can see, the interplay of all above-mentioned sources of variability (see Appendix C at the end of this book) make all mobile and pager reception essentially probabilistic. A success rate value of 100% is never guaranteed.

2.3.2 Noise and Interference

In a radio receiver tuned to a channel, if unwanted radio power present in the channel approaches or exceeds the power of the wanted signal, reception quality suffers. The quality reduction can vary from slight to total obliteration of the wanted signal.

A fundamental law of physics is that all receivers generate within themselves electrical noise (i.e., unwanted radio power). The designer's task is to make this noise as small as practicable, but there is a limit below which reduction is not possible. If the power level of the received signal is comparable with the receiver noise, reception will be impaired. This sets a *noise limit* to the sensitivity of the receiver.

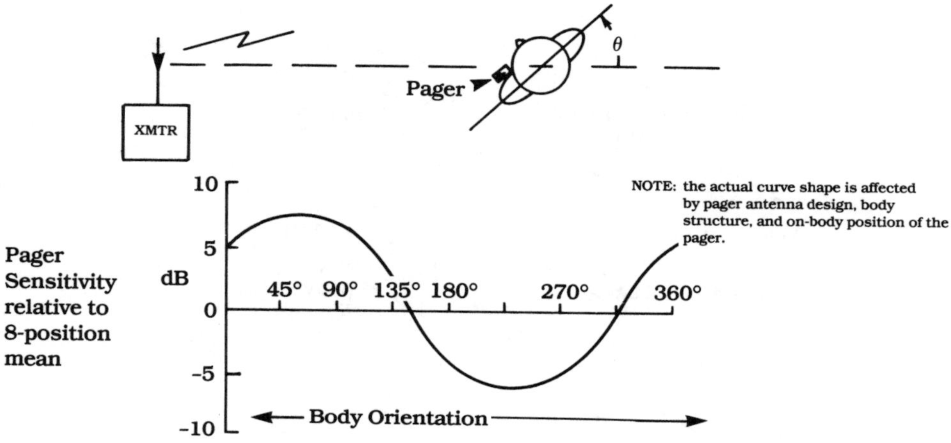

Figure 2.4 Typical pager sensitivity *versus* body orientation.

Unwanted power can arise outside the receiver from another transmitter on the same channel (*co-channel interference*), or from transmitters on other frequencies combining to produce power within that channel (*intermodulation interference*). Moreover, the channel selectivity of a receiver cannot be perfect, and a nearby transmitter tuned to an adjacent channel can interfere. Finally, man-made noise power (e.g., from poorly suppressed engine ignition systems), frequently causes interference. In laying out any radio transmission system, the task of the system designer is to try to arrange that the wanted received power exceeds the unwanted noise and interfering signal powers over some high percentage of the intended coverage area (e.g., 99%).

2.3.3 Radio Coverage and Interference Between Transmitters

The basic service that mobile radio offers is communication with a person, wherever he or she is. The recipient may be out in the open on a high hill or deep inside the lower levels of a building in a valley. As a person travels, his or her reception situation varies enormously.

As already described, the physical laws governing radio propagation are such that to cover the whole of a desired wide area with great reliability usually is not possible from a single transmitter site. Thus, several or many widely spaced transmitter sites are needed, and a wide-area system must be engineered so that these transmissions do not mutually interfere. For instance, Figure 2.5 shows a pair of sites sufficiently separated so that co-channel interference cannot occur. The techniques of wide-area transmission are concerned with the avoidance of interference.

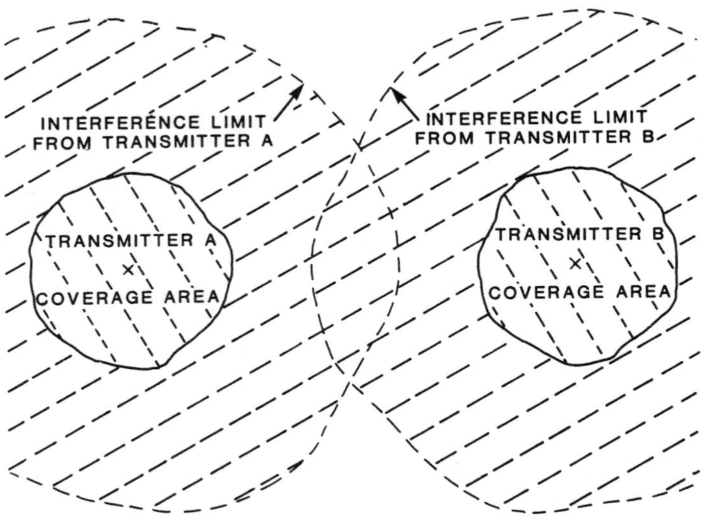

Figure 2.5 Co-channel coverage and interference areas of transmitters.

Where radio fields overlap, there are three principal methods of avoiding interference:

(a) Frequency Separation. Different frequency radio channels are used for adjacent transmitters.

Alternatively, using the same channel frequency for all transmitters:

(b) Sequential Transmission. While any one transmitter is being used, all surrounding transmitters which might interfere are switched off. Thus, the transmitters are switched on in turn (see Figure 2.6).

(c) Simulcast (or quasisynchronous transmission). The modulating (data or speech) signals of all the transmitters in the area are controlled so that they are substantially in synchronism. Figure 2.7 illustrates the potential effects of unsynchronized and synchronized digital signals from two transmitters, interference being possible whenever the two signals differ. The actual degree of interference at any location depends on the relative field strengths of the transmitters at that point. Analog signals, such as speech, are much more difficult to synchronize than low-rate digital ones. Typically, speech signals must be synchronized to within 40 μs, whereas 500 b/s digital signals can tolerate up to 500 μs of mistiming, the latter tolerance being much easier to maintain.

The basic arrangements required for simulcast to cover an area are shown in Figure 2.8. As illustrated, all the transmitter sites are fed by land lines radiating

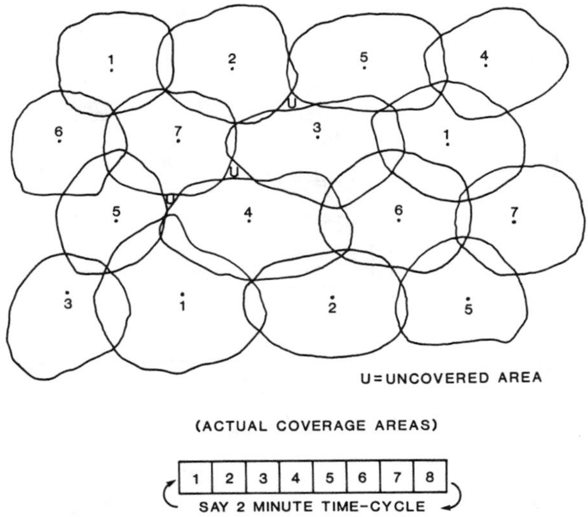

Figure 2.6 Layout of sequential paging transmitters.

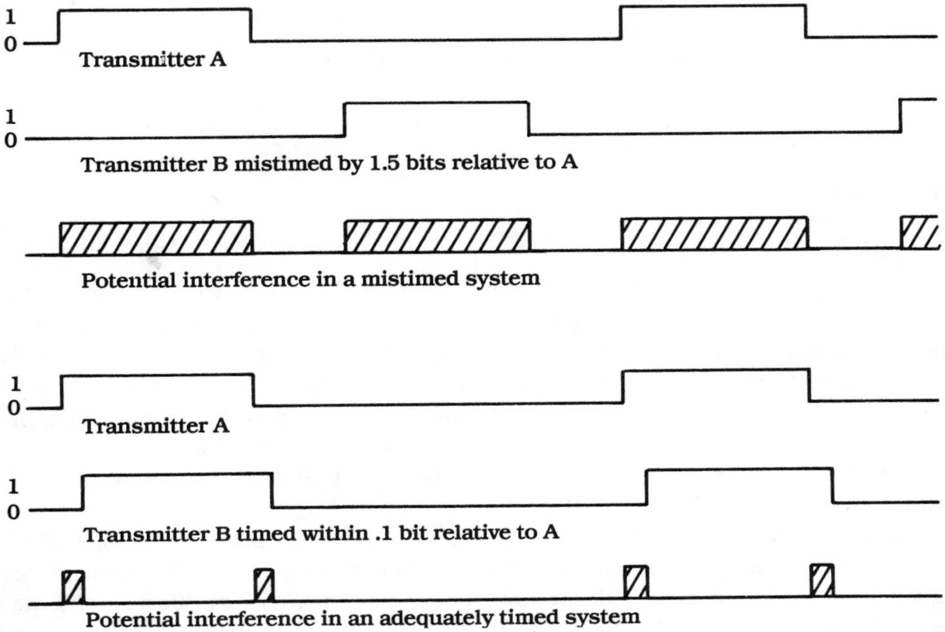

Figure 2.7 Potential interference in a simulcast binary FSK system.

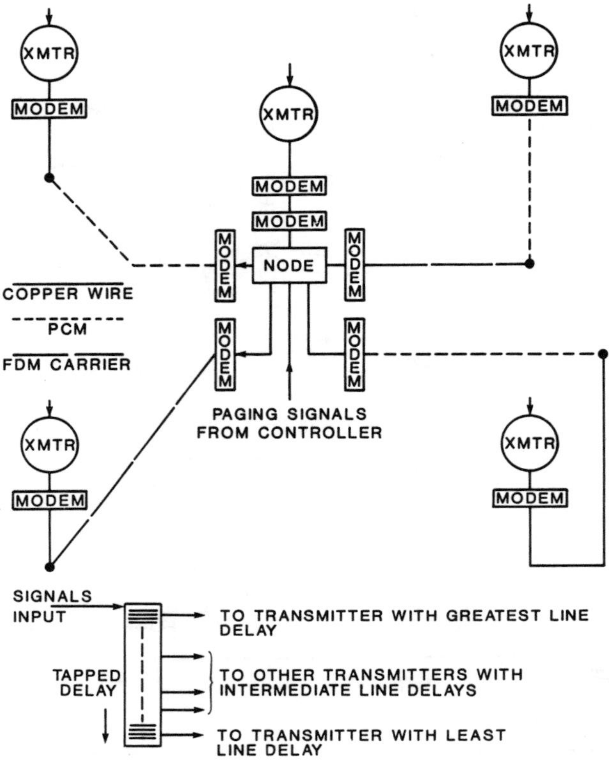

Figure 2.8 Simultaneous paging operation.

from the node. Of course, the transmission delay of these lines varies according to the line length and transmission equipment composing each line. To achieve transmission synchronism, each line is padded with extra time delay so that the total delays (of each line plus its time pad) are all substantially equal. The time pad can consist of a delay line tapped at intervals suitable to cope with the requirements of the transmission codes in use on the system. In this illustration, the tapping is at 50 μs intervals, which is generously appropriate for the CCIR RPC1 transmitted at 512 b/s.

As described, the adjustment of the time padding would be done manually. Various arrangements for automatically performing this adjustment are known (see Appendix 2A). Of course, if all the links to the transmitters have substantially the same delay, as is often the case if radio links can be used instead of lines, the adjustable time pad may be omitted.

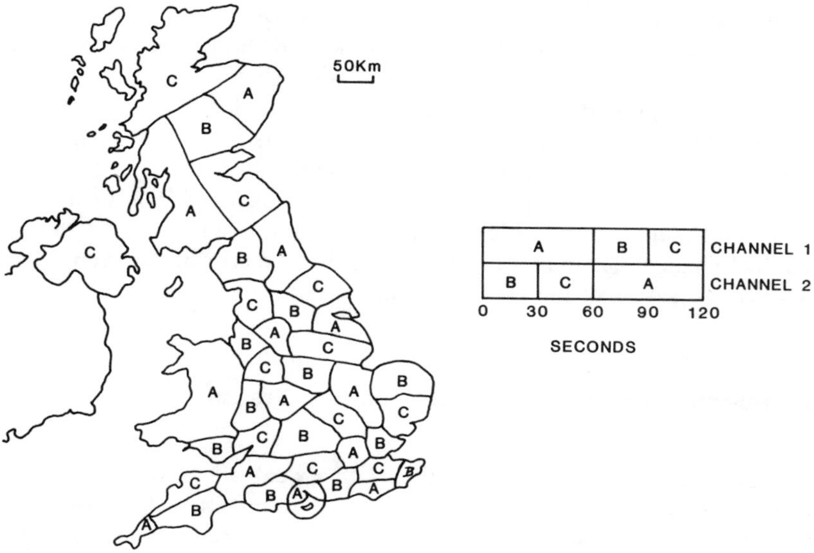

Figure 2.9 Radiopaging zones and time slots.

The foregoing properties and techniques apply to most mobile services. Those that follow are more particular to paging, but may also be found in other services.

2.3.4 Transmission Zones

To achieve frequency reuse, large paging systems are sometimes organized into zones. Each zone is covered by simulcast transmission. The user pays only for the zones in which he or she wishes to be paged, and a user's calls are transmitted in each of the chosen zones.

Figure 2.9 shows the arrangement once used by the British Telecom (BT) national system. There were 40 zones, and each was covered by simulcast transmission from several sites. Of course, the transmissions of the various zones differed due to user zone selection. To avoid interzone interference, the zone boundaries and transmitter sites were carefully planned so that the zones could be divided into three sets, A, B, and C, with the properties that all zones in any one set were nonadjacent and noninterfering. Consequently, by dividing channel time into three slots and allotting each set of zones its own slot, and by switching on the transmitters only in the slot appropriate to that zone, all interference was avoided and the frequency was reused about 13 times. However, about one-quarter of the UK population resides and works in London and its immediately surrounding zones, so the full gain from frequency reuse cannot be realized.

An important property of this air time slotting is that it is geographically infinitely extensible, and thus is particularly suited to the sharing of a single channel across the boundaries of two or more abutting paging systems.

Finally, we can see that provided no time slot exceeds 50% of the complete time, if two channels are available, a single transmitter can be used with remotely switched dual-channel capability at each site. Thus, the cost of two transmitters and two lines per site (to serve the two channels) can be halved.

2.3.5 Comments on Multitransmitter Wide-area Systems

Within the system coverage area, a location with a weak signal from one transmitter is likely to have a strong signal from some other transmitter. Both of the single-channel systems (in Section 2.3.3, b and c) have a marked advantage over a frequency-separated system in that the receivers are always tuned to all of the transmitters. Thus, the transmissions mutually reinforce each other. The transmissions of a frequency separated system offer no such mutual reinforcement.

In a frequency-separated system, to receive throughout the whole area, the receiver must be equipped with automatic channel tuning. (Users are forgetful of manually tuning a pager.) Tunable receivers currently are more expensive, deplete their batteries quicker, and are more bulky than fixed-channel receivers. Moreover, the automatic tuning circuitry constantly must decide whether it should search for some other channel, and communication cannot occur while tuning is in progress. The system design would have to account for this.

The throughput of a simulcast system is equal to the maximum possible for a given data (baud) rate, and is unaffected by the number of transmitters in the system. The system delay needs careful set-up and maintenance, but this can be done automatically. A sequential transmission system is the easiest to set up and maintain. However, for a given baud rate, the paging throughput is inversely proportional to the number of time slots in the transmission cycle. Thus, the achievable efficiency or radio spectrum usage is much less than that of a simulcast system using the same data rate. This is of no significance if the wanted throughput is small.

A system using sequential transmission can be easily changed to a simulcast system as the paging traffic increases. Because the same pagers are used, there is no need to disturb customers at the time of this change. Note that if a paging channel must be shared among a number of paging service providers covering overlapping areas, time division of the channel is essentially similar to that utilized in the sequential transmission scheme outlined above.

2.3.6 Combining Paging with Another Mobile Service

Usually, paging channels are dedicated to one service (although sometimes shared between paging operators). The transmitters can thus be planned and sited to give any desired degree of coverage (e.g., penetration into the heart of selected large buildings). Of course, the full cost of transmitters then is borne by the system provider. In contrast, one can carry paging signals on another service, particularly FM broadcast radio (see Appendix 2B). The transmitters are already in place and no separate ones are needed.

However,

(a) the pagers need to incorporate demultiplexing circuitry to separate the paging signals from those of the main service;

(b) the coverage given by the broadcast transmitter network could well provide a lower probability of paging reception than that desirable for a good paging service; and

(c) careful business agreements with the broadcast (or other service) organization are necessary to maintain the paging service to the satisfaction of the users.

Generally, sharing with a broadcast system is cost effective in areas of low population density if indifferent paging coverage is acceptable.

Another service that may seem ideal for combination with paging is mobile radiotelephone (e.g., cellular; see Chapter 3). A user might be paged for incoming telephone calls and could ring back at his convenience rather than answer immediately. Further, the pager and phone might be combined into one portable instrument. However, to date, successful sharing of these two services on a single transmitter network is virtually unknown. The reasons for this lack of sharing are becoming smaller with the advance of technology. The advances required are:

(1) An increase in the field strength of radiophone systems to allow operation deep inside buildings (but cellular phone systems now tend to operate at high field strengths).
(2) The called user needs to be informed of the caller's number without answering the incoming call. (Modern electronic PSTNs provide *calling number indication,* and this could be included in the paging signals to the wanted radiophone.)
(3) A single instrument would have to be capable of receiving both paging and phone channels (which are not usually in the same radio frequency band). In the paging mode, the phone part of the instrument would be switched off to save energy.
(4) The phone would have to be reduced in size and weight to compete with the

carrying convenience of a pager. (Proposals for "microenergy" phones have been made.)

To date, dedicated single-channel paging systems are much more popular than the shared service possibilities mentioned above.

2.3.7 Paging Codes

Paging calls and similar digital transmissions occur in coded form. The code is chosen not for secrecy, but for ease of transmission, accuracy, and ease of reception (see Appendix 2.C). Originally, audio "tone coding" was used, but this is now obsolete.

Binary coding is becoming prevalent, using *frequency shift keying* (FSK) (i.e., with positive shift of radio frequency representing a binary 0 and negative shift a binary 1). The bits are transmitted in continuous streams or blocks (i.e., *codewords*). These transmissions consist of the information bits and some redundancy bits (also known as *check* or *parity bits*). The function of redundancy is to permit the received information to be at least checked for accuracy, and perhaps for a few errors to be corrected.

Unfortunately, slow movement through a multipath fade can produce a long burst of errors, and the amount of redundancy that needs to be added to permit correction of such a burst is formidable, and greatly reduces the information throughput of a channel. For all those situations within a coverage area that receive a good signal, this massive redundancy is not required and is a waste of transmission time. Therefore, codes must be carefully chosen to balance between being unable to correct a high proportion of received codewords because too little redundancy is included and wasting transmission time because the redundancy is excessive for most situations. The RPC1 code has been subjected to extensive simulations [Mabey *et al.*, 1985] and field tests [Tridgell and Denman, 1984]. Whatever code is chosen, some error bursts will be too large to permit correction, and message repetition is then necessary.

2.3.8 Outline of Pager Technology

Figure 2.10 shows a typical pager design. It is powered by one or two small batteries, which last for several months of normal use in a good pager. The designer aims to achieve great sensitivity together with very low energy consumption.

2.3.8.1 The Front End

Typically, a loop or small ferrite rod antenna is integral. This rather inefficient antenna (i.e., 20 dB less sensitive than a dipole) feeds the *front end,* which selects the

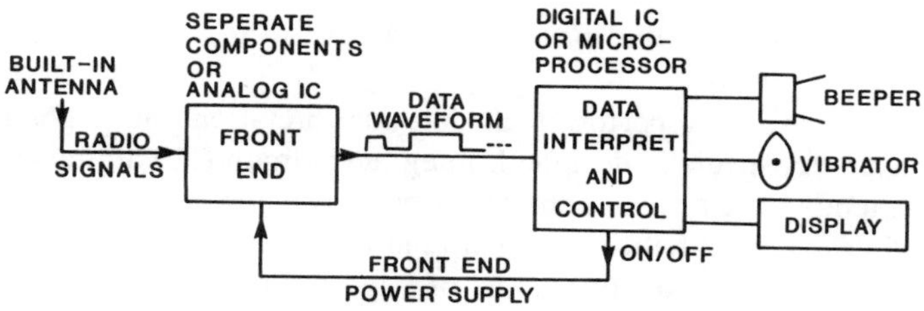

Figure 2.10 Typical pager design layout.

radio channel, amplifies, and recovers the modulation waveform from the radio frequencies. The front end is a major power consumer and is automatically turned off whenever possible to prolong battery life.

To manufacture the complete front end of a single-frequency pager on one silicon chip is now common. Typically, only two factory adjustments are needed to assemble and align such a front end, compared to perhaps seven adjustments for front ends using discrete components. Figure 2.11 shows a single-chip front end, the operation of which is explained in Appendix 2.D.

To date, multifrequency pagers use discrete component circuitry, which makes them both bulkier and more expensive than the single chip solutions. Pager front ends tend to use about 5 mW of power, a low value which makes them rather susceptible to electrical noise. The digital parts of the pager tend to be electrically noisy, and the designer has to be careful that these parts do not detract from the front end performance.

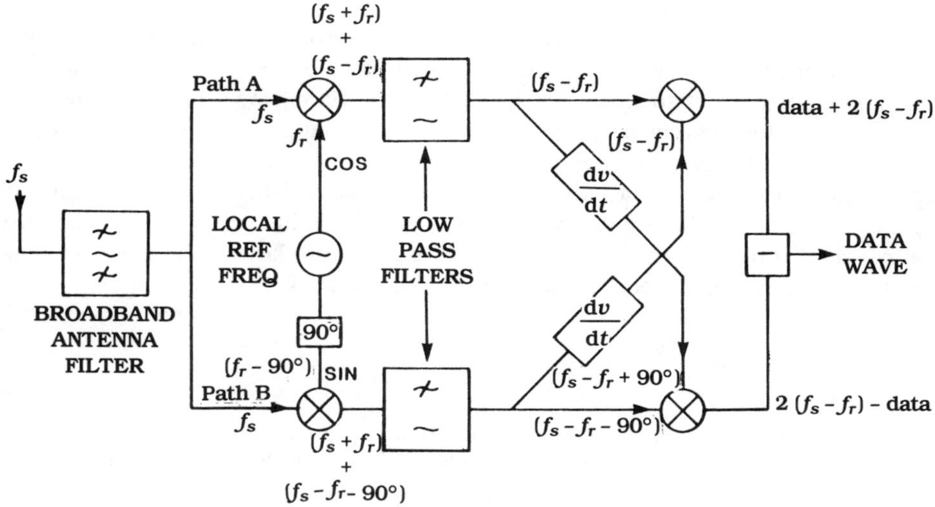

Figure 2.11 Schematic of a direct conversion receiver.

2.3.8.2 Data Interpretation and Pager Control

An alert-only pager usually performs all data interpretation and pager management functions on a single, specially designed, integrated circuit (IC) digital chip. Message display pagers tend to use a small general-purpose microprocessor and a storage chip. The pager address is often held in a small read-only memory (ROM), which can be changed as required (e.g., to suit change of pager ownership).

The functions of the interpret and control part of the pager are:

- to derive bit timing from the recovered data waveform,
- to synchronize with the data frames (and perhaps generate an "out of range" warning if synchronism cannot be achieved),
- to interpret (and perhaps correct) the data,
- to search the corrected data for any of the pager's addresses,
- to generate beep patterns or drive the vibrator,
- to memorize any received calls when switched to the memory mode,
- to display any stored messages,
- to manage all battery energy saving.

2.3.9 Methods of Originating a Paging Call

2.3.9.1 On-Site Systems

Manual Input

The control for small on-site systems frequently consists of a keyboard plus an encoder for the pager addresses and messages (if any). Sometimes a small transmitter is built into the unit to make the base station a complete, stand-alone item. The control operator (who usually also operates the manual phone switchboard) receives a call over the telephone system, keys in the wanted address, and transmits the call. Typically, the pager has at least two addresses, one of which is reserved for emergency calls. When the wanted pager user phones back the operator, she can deliver a message. If the system is a message type, the operator keys in the message as well as the address, and the called user need not phone back.

Automatic Operation

Assuming the on-site phone system is automatic, the paging system might be given a block of phone numbers (e.g., all those beginning with digit 7). The paging system controller is connected to the phone switch, which immediately routes any calls beginning with 7 to the pager controller. Thus, the controller can interpret the remainder of the dialed digits, code them as a pager address, and transmit. One "answer page" phone number is used, and the called user dials that number and is connected to the caller.

Frequently, numeric messages are used as codes for previously agreed text messages (e.g., "123" might represent the message "Duty doctor, go immediately to operating room X to assist"). These prearranged numeric messages can be dialed by the caller as an extension of the dialed pager address, and users only need a numeric message pager instead of a full alphanumeric one. This saves both pager cost and transmission air time and removes the need for an operator.

Of course, the most sophisticated systems also permit entry of text messages from remote keyboards. Generally, any of these automatic facilities are easily handled by a microprocessor within the system controller. The same processor can also be used to log and time the calls, store the messages for subsequent recall, *et cetera*.

2.3.9.2 Wide-area Call Origination

PSTN Interconnection

Public systems usually interconnect with the PSTN. Many PSTNs use *Dual Tone Multifrequency* (DTMF) tone dialing, which permits continued use of the keypad after the call has been connected to the paging controller. Thus, a common dialed number can be used for all paging calls, and the wanted pager address and any numeric message can be dialed after the paging controller answers the call. Various proposals have been made for entering text messages from numeric keypads, but none of these seems to have achieved widespread use.

For PSTNs which require dc dial pulses, the principle is similar to that used in an on-site system (i.e., a block of phone numbers is reserved for the pager addresses, and any dialed number within this block is routed to the paging controller, which interprets the number and codes it as a pager address). If the PSTN automatically provides the caller's number, the paging controller can interpret this as a numeric message to be attached to the call. For other PSTNs, a few manufacturers make equipment which is advertised to detect and translate dc dialing after the call has been set up to the controller. The pulse detection success rate and caller acceptability of this type of numeric message input is not known, but post dialing pulse detection systems have not achieved great popularity.

Text message input still presents a difficulty. It is possible to connect paging controllers to telex and data networks, or PSTN data via data modems, and so allow those with access to keyboard machines to input text messages directly. (A portable personal computer can be used with an acoustic coupler.) Otherwise, text messages are input via a manual bureau, such as a *telephone answering message service* (TAMS).

Composition of a Wide-area Paging Controller

A wide-area paging controller may contain:

- A file of permitted pager addresses with the services and areas allowed for each pager;
- A dialed number file if dialed numbers bear no simple relationship to the pager addresses, and arrangements to alter the dialed number-address relationship for each number;
- Interconnection to the PSTN (and perhaps other networks);
- Encoding subroutine(s) (it is possible to transmit several different paging codes sequentially);
- Batching arrangements for the calls;
- Transmitter power-up and power-down control;
- Paging code output arrangements;
- For linked-controller systems, an intercommunication package;
- A paging traffic statistics package;
- Automatic changeover of faulty lines, transmitters, *et cetera;*
- A fault diagnostics and reporting program;
- A system configuration package;
- A billing output, sometimes in bill format, but more usually as a log of calls for off-line processing.

2.4 THE USES OF RADIOPAGING

2.4.1 Alert-only Pagers

Originally radiopaging resulted in a simple beep to the user. Even in this form, the benefits to users and their business colleagues, families, and friends are considerable. Thus, a person in the main office of a business alerts a mobile colleague or boss carrying a pager to some change of circumstances. The user finds a telephone and makes a call to ascertain the nature of this change and hence reacts accordingly.

Typically, a paging call can result in alerting a businessperson to some change of circumstance while he or she is out of the office, and the commercial importance of the swift reaction that then becomes possible is obvious. Similar remarks apply in other fields (e.g., fire damage limitation). Less dramatically, repair technicians and truck drivers can be saved abortive journeys, cab drivers may collect extra fares, physicians and surgeons can have greater freedom (e.g., to enjoy a game of golf knowing that they can be alerted to emergencies). A typical pager user receives about one call per working day, and each of these is liable to result in significant savings, earnings, or gain in personal satisfaction.

Single-beep pagers all suffer from the limitation that the user must call back to a single point to ascertain the reason for the call. This makes the pager a natural partner for a TAMS bureau because the user can quote the bureau phone number on his business letterhead. A later development that serves the same purpose is the voice message recording service, which automatically calls the pager as soon as message input is completed. Either service solves the problem of staffing a user's response point. The advent of the multiple-address pager allows more than one source of call origination because the user can distinguish between beep patterns, or allows the degree of urgency of the call to be indicated.

2.4.2 Tone-plus-voice Paging

Regarding the conveyance of messages, at first sight it might seem that tone-(beep)-plus-voice paging offers an ideal system to replace the TAMS or an automatic voice-message recorder. However, the utility of tone-plus-voice is limited because:

(a) the transmission time is limited to about ten seconds per call;

(b) the recipient may not react to the beep quickly enough to fully understand the speech;

(c) the speech quality and volume are often insufficient to make the message clear in noisy surroundings; and

(d) the message is liable to be heard by everyone near to the user, and hence must not be confidential.

On a tone-plus-voice system, a high proportion of messages simply say the equivalent of "call the office," which is no more information than is contained in a single beep. Assuming that a tone-plus-voice paging call occupies ten seconds of air time, up to 350 beep-only calls or 25 average text display calls can be accommodated in the same time, depending on the paging code in use.

2.4.3 Display Paging

Display message paging enables the user to recall the message privately and at his or her convenience as often as desired. Thus, display messaging is entirely suitable for persons attending a meeting or in noisy surroundings, and relieves the user of any need to make a note of the message. There are two types of display message pager, numeric and alphanumeric (hereafter denoted as *text*).

The numeric message pager can receive the numerals 0 to 9 inclusive and six other characters including opening and closing brackets, hyphen, space, and an urgency symbol. Thus, as well as the caller's phone number to ring back, many other types of message are possible. One of the most popular is the "standard" coded message, where the user carries a numbered list of all possible messages and the caller sends only the number of the message. On receipt of the call, the user consults his or her message list. More than one standard message can be included in a call if appropriate.

The chief advantages of the numeric display pager are that:

(1) each numeric digit needs only about 60% of the air time that a text character needs;
(2) the caller's phone number may be given automatically as a message;
(3) alternative message input is simple from any DTMF keypad; and
(4) apart from telephone numbers, the coded message is not easily interpreted by eavesdroppers.

The text display pager has obvious advantages to the user over all other pagers. A business owner can include the message pager number in his or her letterhead. Messages of up to about 80 characters can be sent, but because most text pagers can store six or more messages, a follow-on message can be sent if this limit is insufficient. The problem with text paging is the input of the message, and use of a manual bureau is the most general method. However, machine input via telex, PSTN data transmission, or a packet-switched data network are possibilities. Machine input is cheaper and faster than a manual bureau, and can be entirely automatic. For example, some spare parts arrive at a depot and are booked into the computer. The computer notes that they are urgently wanted in the field and generates an immediate message paging call. British Telecom reports that the text paging service has the highest percentage growth rate of all its paging services.

2.4.4 The Benefits of Paging

Some of the qualitative benefits of paging can be derived from these descriptions of methods of use. These benefits might be summarized as given below.

For the user:

(1) A pager is light and unobtrusive to carry.
(2) The paging calls will be received even deep inside buildings (with a well designed transmitter network).
(3) The user can travel without worry about being unable to react to sudden (good or bad) changes in circumstances.
(4) It allows the user to make best use of time while away from base (office or home).
(5) Pager models are available to suit every possible environment, (i.e., noisy, completely quiet (use vibrator pager or pager memory), inflammable atmospheres).

For the user's organization while the user is away from base:

(1) The user can be redirected if circumstances change.
(2) The user's attention can be attracted if his or her advice or support is needed.
(3) The costs and troubles of frequent phone calls by the user back to base are avoided.
(4) The paging service is generally very cheap compared to the benefits it gives.

The qualitative benefits of radiopaging are well recognized. BT has found that commercial sectors subscribing for the greatest number of pagers are distribution, engineering, business services, and construction [Tridgell, 1987]. The leaders in terms of market penetration are public utilities, engineering, transport, education and justice. In the UK, the areas with greatest percentage growth are business services and government.

People who use pagers are in employment requiring a high level of mobility, so that it is difficult to track their movements. They tend to be on call for more than eight hours per day, and the nature of their work demands urgent attention at or redirection to another site, or urgent contact for decision making. Key occupational groups include company directors, proprietors, managers, engineers, salesmen, service representatives, medical personnel, and drivers.

For many situations, the user only needs relatively short-term possession of a pager (e.g., one may be leased by a visitor for the duration of his stay or by an expectant father for a few weeks before his wife gives birth). It is more difficult to provide a realistic estimate of the quantitative benefits of paging. How do we value the extra goodwill of a client satisfied because of the immediate attention given, or the peace of mind of an expectant father who knows he can be contacted as soon as his wife goes into labor? However, the popularity of the service amply demonstrates that the perceived benefits far outweigh its cost.

2.5 THE STATUS OF PAGING

The initial development of radiopaging relied primarily on a cheap and easily available telephone service because of the need for a called user to phone to ascertain the reason for each call. In most industrial and medical complexes, there is a good private telephone system, so private systems are still being installed or expanded in large numbers. For wide-area paging, a cheap and readily available public telephone network is a necessity. Thus "telephone conscious" countries such as the US and Japan have shown the greatest penetration of wide-area paging. The continental status of paging follows in alphabetical order.

2.5.1 America

The US is the largest radiopaging market. Most wide-area systems were started by entrepreneurs and covered one city or town. However, over the last few years, there has been a marked tendency for these relatively small organizations to be gathered into larger ones. Frequently, there are competing organizations covering the same town, and in some cases (e.g., New York and Chicago), competition with entrepreneurial systems is provided by the local telephone company.

For wide-area paging, two types of pagers are popular, the tone-only and the tone-plus-voice. However, the advent of message display pagers, combined with a shortage of paging channels, is making considerable inroads into the tone-plus-voice market for reasons already explained.

US paging systems are fairly typical of those found anywhere else with a modern PSTN. Call input is via the telephone DTMF keypad to a paging controller. Recorded voice confirms receipt of a tone-only call. Synthesized voice reads back to the caller any numeric message input via the keypad, or the PSTN automatically provides the paging controller with the caller's number so that it can be used as the numeric message. Text message calls are input either via a manual bureau or by some form of data transmission. Transmission to the paging transmitters generally is via UHF radio links where frequencies are available because this is cheaper than hired (leased) lines, and simulcast time equalization requirements are eased.

Regarding the choice of paging code (see Appendix 2C), tone-coded pagers are being replaced by binary digital types using either the Motorola GSC or the CCIR RPC1 code, the latter of which is said by some authorities to be dominating the market growth for both tone-only and message paging. There are estimated to be well over two million wide-area pagers in use in the US (*circa* 1988).

2.5.2 Asia

In Japan, paging also covered separated cities and towns. Digital paging was introduced in 1978 with an NEC code, which only permitted 65,000 addresses and was transmitted at 200 b/s (compare to RPC1 with 8 million addresses and transmission rates of 512 or 1200 b/s). This proved insufficient for many cities, and extra channels were allocated. However, Japan recognizes the spectrum-saving virtues of a higher speed code with much larger address capacity, and RPC1 is now being introduced. There are at least 1.5 million pagers in Japan. Parts of Southeast Asia where paging is very well used are Hong Kong, Singapore and South Korea, with the RPC1 code dominating market growth.

2.5.3 Australia

Both Australia and New Zealand have RPC1 paging systems, the latter being a nationwide system. Of course, in Australia, apart from main towns, the population density is too low to support paging. Some form of direct satellite paging system might be a solution to the low population density problem there and in other countries.

2.5.4 Europe

In Europe, the UK leads the paging field. The BT system is said to be the largest in the world [Tridgell, 1987], with over 400,000 pagers and growing at about 3000 per month. Now only the RPC1 paging code (at both 512 and 1200 b/s) is used. The BT system includes five paging controllers linked by data transmission lines. Apart from BT, there are several other competing systems. For all systems, facilities are similar to those in the US, except that DTMF keypads are not in common use and there is no automatic advice of the calling number. Where DTMF is not available for input of numeric messages, the caller can use a wired or acoustically coupled DTMF keypad.

In France and West Germany, the main paging system is *Eurosignal,* which uses tone-coded amplitude modulation, and was designed *circa* 1971. The maximum calling rate that any transmitter can handle is 0.6 tone-only call/s, and there is no display message facility. Moreover, pagers must be manually or automatically tuned. The total number of such pagers is thought to be about 170,000. However, in West Germany city-wide systems using RPC1 are now being allowed as competition. Neighbors Austria and the Netherlands both use RPC1 in their national systems.

In the Nordic countries, paging is a fast growing service. All except Sweden use only RPC1. Finland has developed an interesting system of simulcast over the whole country (see Appendix 2A). In Sweden, the main system is shared with the FM broadcast radio, being invented and operated by the telecommunication administration. However, a competing, dedicated paging channel RPC1 system is now being constructed.

Regarding international paging in Europe, plans are maturing to provide a common paging channel at 466 MHz to cover significant parts of France, Italy, the UK, and West Germany. Users of this system will be able to roam among these countries. The code will be RPC1, transmitted at 1200 baud. In the UK, the channel will be operated by a consortium of paging interests. A much larger pan-European development, known as *ERMES,* is under study. Studies are being conducted in ETSI (see Appendix B) to define a 16-channel system using frequency-agile pagers at about 169 MHz to permit roaming throughout all participating European countries. Market studies forecast about 13 million users by the end of the century.

2.6 RELATIONSHIPS WITH OTHER MOBILE RADIO SERVICES

At the beginning of this chapter, we noted that many of the technical facets of radiopaging were common to other mobile radio services. For instance, the problems of using this very variable transmission medium to cover a wide area, and laying out multitransmitter systems so as to minimize interference, are considered in several parts of the book. The reader is encouraged to become familiar with these concepts.

There are also strong commercial links between radiopaging and the other services. For instance, consider how a portable cellular phone user might react to an incoming call if he or she were walking down a street in the rain, with the phone in a briefcase in one hand, and an umbrella in the other. Alternatively, the user might be inside a heavy building into which cellular service would not penetrate. In the first case, the user would probably deliberately ignore the call, and in the second case would never be called.

Radiopaging can solve both of these problems, first because it does not demand immediate reaction from the called user, and second because it can be economically engineered (due to the large number of subscribers per channel per transmitter site) to penetrate very heavy buildings. A cellular phone user who carried a pager whenever the cellular phone is switched off could be paged automatically by the cellular system when it failed to detect that the phone was switched on. Thus the user's attention can be attracted to an incoming call (e.g., so that the user can find a drier area or a place near a window before answering the incoming phone call).

Likewise, a radiophone service without significant incoming call capability could use radiopaging to provide the alerting facility. The future "phonepoint" service will be an example of such a situation (see Section 2.7.1).

When it shares the broadcast data channel, radiopaging may also have commercial links with both traffic information broadcasting and a radio data system because both of these services, in principle, can also use that data channel and have it multiplexed with the paging signals. Here the link is in the medium rather than in service complementarity. Radiopaging could have applications in route guidance and vehicle location systems. Typically, these show the position of a vehicle on a projected map in a control room. Route guidance messages could be paged to the drivers.

As we can see, many of these mobile services have synergistic links. Due to its simplicity, relatively low cost, and user attractions, radiopaging can be considered as an enhancement to other mobile systems.

2.7 THE FUTURE OF RADIOPAGING

Even with its current facilities, paging service growth will continue to fulfill market forecasts as high as one pager for each individual in 8% of the working population in developed countries. This will be accelerated as methods and systems which encourage user roaming are gradually introduced (e.g., ERMES; see Appendix 2.E). Apart from this, there are a number of possibilities that can all contribute to further technical and commercial development of radio paging.

2.7.1 Combined Pager and Cordless Telephone

Cordless phones for PBXs and the public "phonepoint" systems are making rapid advances in lightness and ease of use. When idle, cordless phones do not register their presence with a base station, so there can be no forwarding of incoming calls to a user. Moreover, the range of a PBX or phonepoint base station is only a few meters. Thus incoming call attempts to the user will generally fail, and another means of attracting the attention of the user will be necessary (e.g., radiopaging). We could postulate that when these phones become as convenient as a pager to carry, then pager functions could be integrated so that the item becomes a phone-and-display-pager. This might become a replacement for current on-site pagers. A *private advanced radio service* (PARS) [Hudson, 1988] which could incorporate this concept is under development (see Section 3.8.1).

2.7.2 On-air Programming of Pager Addresses

Programming a new address into a pager currently requires a physical change to the pager (e.g., by inserting into it a new programmable read-only memory (PROM) at a maintenance center). User cooperation is required to make the change. A function of pagers now becoming feasible is on-air alteration of the contained addresses by specially secure paging calls. For instance, a user could instantly be reassigned to a new team with its own group call address. This would allow, for example, two teams to be dynamically redeployed into three teams, each having its distinct paging group call.

The display pager could be used to receive textual broadcasting of classified information (e.g., share prices). Users would wish to be alerted only to events relevant to their current situation or activity, and thus the number of classes of information would need to be large, and hence each class probably would be narrow. Because of this narrowness, such a facility might not become popular unless users could very simply change the class of information which they received (e.g., rubber share price movements instead of those of oil). If each information classification has its distinct group address, on-air changing of the class of information received by a pager could be automatically achieved as a result of a request from its user (e.g., from a DTMF keypad). (The RPC1 code allows eight million different addresses, and extension is possible.)

2.7.3 Message Input by Speech

Text display paging for individual messages loses potential because of the difficulty that the ordinary person faces in entering the message into the system. Moreover, the information in the message is what the user wants rather than text. Ongoing technical developments may remove the message input barrier and enable the full potential of message paging to be realized. Traffic volumes could be very large because the service then would appeal to such a wide spectrum, particularly in the domestic sphere (e.g., a wife might request her husband to collect some special food item on the way home from business; one could even imagine a child wearing a pager when out at play in the local park so that a mother could call the child to return home). A mass market could develop, which might outstrip the current maximum forecast of pagers for 8% of the working population.

2.7.3.1 Machine Translation of Speech into Text Messages

Voice recognition and interpretation systems are developing. When reasonably reliable voice interpretation and translation into text becomes technically feasible, the message input barrier will be reduced if paging controllers contain a translator so that a phoned message can be transmitted as text.

2.7.3.2 Coded Speech Messages

The techniques of compressed digital coding of speech are rapidly advancing. One problem is that the more the speech is compressed, the greater is the processing needed to perform the coding and decoding. For a given speed of processing, the delay before the reconstituted speech is delivered increases with the degree of compression. A delay in excess of about one second would be inhibiting in a phone call, but for paging of messages a delay of that length would be unimportant. At the same time, paging codes are increasing in speed, and digital storage and processing are reducing in size while gaining in speed. We can postulate that these factors will all combine to give a digital speech message paging service. The received coded speech would be put into pager memory for subsequent recall in private, and could be replayed as many times as the user wished.

Note that text can be considered as a very compressed form of speech. If a pager could contain a text to synthetic voice translator (perhaps relying on phonetic rather than current highly irregular English spelling), the paging code could be the same, regardless of whether the pager gave text display, voice output, or both.

2.7.4 Acknowledgement of Receipt of a Paging Call

A facility which paging lacks is immediate acknowledgement of message receipt by the pager. The volume of information in an acknowledgement is small, perhaps as little as one bit. Therefore, the bit transmission rate of the acknowledgement direction could be much lower than in the message paging direction. In theory, a few bits require less energy and, at low rate, less radio spectrum than do many bits. As little as 1 W for 1/16th s has been calculated to be sufficient transmitted power for acknowledgement in a wide-area system with acknowledgement receivers co-sited with the paging transmitters. Of course, if the acknowledgement receivers could be spaced much closer, even such a modest transmitting requirement could be greatly reduced.

Acknowledgement is not particularly difficult to arrange in a private system, but there is no great need for the facility in those systems because the users are easy to discipline. In a wide-area system, acknowledgement of receipt has been more difficult to arrange for two reasons:

(a) the required pager transmitter power, bulk, and cost were excessive; and

(b) there was difficulty in passing the acknowledgement back through the PSTN, unless the caller hung on to the originating call. Hanging on is not attractive if batching of calls results in transmission delays of up to two minutes in duration.

There are now several technologies that can contribute to acknowledgement paging:

(1) higher speed paging codes (to reduce the transmitting delays and increase traffic handling capacity to allow all calls to be paged over a whole country);
(2) improved methods of simulcasting over a very wide area or whole nation;
(3) input via data packet-switched networks in which the user would not need to hang on waiting for an acknowledgement; and
(4) improvements in transmitter miniaturization.

Market studies must be done before the developments necessary to combine these technologies into a wide-area acknowledgement display or stored coded speech message paging system can proceed.

2.7.5 Paging in Areas with Low Population Density

Finally, the problem of providing paging service to areas with low population density might be tackled. Direct transmission to rural pagers from communications satellites has been suggested to be technically possible if a spot-beam satellite antenna (preferably remotely steerable so that it can be time-divided between a number of areas), carrying only paging signals, could be used. Even if the technical possibility is proved, the commercial feasibility would need to be studied. Nevertheless, we can argue that the penalties of not being contacted are greater in an area of low population than in one of high population, and hence higher charges are justified.

2.7.6 Concluding Remarks

The developments outlined above are unlikely to exhaust the technical possibilities. There is every reason to expect that every household eventually will have at least one pager. Of course, by that time, the pager may merely be an integral part of a "personal digital radio communicator" with worldwide reception ability, rather than a separate item limited to usage on a single system.

REFERENCES AND BIBLIOGRAPHY

CCIR, *Recommendations and Reports of the CCIR, Vol VIII,* ITU, Geneva, 1982,[1] Recommendation 584. (Easy reading.)

Ibid., ITU, Geneva 1982,[1] Report 499. (Easy and instructive reading.)

Ibid., ITU, Geneva 1982,[1] Report 900. (Easy and instructive reading.)

Hudson, A., "The new private advanced radio service, PARS," *Proc. IEE Mobile Radio Networks Colloquium,* March 1988, IEE Digest No. 1988/36. (Easy reading.)

Jakes, W.C., ed., *Microwave Mobile Communications,* John Wiley and Sons, 1974 (Highly mathematical and rigorous. A standard book for cellular radiophone engineers.)

Mabey, P.J., D.M. Ball, and C. Desmarchelier, "The application of CCIR Radiopaging Code No. 1," *IEEE Vehic. Tech. Conf.,* May 1985.

Tridgell, R.H., "Experience with the CCIR Radiopaging Code No. 1," *ITU Telecommunication Journal,* No. III, Vol. 54, 1987. (Easy reading.)

Tridgell R.H., and D. Denman, "Experience of CCIR Radiopaging Code No. 1 (POCSAG) for message display paging," *IEE Mob. Rad. Conf.,* September 1984. (Contains a fuller account of practical message paging field trials than does [Tridgell, 1987].)

The Book of the CCIR Radiopaging Code No. 1, pub. by Radiopaging Code Standardization Group (RCSG), 1986. (Available from British Telecom Mobile Services and other RCSG members. Easy reading. Contains [Tridgell, 1984] and [Mabey, *et al.* 1985].)

APPENDIX 2.A METHODS FOR AUTOMATIC DELAY EQUALIZATION IN A SIMULCAST SYSTEM

The requirement for simulcast transmission is that the modulating signals be as nearly as possible in synchronism at all transmitters. An easy way to ensure this is to use radio links between the control center and each transmitter, but shortage of radio spectrum may make this impractical.

If lines are used for linking a control center to the transmitters, any rerouting by the line provision authority is liable to disturb the line transmission times and the simulcast setup, and such disturbance may not be detected until complaints by pager users start to accumulate. Further, *packet-switched data networks* (PSDN) are now available in many areas, but they cannot be used for conveying paging information to base stations unless the inherently variable transmission times of a PSDN were taken into account. Thus, *automatic equalization* of the delays in a simulcast system

[1]CCIR (and CCITT) *Recommendations* and *Reports* are updated about every four years, and are usually known by the color of the book jacket. While the edition cited here is the 1982 version (CCIR *Green Book*), the latest should always be used unless history is being researched.

is often desirable. Two methods of automatic equalization are described here; the first is for continuous lines only, and the other can include a PSDN.

Method 1: Continuous Lines

The paging signals traverse a tapped delay equipment before being distributed to each line and transmitter. Given that the line delay is T_L and the tapping delay is T_D, the object is to adjust each tapping so that

$$T_L + T_D = T_K \text{ (constant for the system).}$$

Referring to Figure 2A.1, we can see that the signal for any line is looped back at the base site, and that the returned signal is (frequently or continuously) compared in time with the outgoing signal. The loop delay is divided by two to yield the one-way delay, and the measurement result is then used to select the appropriate delay tapping. Although the time comparator has been shown as individual to each line, because lines do not alter in transmission delay frequently (as compared to the time required for the automatic measurement), a single comparator can be time-shared between all the lines.

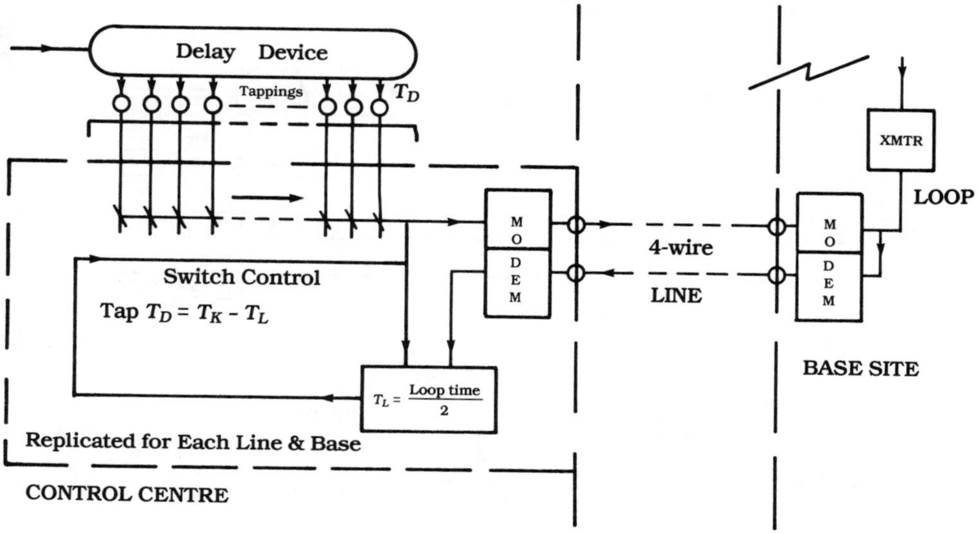

Figure 2A.1 Automatic delay equalization for a simulcast system.

Method 2: On-air Synchronization

The method assumes that, apart from a master transmitter, transmission occurs in bursts (i.e., up to several tens of seconds per burst). Each slave transmitter is switched off between bursts, and is equipped with a receiver that can receive the master transmission only while its own transmitter is off. The start of transmission of each slave is derived from the master by this reception. Transmission of the remainder of the burst is then under control of an accurate local clock at each transmitter.

Figure 2A.2 shows a simulcast area with a master and three slave transmitters. One slave is fed by a physical wire, another by a carrier system, and the third by a packet-switched network. Figure 2A.3 shows that, although the line transmission

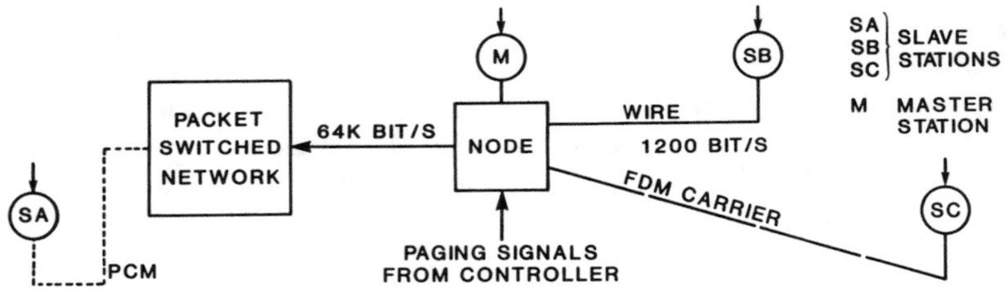

Figure 2A.2 General unequalized transmission system.

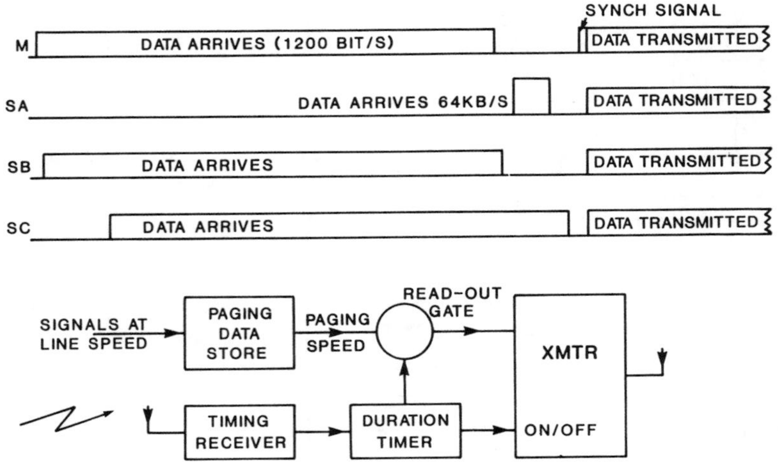

Figure 2A.3 Radio synchronization timing.

rates differ, some paging data in all cases arrives at its transmitter before transmission begins and is stored to await the start of transmission. The master precedes its paging transmission with a broadcast synchronization signal, the end of which the slaves interpret as the moment to begin their own transmissions.

It is possible for one or more of the slaves to act as a submaster and pass on another synchronization signal to further slave transmitters out of reception range of the original master, *et cetera*. In Finland, the whole national radiopaging system is coordinated into a single simulcast area by this method. Because timing is passed along the transmitter chain, there is a significant time difference between transmitters in Helsinki and those in the far north, but because these are too far apart to interfere, there is no adverse effect.

As we note that interference should not occur over more than about 0.25 of a direct binary FSK signal element, if the frequency accuracy of each local clock (including the master) is $\pm A$ to allow for both master and slave inaccuracies, then

$$A < 0.25/2T_B$$

where T_B is the total bits per transmission burst.

For a rate of 1200 b/s and a transmission burst of 30 s, for example, the accuracy required of the transmitter site clocks is about ± 3.5 parts/million.

APPENDIX 2.B
MULTIPLEXING STEREOPHONIC BROADCASTING AND PAGING

Figure 2B.1 shows the standard FM broadcast stereophonic baseband signals with an added data channel (see Chapter 5). The data signals have a rate of 1187.5 bauds, this being time-locked 1:48 to a 57 kHz subcarrier frequency. This subcarrier frequency is phase-locked to the third harmonic of the 19 kHz stereophonic pilot tone. The data channel is recommended by the CCIR. Data is conveyed in 26-bit blocks (16 information bits and 10 redundancy bits), which permit correction of any one error burst up to 5 bits long. One class of data catered for is radiopaging.

A pager for this multiplexed band sweeps continuously across the whole FM

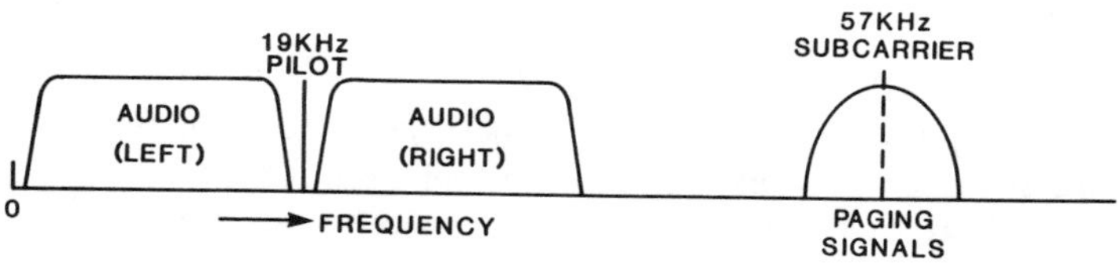

Figure 2B.1 Multiplexed paging and broadcast baseband signals.

sound broadcast band until it finds and locks onto a good channel carrying data and paging signals. The whole FM signal is demodulated, and then the audio channels and pilot tone are filtered to leave only the data signals. These are hence demodulated to recover the 1187.5-baud data.

The information coding can carry several types of data (e.g., road and traffic condition information; see Chapter 5) as well as paging. The unwanted information must be ignored.

APPENDIX 2.C
CODING FOR RADIOPAGING

2C.1 Requirements of a Paging Code

The minimum requirements of a modern radiopaging code, with comments, are as follows:

(a) adequate address capacity (especially in an expanding market);

(b) transmission speed adequate for the intended traffic;

(c) high protection against false calling;

(d) suitability for sequential and simulcast transmission;

(e) inherent methods of battery economy. (Nearly all paging transmissions start with a preamble of fairly long duration. Thus, the front end of the pager can be switched off most of the time, but, at intervals less than the preamble duration, the pager can be switched on briefly to sample the channel. The pager stays on if a signal is detected during one of these samples);

(f) low-cost decoding implementation within a pager (see (i));

(g) adequate sensitivity (not highly important because pager antenna and front end are the dominating factors in pager sensitivity);

(h) some protection against multipath propagation effects, but all codes will still suffer some effects (see below);

(i) acceptable to more than one manufacturer (for wide-area codes);

(j) numeric and text message capability (essential);

(k) adequate protection against false message generation (i.e., less than one false message per year for any normal user [CCIR Report 499; Section 1.3]). (Note that false messages can often be rejected by the user); and

(l) compatibility with other codes (important when a code is introduced to an already working system).

2C.2 SEQUENTIAL TONE CODING

A number of audio tones are transmitted sequentially without a break between the tones. The first tone of each call is the longest and may form the preamble. Each subsequent tone differs from the immediately preceding tone. These differences are easy to detect by audio filters and obviate the need for strict timing, but the detection cost is greater than that for a binary code.

For example, the decimal five-tone system has eleven different tones, one for each of the ten digits and a "repeated adjacent digit" (*R*) tone. Thus, the paging address 12233 would be represented by tones 1, 2, *R*, 3, *R*. In its most densely packed form, the first tone might be of 100–130 ms duration, and the other four tones might be 30–40 ms each, so that the calling rate would be about 5/s.

Tone coding offers some protection against multipath effects because of the comparatively long duration of each symbol. However, tone coding has almost no redundancy and is somewhat prone to false calling, especially in the face of co-channel or intermodulation interference. The five-tone system can be extended up to sixteen different tones, but it is unsuitable for text message transmission.

2C.3 BINARY CODING

2C.3.1 General Description

The addresses and messages are nearly always transmitted in *N*-bit codewords. Some codes use one codeword per address and others use two. One code alters both bit rate and codeword size when proceeding from addresses to messages.

Each codeword includes its own fixed number of redundancy bits, usually according to a *cyclic code* [Peterson and Weldon, 1972], which permits simple decoding with a good degree of error detection. For a given quantity of redundancy (as fixed by the code), usually some is used in the pager for error correction, but the amount then remaining for error detection is thereby reduced, and thus the protection against false decoding is weakened. The amount of redundancy needed to achieve error correction rises very rapidly with the amount of correction chosen. Thus, code selection is a compromise between traffic throughput and redundancy.

The *cyclic block code* [Peterson, 1972] treats the information as a binary number followed by the same number of zeros as wanted redundancy bits. This is modulo-2 divided by a generator number (i.e., binary division without carrying because of simplicity of implementation in digital circuits), and the remainder from this division becomes the number of redundancy bits so that the codeword is an exact multiple of the generator. To detect received errors, the received information and redundancy bits are divided by the generator number. If the resultant remainder is not zero, errors have occurred.

"Hard" or "soft" decision decoding is possible. In hard decoding, a firm decision is made on each bit as it is received, and all the analog information in the signal is then discarded. For soft coding, at least some of the analog information in the signal is preserved as a digital *signal quality statement,* usually one such statement for every received bit [Mabey, 1985]. See [Peterson and Weldon, 1972] for error correction algorithms.

2C.3.2 The Radio Paging Code No. 1

The RPC1 is summarized as an example of a binary code. Comments in parentheses indicate how pager and system design is affected.

The code has a binary format with normal transmission rates of 512 or 1200 b/s using direct FSK modulation. (This bit format permits a synchronization tolerance between neighboring transmitters of 488 or 208 μs for the respective rates. The signaling rates are different from those of proprietary codes, and this permits pager battery saving in systems that transmit a mixture of codes. Direct FSK permits the whole front end to be implemented in a single integrated circuit.)

Each transmission starts with a preamble of at least 576 bits of alternate 0 and 1. (The preamble permits both battery saving and fast bit synchronization.)

The preamble is followed by a string of 32-bit codewords, each with 21 information bits and 11 redundancy bits. For RPC1 codewords, at least the following error protection algorithms are possible:

(a) Using "hard" decision decoding, detect any 5 random errors or any 1 error burst up to length 11, correct any 1 error and detect up to 4 errors or 1 error burst up to length 7, correct any 2 errors and detect up to 3 errors, or correct any 1 error burst up to length 4.

(b) Using "soft" decision decoding, correct any 1 burst up to length 11 including poor quality bits, correct any 5 poor quality bits, or select either according to the quality pattern. (Note that there is no standard method of decoding, this being left to the ingenuity of the pager designer; any good standard leaves as much as possible as designer's options.)

The codewords are transmitted in contiguous batches, each starting with a unique synchronization codeword and followed by 16 information words. (Each batch thus is only 1.0625 or 0.453 s duration according to bit rate. The transmission can then be stopped. Thus, time-slotted multitransmitter systems can be engineered very conveniently. Priority calling within about 1 s can be included in nontime-slotted systems.)

The first bit of each information word is a flag to indicate whether the word is an address or message codeword. (Twenty bits are available for information.)

The last two information bits of each address codeword convey the call function.

(The first 18 bits form part of the pager identity, and each pager thus has four addresses. Typically, these are indicated by four different beep patterns.)

For the transmission of pager addresses, a batch is partitioned into 8 frames of 2 codewords each. (The frame number is part of the address. This multiplies the address capacity of the code by 8 to give over 2 million identities and 8 million addresses, permits further battery saving of about 82%, and reduces the false calling rate by a factor of 8:1. Further methods of battery saving with this code are also known. Batteries may have a very long life.)

Any message immediately follows the pager address. Two types of message format are standardized, numeric and text. (Any potential paging service provider can start with the simplest alert-only service and later may add numeric and text message services, knowing that air time will not thereby be wasted.)

The numeric format packs 5 digits into each message codeword. Each digit can be any of the decimals or space, or hyphen, or an urgency symbol, or a bracket, or a spare character.

The text format utilizes the full CCITT Alphabet No. 5. (Also known as ASCII or ISO 7-bit. This may be very important in the future for linking pagers to "laptop" computers.)

A unique "idle" address word is specified to fill in any part of a batch containing no information. (The synchronization and idle codewords differ by 16 bits, but have a very simple relationship, which can reduce pager implementation costs.)

A *traffic simulation,* [Book of RPC1, 1986, p. 5], assumed 0.1 calls per pager and a mixed system with 50% alert-only calls, 37.5% numeric message calls (10-character average), and 12.5% text message calls (30-character average as found on the BT system). Assuming that 2% of the population had pagers, the results indicated that, for air-time fills of 80 or 90%, a 1200 b/s nontime-slotted system could accommodate 333,000 or 372,000 subscribers, and the system would be appropriate for populations of over 16 or 18 million, respectively. Even for 90% air-time fill, the mean delay between call input and transmission could be as low as 3.6 s.

2C.3.3 The Golay Sequential Code

This description is sufficient only to illustrate some of the differences and similarities between this code and the RPC1. A manufacturer's full description of this code can be found in [CCIR Report 900, 1986].

2C.3.3.1 Addressing

Addressing occurs at 300 b/s using two codewords of 23 bits each (the *Golay sequential code* (GSC) [Peterson and Weldon, 1972]), with 12 information and 11 redundancy bits per word. The first address word is limited to about 1000 combinations so that the total address capacity is approximately 400,000. (A method of increasing the address capacity to 4 million has been described, but a full transmission cycle including messages might then take about 48 s. This could create difficulties in time-slotted paging systems, especially if other codes were also to be transmitted. RPC1 normally can be stopped in about 1 s.)

Each address is introduced by at least 14 half-bits of alternating sense, and the two codewords are separated by another half-bit. This arrangement makes it difficult to confuse the introduction with the first address word, and the first with the second address word. (Synchronization with an RPC1 signal is a little more complicated, but appears to be just as secure. Any cost differential due to this complication is insignificant in digital integrated circuits. In a given transmission time, RPC1 used at 512 and 1200 bit/s could convey 3 or 7.1 addresses, respectively, for every GSC address.)

2C.3.3.2 Message Transmission

For message transmission, the GSC code changes to 600 b/s. The codewords are 15 bits long, with 7 information and 8 redundancy bits per codeword. Two errors per codeword can be corrected using simple hard decision decoding. Eight bit-interleaved message words compose a message block (i.e., 56 information and 64 redundancy bits), which allows correction of any error burst up to 16 bits long (26.7 ms duration). (The GSC code thus is more resistant to multipath fading than RPC1, but the CCIR noted that much user activity is at very low speeds or stationary, and was not convinced that the extra resistance is of high value.)

However, because of the high probability of false corrections, 7 information bits are devoted to a block check-sum. One further information bit indicates whether another message block follows. Thus, a message block contains 48 user information bits and 72 redundancy and control bits. For text transmission, a 6-bit special version of the ASCII code is used. (Compare this to 20 user information and 12 redundancy and control bits per codeword in RPC1. Including synchronization, for any given bit rate, RPC1 conveys 56% more text message information. The usual RPC1 text message uses the normal ASCII code.)

REFERENCES AND BIBLIOGRAPHY (APP. 2.C)

CCIR Recommendations and Reports, Vol. VIII, Report 900, ITU, Geneva, 1986.

Mabey, P.J., "Mobile digital transmission with soft decision decoding," *Proc. IEEE Vehic. Tech. Conf.*, May 1985. (Moderately easy but specialized reading.)

Peterson, W.W., and E.J. Weldon, *Error-Correcting Codes,* MIT Press, 1972. (Very rigorous mathematics, although modulo-2 arithmetic is much simpler than decimal arithmetic as taught in school.)

The book of the CCIR Radiopaging Code No. 1, pub. by Radiopaging Code Standardization Group (RCSG), p. 5, 1986. (See Chapter 2 References and Bibliography.)

APPENDIX 2.D
AN ANALOG SINGLE-CHIP FRONT-END FOR A PAGER

This *homodyne* front-end is effectively a *superheterodyne* (superhet) receiver with a local reference (crystal) frequency, f_r, nominally equal to the received carrier frequency. Thus, the received FSK signal, f_s, is either higher or lower in frequency than this local reference.

Referring to Figure 2.11, we can see that the received signal is processed in two paths, A and B, which are essentially similar. The signal is first fed to multipliers (modulators). The output components of each multiplication are the sum $(f_s + f_r)$ and difference $(f_s - f_r)$ between the received and reference frequencies. The sum frequency, being on the order of 300 MHz, for example, is removed by the low-pass filters. The (positive or negative) difference frequency is in the audio band, and proceeds along the signal paths. (Frequency can be negative, in which case the phase "advances backwards." The effect is quite real, and it is important in this receiver.)

The distinction between the two paths is that the reference frequency of path B lags 90° behind that of path A. This phase difference is retained in the audio-frequency component of these multiplications.

Differentiation has the effect of shifting phase forward by 90°. Thus, differentiating the phase-lagging signal has the effect of bringing it into phase with the undifferentiated phase-leading signal. Differentiating the leading signal brings it into antiphase with the undifferentiated lagging signal. Because of these shifts, the output components of each second stage of multiplication are a (wanted) data signal and an (unwanted) ac wave with a frequency of $2(f_s - f_r)$. Comparing the two paths, the former components are of opposite sign and the latter components are in phase. Consequently, by subtracting one multiplier output from the other, the wanted data signals add and the unwanted ac ones cancel.

The sign of the data output changes according to whether the received signal is above or below the local reference frequency. Thus, the resultant data output represents the signed difference between the received signal frequency, f_s, and that of the local crystal reference, f_r, and the data hence is recovered. The two paths need to be equal in performance if the circuit operation is to be successful, but this can be achieved by careful processing in the manufacture of the integrated circuits.

APPENDIX 2.E
THE EUROPEAN RADIO MESSAGE SYSTEM (ERMES)

2E.1 INTRODUCTION

In 1987, CEPT conceived the idea of instituting a paging and messaging system to cover Western Europe. A study group, called "RES4," was set up to produce the specifications. Later RES4 was incorporated into ETSI and renamed "PS"; it is charged with completing the specification work. Currently, the system is referred to as the *European Radio Message System* or ERMES, but the latter name may not be acceptable in all countries.

At the time of writing (1989), the services and facilities of ERMES have been fully defined, but much work remains to be done on the definitions and specifications of the radio and networking aspects. The action plan is to complete these definitions within ETSI-PS, and for an ETSI project team then to write the specifications. The target date for opening of service is now January 1993.

2E.2 DIMENSIONS OF ERMES

The dimensions of ERMES were derived by considering the forecast number of paging users, the expected calling rates and message volumes, and the number of common radio channels that can be made available throughout Europe. An additional complication is that each country will provide its own system (or systems) on its own radio channels. Studies by CEPT members and marketing consultants have been combined to provide the following dimensions:

(a) The number of users is expected to be about 13 million by the end of this century, and may double before the market is saturated. (*Note:* The total population of Western Europe is about 350 million.)

(b) The area with the greatest traffic density will be where France, West Germany, and the Benelux countries share borders. In that area, at peak hour, the net user traffic could be 20 kb/s, 25 kb/s, and 15 kb/s, respectively.

(c) A block of eight 25 kHz channels at 169 MHz is designated for 1992, to be followed later by another similar block (i.e., 16 channels total). Of course, not all of these channels will be available at the border in any one country.

(d) From (b) and (c), the peak net user bit rate per channel should be 3.75 kb/s, although a rate as low as 3.0 kb/s might be acceptable. Allowing for error checking bits and system overheads, the eventual channel transmission rate should be in the range of 5.0 to 6.5 kb/s.

(e) Up to 5% of users will ask for roaming. Not all of these will have their roaming facility simultaneously activated.

2E.3 ERMES USER AND CALLER FACILITIES

2E.3.1 Roaming

Any user will be able to specify the areas in which it wants its calls to be transmitted for a defined period, and the pager there will adjust automatically to this. Additionally, should the user so wish, the caller will be able to nominate the desired area in which the call is transmitted. As a result, the pager must be able to select the appropriate channel (or channels) wherever the user roams.

To avoid possible loss of calls while the user is in a coverage overlap situation, any call transmitted in an area adjacent to the user's home area will also be transmitted in the home area. If there is more than one network covering the chosen roaming area, the roaming network used will be chosen by the home area's network operator.

2E.3.2 Types of Paging Call

Ascending categories of calls are tone-only, numeric message (maximum 20 digits), and text message (maximum 400 ASCII characters). The highest category that can be received by a pager is a designer's option, but then the pager must also be able to receive every lower category.

The remaining category of call includes a transparent data message (maximum 4000 user bits). Because there is no standard format, the pagers for this category will be special. The category could be important for dealing with paging developments (e.g., highly compressed digital speech).

2E.3.3 Call Controls

Generally, networks have the option of which call controls they offer. Authentication and legitimation will be used to ensure that call control does not fall into unauthorized hands.

There are three call priorities: Priority 1, which is transmitted within the home area within one minute; Priority 2, with mean busy hour transmission delay of less than two minutes; and Priority 3, which has no defined transmission delay. Also,

- With user permission, a network may provide a message user directory.
- The caller may be informed of the call charge.
- The caller will receive an acknowledgement of each call made.
- The caller may be informed of the times when his or her calls were transmitted.
- Users may bar calls for a specified time.
- Both caller and user may request that transmission of calls be deferred until a specified time.
- A user may divert his or her calls to another pager.
- Calls may be confined to members of a closed user group.
- Generally, the caller will pay the call charge, but a user may opt to pay instead.
- A call may be repeated automatically after five minutes.
- The network will number every message call, and pagers may indicate if any

message number is missed; networks may store the messages for up to 24 hours for retrieval by the user, or may offer to transmit the latest message number.

- A message encryption service may be offered by using special pagers.
- Calls may be to individuals, to previously specified groups, or multiaddressed.
- A call may be transmitted with an "urgent attention" indication.
- A network may hold a list of standard messages to which the caller can refer by number.

2E.3.4 Pager Facilities

Each pager will have capacity for at least eight addresses (or *radio identity codes,* RICs) and corresponding alerting signals.

A "silent mode" can be activated by the user, but will be overridden by any urgent message. In the silent mode, a tone-only pager will be able to store a call on each of its activated RICs.

Normal message pagers will store at least ten messages of each type that they can receive, and will also indicate if a message is repeated. The storage properties and capacity of transparent data pagers are not defined.

All pagers will indicate the presence of stored calls, and also whether the call memory is full. If the message memory is full, the earliest message will be discarded in favor of an incoming one.

Additionally, pagers may offer various message manipulating and reading facilities, an "out of range" indication, a lost message indication, date and time of call indication, and remote programming of RICs.

2E.4 SOME TECHNICAL DETAILS

The incomplete status of the scheme has already been mentioned. The following details can be given.

2E.4.1 Channeling

An early decision in RES4 was that all pagers would have the same roaming capability. Consequently, all pagers will provide multichannel reception.

Two schemes were considered:

(a) Local channels would be provided as necessary for local calls. Additionally, a single, time-shared channel would be provided throughout Europe for transmission of roaming calls. Pagers would have two channels only (i.e., their "home" local channel and the common channel). This simple arrangement required only an extra crystal and crystal selector in the pager.

However, commercial and political difficulties were said to make operation of the common channel a dubious proposition and the scheme was abandoned.

(b) The decision was made that all pagers should be capable of selecting any one of 16 contiguous channels. A synchronized time cycle of about 7.5 s, consisting of 16 equal periods, will be established throughout ERMES. Pagers are divided into 16 groups, and calls for any one group start only in their time slot on that pertinent channel.

Pagers situated within range of their home area will select the home channel. Pagers not within range of their home area will, in turn, select each channel for the duration of their group period of that channel. The scheme requires inclusion of a frequency synthesizer in each pager to perform the channel selection, which will significantly increase pager costs, at least initially.

2E.4.2 Modulation

Several types of modulation have been considered. By taking into account the need to avoid adjacent channel interference and to be able to use quasisynchronous transmission, a transmission rate of about 6 kb/s was determined to be too high for any two-level type of modulation. Eventually, four-level FSK was chosen.

2E.4.3 Quasisynchronous Timing

Simultaneous transmissions from adjacent cochannel transmitters will be synchronized to within 10 μs. Transmitters will be sited so that, at a pager, the timing difference between received signals of about equal power from these adjacent transmitters will not exceed 50 μs.

2E.5 CONCLUSIONS

The ERMES scheme is a very ambitious one. Whether these ambitions prove to be commercially viable remains to be seen. However, there is no doubt that the volume of paging and messaging data will increase greatly over present levels. Because of the limited radio spectrum available, schemes with high transmission rates will certainly be needed to meet the requirement.

Chapter 3
Cellular Radio

J. Walker and B.R. Gardner
Racal Research Ltd.

3.1 INTRODUCTION

In this chapter, we deal primarily with *cellular radio,* including the current analog systems and a "second-generation" digital system; for comparison, we will also mention other types of mobile radio systems.

The telephone has long been important in modern living, but its use has been constrained by the connecting wire. The advent of mobile radio telephony, and particularly cellular radio, has removed this restriction and produced explosive growth in mobile communication throughout the western world. See Table 11.2 and 11.3 (Chapter 11) for details of market sizes, penetration statistics, and numbers of users. The cellular telephone is now an essential element in business, saving time and money in many situations: the salesman ensuring that his next customer is actually in his or her office; the manager redeploying his service engineers to cope with a new crisis; the change of plans caused by an unexpected traffic jam. (Whether traffic jams will continue to be quite so unexpected in the future is the subject of Chapter 5.)

The cellular radio user in a car (or on the train or in the street) picks up a handset, dials a number, and immediately can talk to the person called. Behind this simple act is an enormous array of technology. Figure 3.1 gives an idea of the complexity of the cellular radio system structure, and Figures 3.7 and 3.12 show the complexity of the transceiver units. We show in this chapter how the needs of the subscriber are met by the cellular radio engineer working within economic constraints and the laws of physics and radio propagation.

3.1.1 Brief History

Until the advent of cellular radio systems, a car telephone was virtually impossible to acquire; the reason was that the limited number of communication channels al-

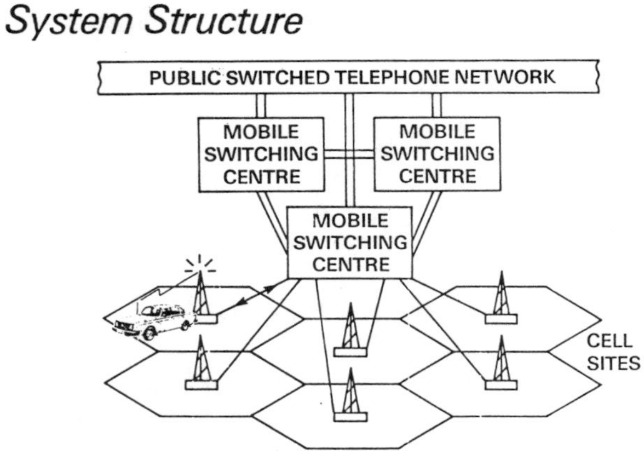

Figure 3.1 Cellular system structure: Base stations at the center of cells are connected by landlines to telephone exchanges ("mobile switching centers") owned by the cellular operator, which are themselves connected to each other and to the PSTN.

located to radiotelephones would support only a few simultaneous conversations, and therefore only a small number of subscribers. Hence, radiophones were a most desirable accessory. Consequently, ways were sought to remove this bottleneck. According to Young [1979]: "The cellular concept and the realization that small cells with spectrum reuse could increase traffic capacity substantially seem to have materialized from nowhere, although both were verbalized in 1947 by D.H. Ring of Bell Laboratories in unpublished work."

Trials in the US, first of all in experiments at Bell Laboratories [DiPiazza *et al.*, 1979; Huff, 1979], demonstrated that the cellular principle would work, and cellular radio entered the public domain. A detailed history of mobile and cellular radio development in the US will be found in Calhoun [1988]. A commercial service (NAMTS) opened in Japan in 1978 [Makitalo, 1978], and in the Nordic countries (the Nordic Mobile Telephone service (NMT)) in 1981 [Bergqvist 1989]. Norway, Sweden, Denmark, and Finland still have the highest "penetration" (number of cellphones per capita—see Table 11.2). The *Advanced Mobile Phone Service,* or AMPS [Young, 1979; Macdonald, 1979], began commercial operation in the US in 1983. AMPS is not a nationwide system; licenses have been granted on a city-by-city basis. A very significant step forward was the introduction of the *Total Access Communications System* (TACS, based on AMPS) in the UK, with Racal Vodafone and Cellnet as competing operators of rival, but technically compatible, systems. The common technical standards were devised by joint committees of British Telecom, Racal, and Cellnet personnel.

So far, an internationally agreed cellular radio system has not emerged. Several broadly similar but incompatible systems have been developed, with some countries opting for one and some for another (see Table 3.3). Furthermore, where different countries have adopted the same basic system, they may have chosen different frequency bands. This means that, in general, the user is unable to take a cellular phone abroad and still use it (the Nordic countries being an exception in having all adopted a compatible system and the same frequencies). In Europe, this has been regarded as a major disadvantage and will increasingly become so in 1992, when the trade barriers are due to be dismantled and Europe will become one large free market. CEPT (the committee of European PTTs) has therefore taken the opportunity to specify a second-generation digital system that will be used throughout the 16 CEPT member countries. The specification has been generated by cooperation between all participants, coordinated by CEPT's *Groupe Special Mobile* (GSM) permanent nucleus located in Paris. The system is known as the pan-European digital cellular radio system (or, more frequently, the GSM system), and is described in Section 3.6.

3.2 BASIC PRINCIPLES OF CELLULAR RADIO

3.2.1 Short Range Transmitters, Cells and Frequency Reuse

Researchers at AT&T's Bell Laboratories realized that if relatively low-power transmitters were used so that the signals propagated only over short distances, the same radio frequencies could also be used again with minimum interference in nearby areas, yielding *frequency reuse*. This is the first key feature of cellular radio—the division of an area into small cells, each served by a low-power transmitter, with frequencies being reused nearby—allowing a much higher number of mobile subscribers (Figure 3.2).

To minimize interference among users, there must be physical separation between cells that use the same frequencies. This is achieved by forming groups of cells into clusters which between them can use all the available radio channels. A typical cluster size is seven when omnidirectional antennas are used at the base station. For convenience, imagine that the cells are hexagonal in shape, and fit neatly onto a two-dimensional map, as is shown in Figure 3.3, where the seven cells of each cluster are lettered A to G. Set A is allocated to the central cell of the cluster, and sets B to G are for the surrounding cells. The nearest cell that uses the same set of frequencies is spaced at a distance of $4.6R$, where R is the radius of a cell [Young, 1979]. This spacing is called the *mean reuse distance*. By repeating these seven-cell clusters, a conurbation or a whole country can be covered (Figure 3.2).

Although a seven-cell repeat pattern has been described, other cluster sizes are possible, such as four or twelve [MacDonald 1979; Appleby and Garrett, 1985]. In reality, the cells are not hexagonal, but have an irregular shape, determined by factors such as the propagation of the radio waves over the terrain, obstacles, (e.g., tall

Figure 3.2 Cellular radio base stations in southeast England (schematic): The diagram, taken from the early days of cellular radio in the UK, shows how the system operator (Racal Vodafone) develops a "cell-plan" to cover the country, including a greater density of smaller cells in urban areas. Two levels of cell-splitting are shown, the smallest cells being "corner-excited."

buildings), and the constraints on the siting of base stations imposed by geography. Complex computer programs are used to predict the effects of propagation and thus can choose the best sites to achieve optimum coverage.

There are also some rather subtle features related to allocation of channels to cells. First, the more channels a cell has, the more subscribers it can support; hence, a cell in an urban area may be allocated more channels than a nearby cell serving a rural area. Second, the relationship between the number of channels and number of subscribers supported is not linear; doubling the number of channels will more than double the potential number of subscribers, at least for the number of channels per cell that we find in cellular radio systems.[1]

Finally, we point out that "line-of-sight" communication does not usually apply in cellular. To receive only a line-of-sight signal is, in fact, unusual. More often,

---

[1]The number of subscribers that can be supported by a particular number of communications channels is the subject of telecommunication traffic engineering. The theory is beyond the scope of this book, but is covered in detail in Bear [1980] and in Siemens [1981]. However, some representative figures can be quoted. At 2% blocking probability (assuming the cellular system is designed so that 98% of call attempts are successful), increasing the number of channels from 20 to 40 increases the telephone traffic capacity from 13 to 31 erlangs (an *erlang* being the unit in which telephone traffic is measured). We also point out that there is not a fixed maximum number of subscribers. The quality of service decreases significantly as the number of subscribers in a cell increases beyond a certain limit.

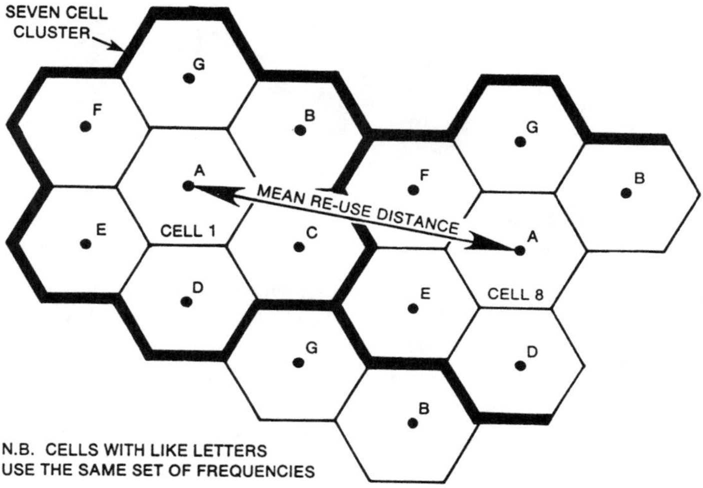

Figure 3.3 Cells based on a hexagonal grid: Base stations are at the center of cells. The radio frequencies allocated to the system operator are divided into seven sets, labeled A to G. The frequencies used in cell 1 (set A) can be "reused" in cell 8, which is sufficiently far away for the interference to be minimized. The separation is known as the "mean reuse distance," and is $4.6R$, where R is the radius of the hexagon. Other "repeat patterns" using different cluster sizes (4-cell clusters, 12-cell clusters) are possible.

the received signal is a combination of direct signals, diffracted signals, signals reflected from buildings, geographical features, and other vehicles; this phenomenon of multipath reception is covered in more detail in Section 3.3.5.

3.2.2 "Handover" Between Cells

When the user's transceiver is switched on and within range of a base station, a link can be established via the base station to a cellular radio telephone exchange, usually called a *mobile switching center* (MSC), and then into the PSTN. For calls between cellular subscribers, however, to connect via the PSTN is not necessary.

As the mobile user approaches the cell boundary, the signal strength falls. This is detected by the cellular system, which then switches the radio link to a base station in an adjacent cell; this process is called *handoff* or *handover* (Figure 3.4). It happens automatically under system control; the user notices nothing, or, possibly, a slight break in the conversation, lasting only a third of a second. Handover is the second key feature of cellular radio (see Table 3.1) and is covered in more detail in Section 3.3.2.

Figure 3.4 Hand-off: As the mobile subscriber moves from cell 1 to cell 2, his channel frequencies will be automatically changed from the set *fx* to the set *fy*. This is known as "handoff" or "handover" and is under system control.

Table 3.1
The Key Features of Cellular Radio

1. Frequency reuse with a "cellular" arrangement of low-power, limited range transmitters.
2. "Handover" of the call from one cell to another as the vehicle moves.
3. Cell-splitting to increase capacity and number of subscribers.

3.2.3 Cell-Splitting to Increase System Capacity

When the number of subscribers increases and approaches the maximum that can be served by a cell, the cells are split into smaller cells, each one having in principle the same number of channels as the original large cell (Figures 3.3 and 3.5). Each cell, therefore, is able to support the same number of subscribers as the original large cell (although there is some loss of capacity in the cells not split). Also necessary is to reduce the power output of the base station transmitters to minimize co-channel interference (i.e., interference between users in nearby cells employing the same frequency). By this process of cell splitting, the potential number of subscribers can be increased without the need for extra bandwidth. Of course, more base stations will be required, but the cost of the additional infrastructure is offset by the increase in subscriber revenue.

In busy urban areas where cell sizes are small, co-channel interference can be a serious problem. This can be minimized by using directional antennas at the base station. Typically, three or six antennas are used, each covering an angle of 120° or 60°. Assuming three antennas, each base station effectively serves three cells and the cluster size is increased from seven to 21. The use of directional antennas reduces

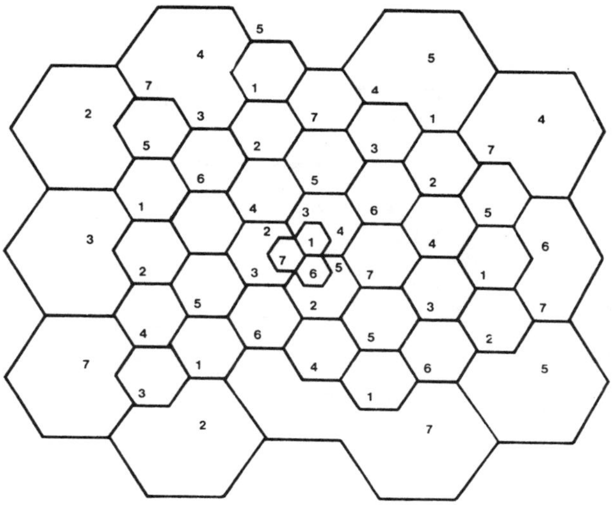

Figure 3.5 Two levels of cell-splitting: The capacity of a cellular system can be increased by "cell-splitting," each smaller cell being allocated the same number of channels as a larger cell and therefore being able to support the same number of subscribers. As the smaller cells become "full," they too can be split, although there are economic and practical factors which limit this process.

infrastructure cost because fewer base stations are required; this, in turn, eases the difficulty of acquiring suitable sites for the base stations. This pattern is shown in Figure 3.6.

Once these smaller cells have as many subscribers as they can support, further cell splitting can take place, as shown in Figures 3.2 and 3.5. Thus, the third key feature of cellular radio is *cell splitting*. Usually, the cell radius is halved and new base stations are set up mid-way between the original base stations [MacDonald, 1979]. Thus, one cell is split into four smaller cells. (The new cells overlap into adjacent larger cells, rather than following exactly all of the old cell boundaries, as shown in Figure 3.5.) There are economic and practical constraints to the splitting process; system cost increases as cell size decreases. In practice, the smallest cells have sizes of about 2 km diameter in urban areas; rural cells are typically 15 km in radius. Also, a difficult engineering task is to have cells abut or overlap so that there are no gaps in the coverage area, but with minimal interference.

3.3 CELLULAR ENGINEERING

Having covered the basic principles of cellular radio systems in a rather simplistic way, we will now give more detail about how cellular systems are engineered. (This

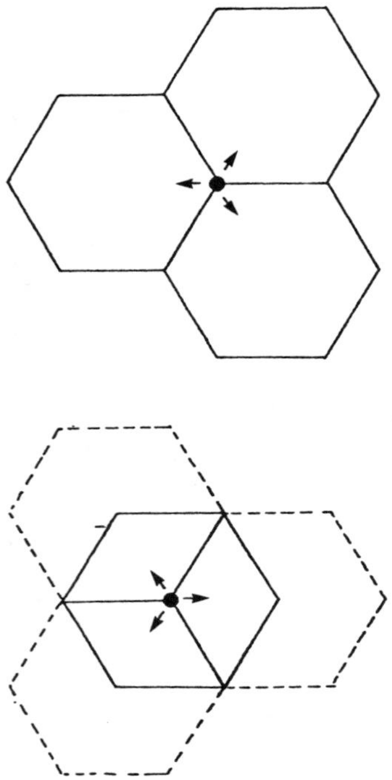

Figure 3.6 Area coverage by 120° directional antennas: A single base station can serve three cells rather than one if directional antennas are used.

discussion is based on TACS and AMPS; other systems may differ in detail.) The basics of noise, interference, and multipath have been covered in Chapter 2 and will not be repeated here. However, there are some techniques which can be used to overcome these effects, and some consequences of them that require explanation.

3.3.1 Signaling

Cellular communication channels are *full-duplex;* each call uses two different frequencies, one for communication from base to mobile, the other from mobile to base (see Table 3.2). Some channels, the *control channels,* are dedicated to signaling and control, such as for call setup. Each base station is allocated one control channel; there are 21 control channels in total. The remaining channels are used for voice transmission.

Each base station radiates a signal continuously on a control channel. When mobile equipment is switched on, it searches across the channels for the control channel with the greatest signal strength. When this is found, the mobile will "register" with the system via this channel (usually, but not always, the closest base station). This operation consists of transmitting its unique identity to the base station, which in turn passes it on to the network switch (MSC). The network now knows in which traffic area that mobile is and accordingly can route any incoming telephone calls. When a mobile is in contact with a base station, the fact is made known to the user by the illumination of a "service available" indicator. When the mobile user makes a call, the initial contact with the base station is via the control channel. The signaling is done by using FSK data, which occupies the whole control channel. The base station will then allocate an unused voice channel, and both base station and mobile will switch to this voice channel for the duration of the call.

For "supervisory" purposes, two types of audio tone are used [Fluhr and Porter, 1979], both outside the voice band in the voice channel. The first is the *supervisory audio tone* (SAT), used while a call is in progress. In AMPS and in TACS, three frequencies are used, at 5970, 6000, and 6030 Hz, although only one of them is used at any particular base station. The base station transmits the SAT tone, which is transponded back by the mobile unit while the call is connected. Loss of SAT tone for more than five seconds means that the receiver is in a deep fade, the signal is weak (as at a cell boundary), or there may be interference, and the call hence is disconnected. The use of three SAT tones effectively increases the cell repeat pattern from 7 to 21, from a signaling point of view, because nearby cells with the same set of frequencies will have different SAT tones. A second supervisory tone, the *signaling tone* (ST), which is at 10 kHz in AMPS and 8 kHz in TACS, is used for several purposes; for example, it is sent by the mobile at the end of a call to indicate "on-hook."

3.3.2 Handoff

The base station monitors the signal strength of the mobile throughout the call. If the signal strength drops below a certain level, the base station assumes that the mobile is beginning to move out of its cell and initiates the procedure for a "handoff" (or "handover"—the terms are used interchangeably). The MSC requests that the base stations covering adjacent cells monitor the signal strength on the traffic channel being used. The mobile is assumed to have moved into the cell corresponding to the base station with the strongest signal. This information is passed back to the MSC, which decides when a handoff (Figure 3.4) is necessary. The MSC then sends a retune command to the mobile. This signaling during a call may be achieved by interrupting the voice signal for a short period to allow the data to be sent across, so-called "*blank and burst*" signaling. The mobile retunes to the new frequency and sends control information to the new base station. Meanwhile, the MSC reconfigures the land-line circuits.

The break in voice transmission of about 400 ms during a handoff is barely perceptible to the user, and does not usually cause any loss of intelligibility. (For data, the break is a different matter, and *Automatic repeat request* (ARQ) is necessary to recover the data lost during handoff; see Section 3.5.2.) In the GSM system (Section 3.6), no break in voice transmission is required for routine signaling, although breaks in voice transmission may be used when urgent messages are to be sent (e.g., handover). One possible problem occurs if a call is set up on a base station other than the nearest. This may occur because of the vagaries of radio propagation. Then, the mobile can move into a cell area that is not covered by base stations which are adjacent to the active one. When this happens, the call may be lost.

The GSM system (Section 3.6) has been designed to overcome this problem. The mobile monitors base stations within range, and can identify them uniquely. The information on signal quality from these measurements is then passed back to the MSC, which makes the decision on handoff. The base station to which handover can be made is thus not constrained by a preset structure, but rather can be chosen, depending on prevailing conditions.

3.3.3 Call Sequence: How a Call is Set Up

The operations required of a cellular radio system are best shown by considering the sequences of events—albeit simplified—which occur in various situations.

When the cellular transceiver is switched on, it scans the control channels, selects the strongest, and sends a signal to "register" its identity and location to the system. The signal is then locked to one of the control channels; furthermore, the system knows in which traffic area the transceiver is to direct a call to it. The system can also request a *reregistration*. For example, "mobile stations" are compelled to reregister when leaving a traffic area.

Mobile Originated Call

The caller enters the dialed digits. Contrary to a "normal" telephone, the complete number is keyed into the cellular phone "locally" before contact is made to the network. Also, the number is displayed and checked by the user before the call is made. When the user presses a "send" button to "originate" the call, the mobile sends the stored digits and its "mobile identity" to the base station, which passes them on to the MSC. The MSC sets up the call into the PSTN as normal, and allocates a voice channel. When the called party replies, the voice connection is completed.

Mobile Terminated Call

Assume that a calling party has been routed by the PSTN to the MSC which is the "home" center for the mobile to be contacted. The MSC converts the dialed number into the mobile's identification number and broadcasts a paging message over a subarea of the MSC coverage, known as a "traffic area." The mobile (provided that it is switched on) recognizes its page and responds on the access channel that it has previously selected on registration. The MSC selects an idle voice channel and informs the mobile, which then tunes to this channel. The MSC transmits a data message to the mobile to indicate an incoming call, and consequently the mobile phone rings. The MSC then also transmits a ringing tone to the calling party. When the mobile answers, the ringing tones are removed and the voice connection is established.

3.3.4 Co-channel Interference

The optimum channel reuse pattern is governed by the strength of signal received at a mobile from its base station relative to the signal received from a distant base station using the same channel. The degree of interference that a radio can tolerate on the same channel (co-channel interference) is a function of the type of modulation employed. In general, the signal strength decreases as the distance from a base station increases. The decrease is not uniform, being dependent on geographical and other factors [Okumura *et al.*, 1968; Jakes, 1974; Lee, 1982], but for a simple example we can approximate the decrease to an inverse nth power law. The interference that a radio can tolerate is expressed as a threshold ratio of carrier power (wanted signal) to interference (unwanted signal) power (C/I) in dB. For the system to operate, we must have the carrier-to-interference ratio greater than the threshold ratio:

$$C/I > C/I_{\text{threshold}}.$$

For analog frequency modulation systems, as used on AMPS and TACS, the value of C/I for satisfactory operation is around 17 dB [MacDonald, 1979], and a gap must be left before the same channel can be reutilized (Figure 3.3). For the modulation systems which will be used on the pan-European digital system, the C/I required is nearer 9.5 dB. Thus, a smaller gap is needed and the channels can be reused more often, thus increasing the system capacity.

The use of directional sites also changes the situation. In this case, the directivity of the antennas must be taken into account. In practice, channel planning is a complex task that needs to take into account the detailed terrain of the area in question. Computer analysis is used extensively to help in this task.

3.3.5 Multipath Interference

Multipath interference is a phenomenon that has a significant effect on a number of aspects of a cellular radio system. The effect occurs when the signal takes a number of different paths from the transmitter to the receiver due to reflections from a variety of sources, such as hills, buildings, and nearby vehicles. Depending on the time delays involved, the two (or more) signals may reinforce each other or tend to cancel, resulting in large variations in received signal strength (Figure 2.3, Chapter 2) and even in total loss of signal. The effect is dependent on the frequency of the signal and the distance that the interfering signals have traveled. It will therefore be different from point to point. At 900 MHz, for example, the distance between peak and trough may be as short as 15 cm.

The fading will not be noticeable to the user unless the signal fades are sufficiently deep to approach the system noise level, in which case multipath may manifest itself in fluttering or fading of the signal being received in a moving car. The effect can be described statistically, and the system designer allows a margin (as indicated in Section 3.3.4) such that the signal is only likely to fade below the system noise level for a small fraction of the time. The human ear is tolerant of small interruptions in speech and, if sufficient margin is allowed, the effect of fading may be largely unnoticed.

However, although speech is tolerant of short interruptions due to fading, data signals generally are not. This is discussed further in Sections 3.5 and 3.6.

3.3.6 Diversity

The fading due to multipath interference depends on the relative positions of the transmitter and receiver, and may change in nature over a fraction of a wavelength (i.e., over a few centimeters at the frequencies of interest). Thus, two antennas less than a meter apart will experience different fading patterns, such that when one is receiving a weak signal, the other may be receiving a strong signal. Selection of the strongest signal at any instant will then reduce the incidence of fading. This is the principle of *space diversity reception.* Diversity reception is difficult to use at the mobile because there is usually insufficient space to mount two separate antennas—especially on a hand-held portable! For this reason, diversity is only used at the base station. The use of diversity reception means that the mobile needs to transmit relatively less power to achieve a given quality of link.

3.3.7 Architecture of an Analog Cellular Radio Transceiver

The major components of a TACS cellular radio transceiver are shown in Figure 3.7. The received signal from the antenna is fed to a "*diplexer,*" a high performance

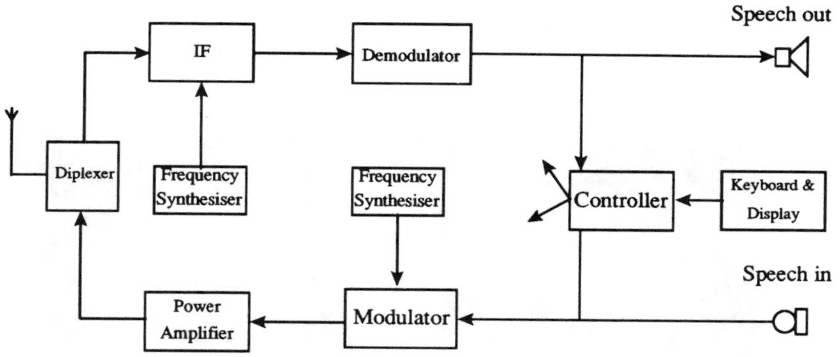

Figure 3.7 Architecture of a TACS cellular radio transceiver.

filter that acts as a selective filter for the receiving signal and the transmitting signal. Because the receiving and transmitting signals are in separate frequency bands, the passbands of the filters are designed to minimize the level of the transmitting signal which is coupled into the receiver.

The received frequency is then downconverted using the signal from the frequency synthesizer to an *intermediate frequency* (IF) for amplification and filtering to reject signals on adjacent channels. Only one IF block is shown. In practice, the signal may be converted to several different IFs, the filtering and amplification increasing at each stage. The signal then goes to the demodulator, the output of which is either data that goes to the control system, or analog speech that can be amplified to drive a loudspeaker. On the transmitting path, the analog speech is modulated directly onto the final output frequency and transmitted to the antenna via the power amplifier.

Note that two frequency synthesizers are shown because transmitting and receiving paths are needed simultaneously. The need for two synthesizers can be eliminated if the IF is chosen to be the same as the spacing between the transmitting and receiving frequency bands (typically 45 MHz).

3.3.8 Subscriber Equipment

The user sees a relatively conventional telephone handset operated in substantially the same way as a "normal" telephone. The only noticeable differences are that a "no service" light may be illuminated when the equipment is not within range of the system and a "send" button may need to be pressed after dialing the required number. There are three basic types of cellular radio subscriber equipment (see Figure 3.8):

a) 

 b)

 c)

Figure 3.8 Types of cellular telephone: (a) Mobile (vehicle-mounted) cellular telephone; (b) Hand-held portable cellular telephone; (c) Transportable cellular telephone. (Source: Racal Vodafone.)

- The vehicle-mounted version takes its power from the vehicle battery and can therefore be operated at the highest transmitting power level.
- The small self-contained hand-held portable equipment runs on internal batteries; transmitting power is therefore somewhat limited, which may result in restricted range, making communication difficult in large rural cells. Batteries may also need frequent recharging or replacement. The user of the hand-held portable is thus likely to be a pedestrian in a large city, where ranges to the base station are small and access to new sets of batteries may be relatively easy. Such units are also frequently used on trains.
- Intermediate in size are the transportable units, which have larger batteries, greater power, and consequently greater weight. Possible users include construction workers with the phone placed in a temporary office or carried around a motorway construction site, and doctors visiting patients at home. Transportables are also found on boats in rivers or coastal waters.

3.4 EXISTING CELLULAR RADIO SYSTEMS AND THEIR CHARACTERISTICS

Unfortunately, a worldwide standard cellular radio system has not yet emerged. There are several different cellular radio systems currently in existence, usually identifed by their initials. Some examples are the original AMPS from the US, the UK TACS, and the Scandinavian NMT. (See Table 3.2 for some of their characteristics.) The systems chosen by different countries (presumably the ones best suited to national circumstances) are shown in Table 3.3. Some countries have their unique development, although there are trends toward standardization, particularly in Europe with the GSM system (see Section 3.6).

One problem of standardization is the choice of frequency bands, which has been determined largely by availability within a particular country or region, subject to the propagation characteristics being reasonable.

NMT was set up to use frequencies of 450 MHz, a band available in Scandinavia, and was designed for the largely rural nature of the Nordic countries, which meant good propagation characteristics and hence good coverage of large areas, and large cell sizes. AMPS, however, was designed for use in cities, and higher frequencies around 850 MHz were chosen, leading to smaller cells and higher capacities; channel spacing is the US standard of 30 kHz. In a smaller country with a high population density, such as the UK, high capacity was important; hence, the TACS system was based on AMPS, which was a mature system with proven technology; a higher frequency band at 900 MHz was allocated, and the smaller channel spacing of 25 kHz, a European standard, was adopted. A later version of NMT, NMT-900, also uses the 900 MHz band, but has additional facilities to increase capacity still further, as well as being designed particularly for hand-held portables.

Table 3.2
Characteristics of Different Cellular Radio Systems

Name	*Began Operations*	*Channel Width (kHz)*	*Frequency (MHz)*	*Number of Channels*	*Characteristics*
NAMTS	1978	25	870–885 b-m 925–940 m-b	600	Increased to 1000 channels?
NMT-450	1981	25	453–457.5 m-b 463–467.5 b-m	180	Low channel capacity. Good radio coverage. Suitable for rural areas.
AMPS	1983	30	825–845 m-b 870–890 b-m	666	City-based. Higher capacity than NMT, but smaller cells.
C-450	1985		451.3–455.74 m-b 461.3–465.74 b-m		
TACS plus ETACS	1985	25	890–915 m-b 935–960 b-m 872–888 m-b 917–933 b-m	1000 plus 640	50% greater capacity than AMPS, but smaller cells.
NMT-900	1986	12.5	890–915 m-b 935–960 b-m	1999	Designed for cities. Caters to hand-held portables.
GSM	1991		890–915 m-b 935–960 b-m		Digital. ISDN capability. CEPT standard.

NOTES:
1. m-b and b-m indicate mobile-to-base and base-to-mobile, respectively.
2. ETACS means extended TACS. Some additional frequencies have been made available to the TACS system, outside the original band, to ease congestion, particularly in London.
3. Sources include Makitalo [1978], Bergqvist [1989], Wickham [1988], and Balston [1989].

A chapter such as this cannot cover all of the current cellular systems in detail (see the end of the present chapter for Further Reading); emphasis will be placed here on one second-generation system becoming available (the GSM digital system, Section 3.6) because it contains all the features of the existing systems and more. In this way, a comprehensive view will be available to the reader. However, we will first consider additional facilities and services that have been developed for the cellular radio user.

Table 3.3
What Countries Use Which Cellular Systems

System	*Country*
AMPS	Australia, Canada, Hong Kong, New Zealand, Thailand, United States.
C-450/NETZ-C	Federal Republic of Germany.
GSM	The CEPT Countries (Table 3.5)
NAMTS	Japan, Kuwait.
NMT-450/NMT-900	Austria, Belgium, China, Denmark, Finland, France, Iceland, Indonesia, Luxembourg, Malaysia, Netherlands, Norway, Oman, Saudi Arabia, Spain, Sweden, Switzerland, Thailand, Tunisia, Turkey.
Radiocomm 2000	France.
RMTS	Italy.
TACS	China, Republic of Ireland, Hong Kong, United Arab Emirates, Malta, United Kingdom.

NOTES:
1. Although different countries may have chosen the same cellular system, they may not use the same frequency bands. So, cellular phones are likely to work only in their country of origin. The exceptions to this are the Nordic countries, which have common NMT systems; AMPS in the US and Canada; and the GSM system in the CEPT countries.
2. Sources include: Wickham [1988].

3.5 ADDITIONAL FACILITIES AND SERVICES ON CELLULAR RADIO

Like ordinary telephones, cellular radios can be used for things other than voice telephony, for example, data transmission (see Section 3.5.3) and facsimile. Other features, some of which are available on the PSTN, are particularly important in the mobile environment; for example, recorded message-taking and answering machines, and especially hands-free operation (see Section 3.5.1). Cellular radios also have as standard some features not available on the PSTN, or at least not on basic telephones (see Table 3.4).

3.5.1 Hands-Free Telephones

When driving a car, making or receiving a telephone call should cause as little disruption to the driving task as possible. The most recent edition of the UK's "Highway Code" asks drivers not to use a hand-held microphone or telephone handset on the move, except in an emergency. The code recommends, by implication, the use of

Table 3.4
Mobile Telephone Facilities*

Facilities	
• NO LOCATION PROBLEMS Your national number locates you anywhere in the coverage area. • DIAL IN HANDSET Dial and handset combined in a single unit. • LIQUID CRYSTAL DISPLAY Clear display with minimal power consumption. • ILLUMINATED KEYBOARD AND DISPLAY Keyboard and display can be read easily under all light conditions. • ONHOOK DIALING Leave handset on its cradle until the called party answers. • SPEED DIALING Conveniently store and instantly recall up to 30 frequently dialed numbers. • ELECTRONIC LOCK (4-digit code) Prevents unauthorized calls being made. • RECEIVED CALL INDICATOR Indicates that you have been called while away from the vehicle. • DTMF SIGNALING May be used to gain access to data systems using DTMF signaling.	• DISPLAY FOR FIRST 6 DIGITS All 16 digits can be read in 10-digit display on simple key operation. • ALERT External alarm on the vehicle is activated when radio is called. • HANDS-FREE OPERATION Conversation without lifting the handset. • LAST NUMBER REDIAL Repeat last number without redialing. • OUTGOING CALL BARRING Prevents unauthorized use of your phone. • CALL TRANSFER Enables calls to be transferred to another number automatically. • ALARM CALL The phone will ring at the arranged time with an appropriate recorded message. • CONFERENCE CALLS For three way telephone conversation. • CALL WAITING Indicates that another incoming call is waiting.

**Source: Racal Vodafone.*

---

hands-free telephones (HFTs)—a microphone fitted in a convenient place in the car, for example, on the sun visor, and a loudspeaker on the dashboard.

This solves the problem of holding the handset during a conversation, but not the problem of dialing the number "hands-free." One solution is *voice dialing,* based on *speech recognition,* also called DVI (*direct voice input*), in which the telephone recognizes the numbers that the driver speaks. Unfortunately, DVI is a major problem that has not been satisfactorily solved anywhere, certainly not in the difficult and noisy acoustic environment of a motor vehicle. Nonetheless, some DVI cellphones have been produced. Neither is implementing an HFT satisfactorily an easy task in the noisy environment of a moving motor vehicle. One attempt to do so is described by Noble [1988], on which this section is based.

The source of the major difficulty in the design of an HFT is the effect of the acoustic path from loudspeaker to microphone. With the sending and receiving gains adjusted for normal listening levels, some of the original sound from the loudspeaker feeds back into the microphone (i.e., the audio path loss from loudspeaker to microphone is only about 6 dB). This figure is much worse than that of an ordinary handset, and gives a poor margin against instability ("howl round") in cases where the telephone circuit produces reflections of the acoustic signal (either from within the network or by an HFT at the far end). Also, if the one-way delay of the telephone unit exceeds about 25 ms, any audible delayed echo is objectionable to the user and can lead to major problems in talking. (This is not normally a problem in the PSTN or in analog cellular systems, but may be troublesome in digital systems (see Section 3.6 below).)

One solution to this is to include automatic detectors that switch in attenuation to make the conversation "one-way" (Figure 3.9(a)). When the caller at the far end is talking, this is detected and the sensitivity of the microphone is effectively reduced. In a similar manner, when a signal is detected on the microphone, the signal to the loudspeaker is attenuated. This can eliminate the possibility of "howl round," but makes the conversation more difficult because both parties cannot speak simultaneously.

To maintain loop stability, the minimum attenuation between microphone and loudspeaker is roughly 31 dB. Existing hands-free units for the car provide about 40 dB (excluding the acoustic loss), and achieve this by means of switched atten-

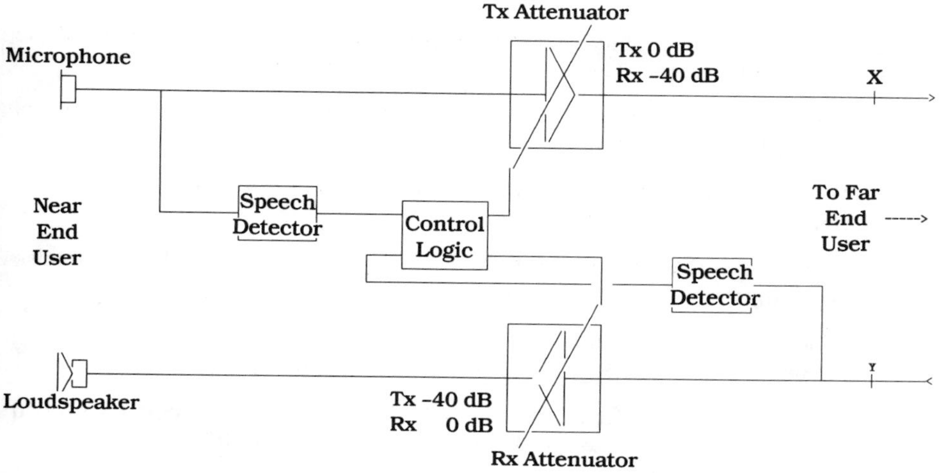

Figure 3.9 Hands-free mobile telephony: (a) A block diagram of existing "switched attenuation" hands-free units. When speech is detected, one attenuator is switched on, and the other is switched off, so that either the transmitting or the receiving path is blocked, and communication is effectively half-duplex.

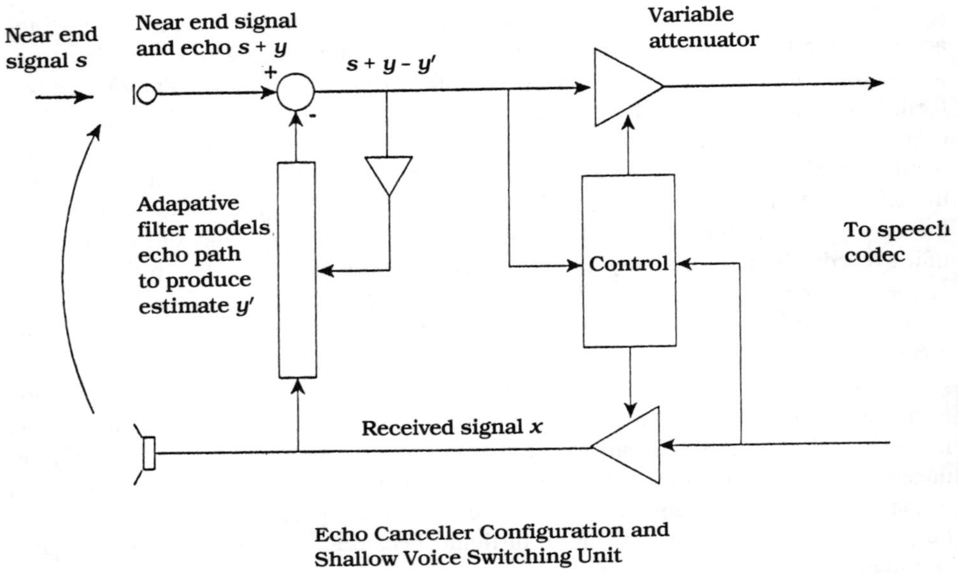

Figure 3.9 (b) A block diagram of an adaptive echo canceller. Attenuation can be less drastic, since any echoes that could cause instability are cancelled adaptively, and a full-duplex link is possible.

uators (Figure 3.9(a)). The units incorporate speech activity detectors to control the switching of the attenuators. The problems with such systems are:

(a) Confusion arises because operation is effectively half-duplex, and users cannot interrupt each other.

(b) Beginnings of words are clipped due to the finite time taken to detect activity and switch the attenuators.

(c) There is insufficient volume from the loudspeaker in high background noise conditions due to the requirement to maintain circuit stability.

(d) When neither direction is active, the system must either "float" with both attenuators at −20 dB, or give precedence to one of the users.

(e) The far-end user suffers from the vehicle noise in his ear being switched out whenever he or she speaks, and this can be very disconcerting.

(f) Noise in the vehicle can cause the transmitting path to open intermittently, and may lead to bursts of noise at the far end.

In hands-free to hands-free operations, the above problems are potentially doubled, and independent control decisions within each hands-free unit can also lead to conflicts. All of these problems combine to produce a system having unsatisfactory performance, particularly when the noise level in the vehicle is high.

Adaptive Echo Cancellers

Most of the problems with current hands-free units would disappear if the required attenuation could be achieved without switched attenuators. Noble [1988] and Burnett *et al.* [1988] report investigations into the use of a digital adaptive echo canceller to reduce the amount of switched attenuation by subtracting any echo, thereby maintaining full-duplex operation.

The basic configuration of an adaptive filter to perform the acoustic echo cancellation is shown in Figure 3.9(b). In theory, the filter models the acoustic path from loudspeaker to microphone, and in the absence of any near-end signal at the microphone, the output from the "summer" will be zero. In practice, this condition is never achieved, the main problems being:

(a) The acoustic path in the vehicle is time varying (people move around) and the filter must track these changes.

(b) When the far-end user stops talking, the adaptive filter has no signal with which to track the changing acoustic path.

(c) The car environment can be noisy, and on a motorway a signal-to-noise ratio (S/N) of 0 dB at the microphone is not unusual [Crown *et al.*, 1987].

The accuracy with which acoustic path changes can be tracked is degraded by near-end noise. The same comment applies when the car user interrupts the far-end user ("double-talk"). Near-end signals (speech and noise generated in the vehicle) in effect constitute interference at the microphone that is uncorrelated with the signal that it is desired to cancel. The signal to interference ratio (S/I) at which the adaptive filter must work can thus be far worse than 0 dB. The filter must cancel as much as possible of the echo (i.e., that part of the microphone signal correlated with the loudspeaker signal), but at the same time not affect the near-end speech.

Performance of the Adaptive Echo Canceller

A real-time implementation using the LMS (least mean square) adaptive algorithm was developed and the experimental system was installed in a Ford Transit van [Burnett *et al.*, 1988]. The Transit formed a particularly harsh environment for the tests because the vehicle is noisier than passenger cars and the interior has a long reverberation time. The attenuation achieved in the test vehicle was around 17 dB, increasing to 25 dB in a test chamber with a shorter impulse response. For use in the

GSM system, this echo canceler fails to satisfy the specification on echo attenuation for the case of a hands-free terminal. For terminals equipped with a handset, the specification could probably be met with the existing system. In any case, adaptive echo cancellation means that less switched attenuation is needed, which alleviates some of the problems discussed above.

3.5.2 Data

Cellular radio services were originally conceived as voice services only. However, in common with the rise of data services on the fixed telephone network, the use of data on the cellular radio network is increasing, as it is on *private mobile radio* (PMR); this is readily understood when the benefits are considered. With an auto-answer modem, the message can be received without the user intervening; he or she does not need to be in the vehicle or, alternatively, he or she is not distracted from the driving task. In contrast with a conversation, there is confidence that the information transferred is accurate and, for a similar amount of data transferred, the data call will take less airtime than the voice call and will therefore be cheaper.

3.5.2.1 Case Studies Of Cellular Data Usage

This section is partially based on preliminary results from a user requirements study of data over cellular radio carried out by the Human Sciences and Advanced Technology (HUSAT) unit at Loughborough University, in collaboration with Racal, as reported in Cole *et al.* [1988]; other applications are listed in Table 3.5.

Field engineers at CASE Communications, a UK manufacturer of data communication equipment, use portable computers to keep records of spares, fault histories, and past diagnoses, as well as to access the company's internal network via the PSTN for telex, electronic mail, and database applications. Use of cellular radio instead of the PSTN has brought the additional advantages that the engineer can be contacted more easily and is not reliant on the customer's telephones. New applications under investigation are downloading of data (and possibly software patches) to the customer's equipment via the engineer's personal computer. Reporters working for a broadcasting organization use cellular data equipment to compose and download reports. There are plans to provide database access so that a journalist interviewing a politician will have on-line access to records of recent interviews with politicians from rival parties.

Several organizations, including a shipping company, an oil company, and a public utility, have successfully investigated cellular data transmission for offshore telemetry and data communication applications, including database access by maintenance engineers and more efficient stock replenishment. Similiar applications have been reported in the US, involving field maintenance managers, realtors transmitting

Table 3.5
Existing Mobile Data Applications

Application	*Communications*	*Stage of Development*
London Taxis	Private Mobile Radio	£1 million investment in on-board computer terminals to replace voice communication. Aims to expand ordering systems capacity by enough to double its fleet to 5000 in 2–3 years. Vehicle location in future.
Aircraft Public Telephone	Satellite	1st link up July 1988. Have agreements for worldwide service.
Fire Service Incident Information	Paging	Transmission of real-time information on incidents. Network of 100 senior fire service officers; 250 in near future. Cost £80 thousand.
Fire Service	Radio	Full mobile data system worth £1.25 million. Uses a liquid crystal display for the display of road maps, text, plans, *et cetera*. Interfaces to existing computer resources.
Electricity Board	Radio	Pilot scheme. Secure data. Day's jobs sent to the vehicle for printout or display, eliminating need to check in at the depot.
Customer Service	Cellular	Europe's most advanced field communications system. Allows direct communication with 450 field service engineers in their vans.
Telex	Cellular	Messages sent to a fixed base station which connects to the telex network.
Taxis	Radio	Data broadcast.
Sea and Air Transport	Private Cellular	750 member companies.
Motoring Organization Breakdown Services	PMR	New computerized breakdown Operations center covers Western Home counties from Bedfordshire to Hampshire. 350 mobile data terminals to be fitted. Will link to the base computer via a two way radio and provide vehicle breakdown details including the driver's name, car type, registration number, and location.
Vehicle Tracking and Location	Dedicated Radio Network	Track progress of a vehicle from the control room to within 50 meters and to display this information on a full

Table 3.5 (cont'd)

Application	*Communications*	*Stage of Development*
		color map display. Presently available in London. 50% of UK within a year. Vehicle position data systems are continually transmitted to a central control room where they are displayed on a digitized map using standard ordnance survey grid coordinates. Users can have other information transmitted as an optional extra. Includes an emergency facility by which the driver can alert the control room in the event of an accident, hijack, or attack. Possible for users to integrate the service into a total computerized dispatching or command and control system.
Motoring Organization Breakdown Services	PMR	CARS (Computer Aided Rescue Service.) £7.5 million contract. 5 computer centers are linked by high-speed data links to 16 further dispatch centers and a central membership mainframe. The system is used to answer distress calls, validate membership entitlements, and to locate the member precisely using a nationwide computer based gazeteer. Information is then broadcast to one of 1900 patrols by a mobile radio data link. Patrols report in data form the symptoms of each breakdown attended and supervisors will interact with the system, checking patrol job loadings and activities as well as transmitting data messages.
Dispatch System	Radio	250 vehicles serving a 60-mile radius around London. Each has a small printer in the cab linked to the van's radio. Used to pick up details such as name, address, floor of building, and telephone number. Voice takes about 1.5 minutes of dialog to provide this information. Data takes

Table 3.5 (cont'd)

Application	*Communications*	*Stage of Development*
		4 seconds. System to cover the whole country. On radio details can be misheard. Saturation of the channels also means that voice messages can take a long time to set up. Data will ease this congestion.
Monitoring		Monitors personal safety and can raise the alarm or call for personal assistance. A base station is programmed to broadcast to mobile stations at regular intervals. If the carrier of the outstation does not respond in a set time, the alarm is raised.
UK Police Force	PMR	Terminals fit in standard radio aperture. Full alphanumeric display and function keys.
Air Travellers	Satellite	Satellite based telex and data transmission from aircraft.
Taxi Company	PMR	Doubled its business within a year with only 10% more taxis.

Source: Husat Research Group, Loughborough University, UK.

real estate information, fire services accessing chemical databases, and an ambulance service transmitting EKG telemetry data [Fontana, 1987].

Most of these applications have been trials, so the user's main concern has been reliable data transmission and solving technical or interface problems. The portable computer and the communication link are used to emulate a terminal connected to a mainframe. In some of these examples, there is a move to more local use of the portable computer, but the end-user is still involved in making the necessary connections to a central computer to transfer data and run appropriate applications. One reason for this is the need for the end-user to participate in security actions with the central computer facilities (such as logging in by using a password).

3.5.2.2 Data Requirements

Working in the "mobile office" environment with a portable personal computer over a cellular telephone link imposes a number of requirements on the system architecture.

A major consideration is sharing the cellular link between data transfer and voice communication. This means that the personal computer may not be permanently attached to the cellular link. The computer may also be in a location where the cellular network cannot be used. Thus, the central computer cannot be sure of making a connection with the portable computer at any particular time. The best solution to this problem is to have the central computer wait for the portable computer to initiate a connection, and then to transfer data. The current UK cellular radio system has a fairly low data rate (1200 to 2400 b/s with an error correcting modem), so data is transferred in reasonably small amounts. A store and forward messaging system is needed to meet these requirements. The CCITT X.400 Message Handling System (MHS) is an obvious choice. The 1988 version of X.400 contains a *message store* [CCITT, 1987], which compensates for the two drawbacks of the portable computer; namely, (a) it has limited resources, particularly for file storage, and (b) it is not permanently connected to the network.

3.5.2.3 Implementation of Data over Cellular Radio

The provision of data services as an "add-on" to a mobile phone presents a number of problems to the cellular radio engineer. One is that no standard interface exists to connect data services to a mobile phone. Usually only the handset connection is available for connection of external equipment, and this interface will vary from manufacturer to manufacturer, resulting in different equipment to provide data services for each make of phone. There are more practical problems to overcome in providing satisfactory data-voice switching and auto-dial–auto-answer services.

While standard PSTN modems can be demonstrated to work on occasion over a mobile phone network, they do not give reliable data transmission. They cope with errors by request for a repeat transmission of the data (ARQ); techniques such as this can be tailored for high throughput rates with minimal errors. However, the vagaries of the radio path such as Rayleigh fading (Figure 2.3) and the specific features of the mobile network such as handover and "blank and burst signaling" would result in many requests for retransmission, and the data throughput would be very low. Hence, the modems developed specifically for cellular radio incorporate *forward error correction* (FEC) as well. The fades are typically of 1 ms to 10 ms duration, while "blank and burst" and handoff last for 100 ms and 400 ms, respectively. The former can be corrected by FEC, whereas the latter demands ARQ. Of the existing V-series modems, the one best-suited to the cellular environment is "V.26 bis" [Jarvis, 1985; Frazer, Harris and Munday, 1986].

Note that some of these interruptions in the signal can occur even when the vehicle is stationary—including handoff—so standard PSTN modems are unsuitable, even if data is transmitted only when the vehicle is not moving. In addition to error correction, a layered protocol is necessary to establish and manage a data link for effective information transfer. The one developed by Racal in the UK is cellular data link control (CDLC), which is based on the CCITT X.25, level 2 (HDLC) protocol (see Figure 3.10 and Table 3.6) [Jarvis, 1985, 1986; Frazer, Harris and Munday, 1986]. CDLC is manufacturer-independent and application-independent. Neither is it locked into V.26 bis; other modems could be used in the future.

In the US, a similar but proprietary solution has been developed by Spectrum Cellular of Dallas, Texas—the "bridge" and "span" (Fontana 1978). As with CDLC, ARQ, and FEC are used. The *bridge* connects to the cellphone handset socket in the mobile; the *span* is at the land-based end. Both can communicate with another bridge or span, or with a standard 212A or 103 modem (in which case the error protection protocol is not used). As well as data, facsimile can be transmitted over cellular, and FEC and ARQ techniques have again been used to improve transmission quality [Furuya *et al.*, 1987]. When suitable portable facsimile machines become

8 - bit leading sync sequence	8 - bit address field	8 - bit control field	Information field n bits (typically 1000 bits)	16-bit frame check sequence	8-bit sync sequence

48 - bit leading sync sequence	8 - bit address field	8 - bit control field	Information field 8n bits (up to 504 bits)	16-bit frame check sequence

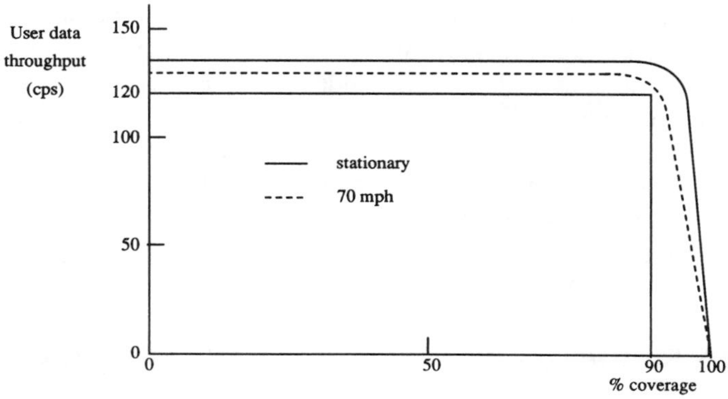

Figure 3.10 The cellular data link control (CDLC) protocol. (Source: Racal Vodata.)

Table 3.6
The Major Features of CDLC (after Jarvis [1986])

Physical	V.26 bis channel standard
Link	ISO procedures ISO error detection Forward error correction Selective block repeat 2-wire and 4-wire working 0.02-bit error rate capability Hand-off and blanking resistant
User	Asynchronous interface Viewdata-compatible data rate True or pseudo full-duplex RS232C/V25bis standards

available, mobile fax is likely to experience the explosive growth seen recently in the land-based office.

3.5.2.4 Cellular Network Data Services

The provision of a special modem for the radio path, essential though this may be, presents a further problem to the user: he or she must have a similar modem at the other end to decode the signal. Network operators (Racal Vodafone in the UK, Ameritech in the US) [Fontana, 1987] have overcome this difficulty by providing a corresponding modem at the network switch. This means that the data user can employ his existing fixed network modems to communicate with the mobile without re-equipping his fixed sites with cellular modems (see Figure 3.11).

Data usage for the mobile user is increasing, and this is readily understood when the benefits are considered. See the first paragraph of Section 3.5.2 above. The disadvantage is that more infrastructure (computers, *et cetera*) is necessary to prepare and receive the messages. However, with the onward march of the electronic office, this disadvantage is gradually disappearing, and we may expect to see more users of mobile data in the future.

3.6 THE PAN-EUROPEAN DIGITAL CELLULAR SYSTEM: GSM

Although the cellular concept was a major breakthrough in the provision of mobile communication services, allowing many more users to share the available spectrum,

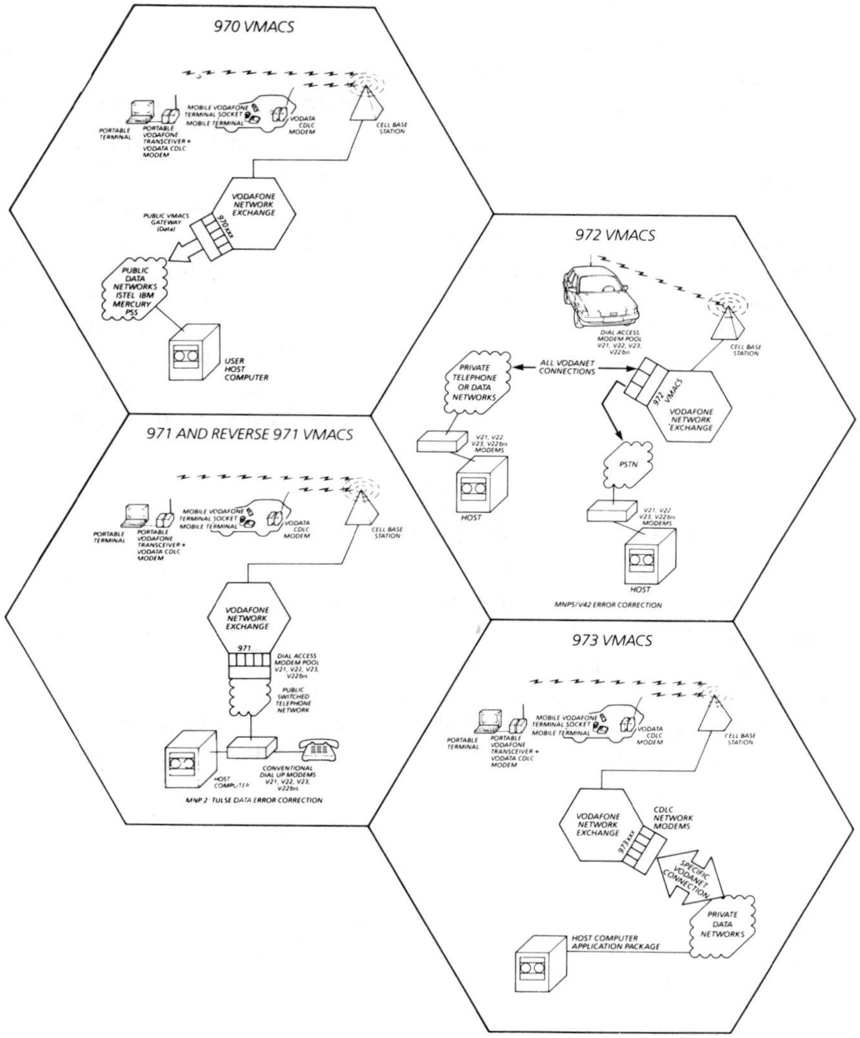

Figure 3.11 Vodafone mobile access communications service (VMACS): VMACS is a service operated by Racal Vodafone on the UK cellular radio network to provide mobile data users with a convenient and efficient method of connecting with host computers.

there are still problems of congestion, and existing systems, though using the same principles, are incompatible. Consequently, CEPT decided to propose and set up a pan-European cellular radio system to accomplish the following:

- to have an integrated European system,
- to increase available cellular radio capacity,
- to take advantage of increasing power and decreasing cost of digital electronics.

The CEPT countries (which are therefore the countries in which this new system will be available) are listed in Table 3.7. The system is usually known as "GSM," after the "Groupe Special Mobile" subcommittee of CEPT that developed the standards to which the system must conform.

Service will begin on the GSM system in 1991. It will allow users to make a call or be contacted on their mobile phones anywhere from the south of Italy to the north of Norway. GSM will not replace existing systems—at least not initially—but will operate in parallel with them. (A TACS or NMT telephone set will not work on the GSM system, nor *vice versa,* but subscribers on these systems can obviously make telephone calls to GSM subscribers just as easily as they can to PSTN subscribers).

3.6.1 The Advantages of the GSM System

GSM will provide a number of additional benefits over existing cellular systems: the voice quality will be better; privacy will be provided via encryption; the data services will be an integral part of the system; and the economies of scale will eventually provide cheaper equipment. Also, "smart card" access will be possible (i.e., you can use someone else's GSM phone, but have the call billed to yourself by plugging your plastic "smart card" into the phone). From the cellular radio engineer's point

Table 3.7
The CEPT Nations

Austria	Netherlands
Belgium	Norway
Denmark	Portugal
Finland	Spain
France	Sweden
Ireland	Switzerland
Italy	United Kingdom
Luxembourg	Federal Republic of Germany

of view, there are also a number of benefits: the system capacity is higher; the mobile participates in the handover process, making it more reliable; and the complexity of the base station is reduced.

The perceived advantages of this system over the first generation of systems are shown in Table 3.8. These benefits are significant; to achieve them, the GSM system is radically different from the systems described above. The major differences are: the signal is transmitted in bursts (time division multiplex); the speech is digitally encoded to a low data rate and transmitted in a very efficient modulation format; error correction is used; encryption provides privacy; interference may be reduced by frequency hopping, transmitting only when someone is speaking, and controlling the power of the transmissions. These differences, and how they are beneficial to the system, are described in more detail below.

One way to achieve a system of high capacity is to pack the cells closely together so that the frequencies are used more often in a given area. This requires a modulation system that is highly resistant to co-channel interference. Also necessary is to use as many radio channels as possible within the allocated spectrum, which means that the frequency separation between channels must be small. The GSM system uses GMSK, which has very good performance in this respect.

The smaller cells mean that handovers occur more often. Thus, the system is designed so that the probability of successful handover is indeed very high. This is achieved by the following strategy. The mobile monitors the frequency spectrum for nearby base stations during a call and reports the quality of the signal to the MSC via the base station that is handling the call in progress. This means that either the mobile must have two receivers or the receiver is time-shared between the call and the monitoring. In the interests of keeping the cost and size of the mobile low, the latter approach was chosen. To reduce cost still further, the mobile does not transmit while it is receiving, but transmits and receives in alternate short bursts. This re-

Table 3.8
The Advantages of the Second-Generation Digital GSM Cellular Radio System over First-Generation Analog Systems

- Economies of scale for equipment manufacture.
- Integration of voice and data.
- Improved system performance in parameters such as voice quality and handover.
- Automatic international roaming so that a user can make calls on his or her phone in any country as readily as his or her own.
- "Smart-card" access, permitting multiple users of the same cellphone, but with individual billing.
- ISDN compatibility.
- "Sleep mode"—longer battery life in standby mode.

moves the need to prevent the transmitted signal from going into the receiver and thereby affecting its operation. Some circuitry of nature similar to a diplexer (see Figure 3.8) is still needed to prevent transmissions from nearby mobiles from affecting the receiver, but the performance requirement is less stringent.

To control the quality of speech, digital encoding is used. This means that error correction bits can be added to the signal of the radio channel and a perfect replica of the transmitted digits representing speech can be obtained at the receiving end, unless the interference is so high as to cause catastrophic failure of the link.

The transmission of the speech as digital information also means that data—an increasing requirement—is easily sent across the link. In addition, it means that the majority of the circuits, apart from the RF components, can be implemented in digital integrated circuits, which are rapidly reducing in price, promising cheaper equipment for the user in the future.

3.6.2 The Architecture of a GSM Transceiver

The incoming signal is picked up at the antenna and passed to the Tx/Rx block. In principle, this block could be a switch because the radio is never receiving and transmitting at the same time. However, filtering is needed on the input to eliminate signals from different transmitters in other frequency bands. This block will normally be a diplexer, as in the TACS system, but in this application a lower performance component is acceptable (Figure 3.12).

The signal then is down-converted, using the signal from the frequency synthesizer to an IF for amplification and filtering to reject signals on adjacent channels. Only one IF block is shown. In practice, the signal may be converted to several different IFs, the filtering and amplification increasing at each stage. Note that, again, because the equipment does not transmit and receive simultaneously, only one frequency synthesizer need be used, time-shared between receiving and transmitting. When the signal has been sufficiently amplified, it can be digitized and applied to the equalizer. This is a high-speed computation block, which removes the distortions caused by multipath from the signal and demodulates it (i.e., turns it into data bits). Following this, the data bits are decrypted, deinterleaved, errors are corrected, and the resultant data go either to the control system, data interface (for external equipment such as personal computers), or speech decoder.

On the transmitting path, the speech signal is encoded, error correction bits are added, and the resulting data are interleaved and encrypted. The data are modulated onto the final output frequency and amplified to a suitable power level for transmitting.

3.6.3 Time Division Multiplex

These considerations result in a *time division multiplex* (TDM) structure in which eight signals are combined on a single RF channel having an overall bandwidth of

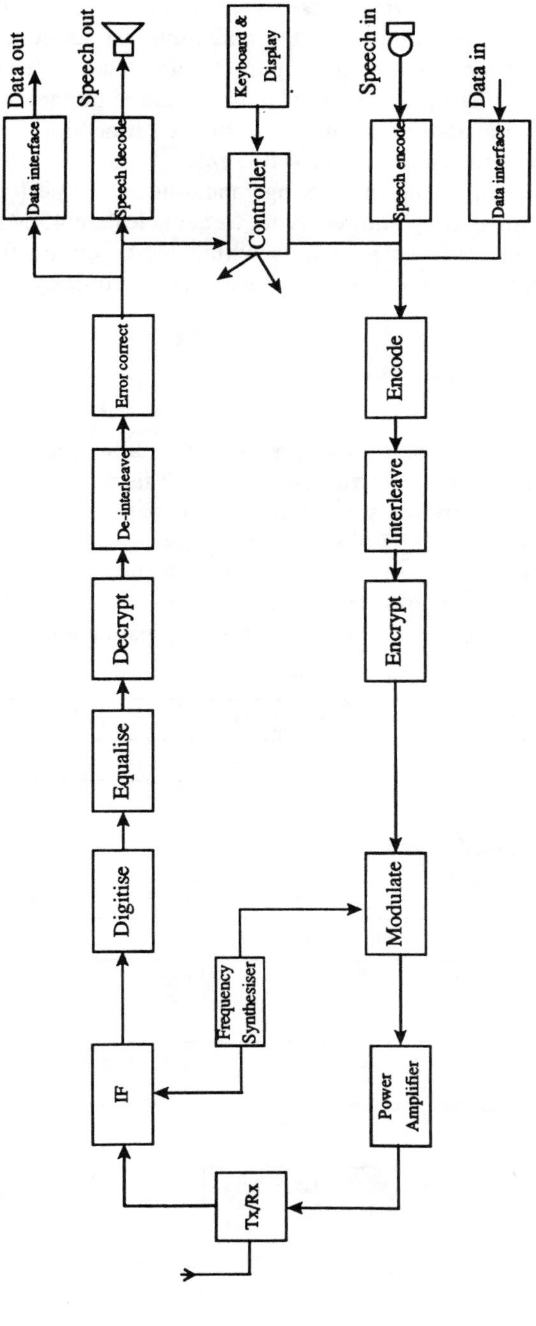

Figure 3.12 The architecture of a GSM cellular phone.

200 kHz. That is, a set of eight time slots has been chosen. A mobile transmits on one time slot and receives on another; of the remaining six, three are available for the mobile to switch between modes (see Figure 3.13(a)) and three are available for monitoring. This means that on any radio channel, eight simultaneous calls can occur, the eight mobiles transmitting in turn, one in each time slot. Each time slot is 576 μs long, so the pattern repeats every 4.615 ms.

Transmitting, receiving, and monitoring functions are all performed on different frequencies. The transmitting and receiving frequencies are 45 MHz apart, while the monitoring frequency may be anywhere within the receiving frequency band. Thus, the radio is naturally "hopping" from one tuned frequency to another, three times within 4.6 ms.

Time-slot Structure

Each of the eight time slots consists of a burst of 148 data bits, as in Figure 3.13(b). Note that the bits do not completely fill the time slots; there is a gap at each end to form a "guard zone" to allow for some timing error between time slots. The data consist of two sets of bits, each 57-bits long; two sets of "tail" bits, each three bits long; a training sequence of 26 bits and two "stealing flags" of one bit each.

The training sequence is used by the receiver to estimate the distortion on the radio channel, which will be mainly caused by the multipath phenomenon. The train-

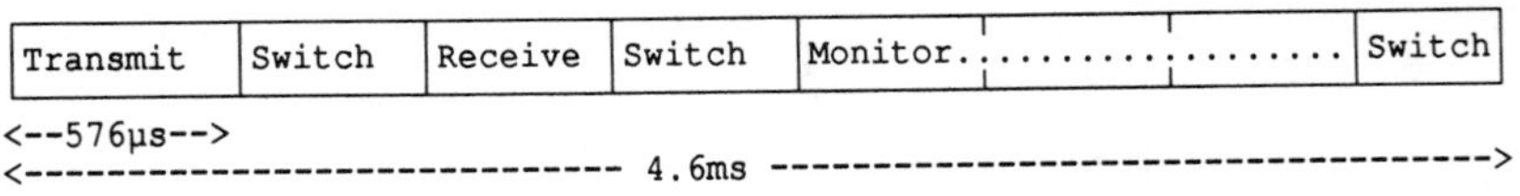

Figure 3.13(a) GSM TDM signal structure.

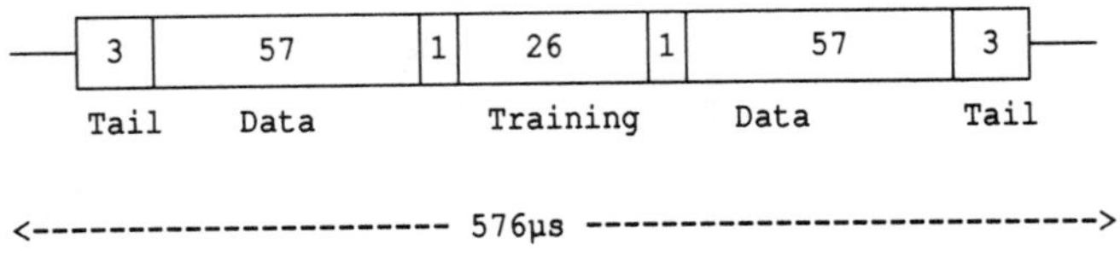

Figure 3.13(b) GSM TDM time-slot structure.

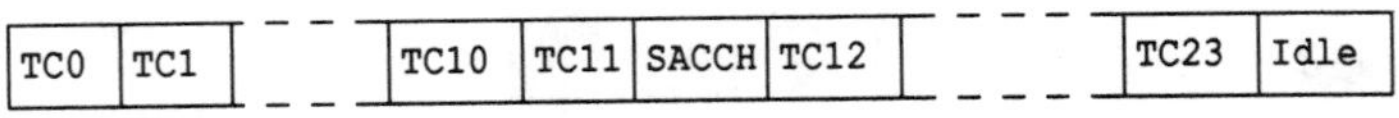

Figure 3.13(c) Time-sharing of the GSM traffic and control channels.

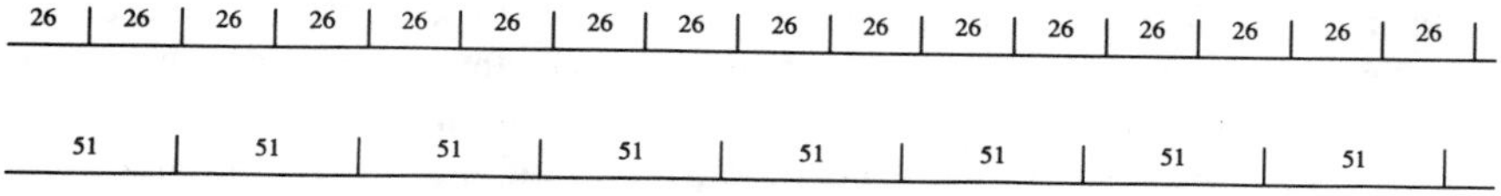

Figure 3.13(d) "Sliding" of the different repeat patterns: How the 26-frame repeat pattern on which a phone will be conducting a call, "slides past" the 51-frame repeat pattern of the broadcast information on another cell. This allows monitoring of any of the information on another cell.

ing sequence is a known pattern of bits. Thus, by measuring the distortion on this bit pattern, the distortion on any pattern of bits can be estimated. One algorithm which can be used is explained in simple terms as follows. If four bits are considered, all possible combinations of these four bits produce sixteen possible patterns. These sixteen patterns are then distorted by the same distortion that was observed on the training sequence to produce sixteen waveforms, which can be compared with the received waveform. The best match between the received waveform and one of the estimated waveforms gives the most likely pattern of four bits that were received. The waveform is then shifted by one bit and the matching exercise is repeated. By working out from the center, one bit at a time, the most likely sequence of received bits can be assessed. The matching is thus performed over four bits, although a decision can be made about only one received bit at any shift. Because of this, there is a problem at the end of the process so that the three known "tail" bits are provided and at the last matching of four bits, only one unknown bit has to be decided upon. Because the process works out from the center, "tail" bits have to be provided at both ends. The two single "stealing flag" bits are used to signal to the receiver whether the data contains normal information (e.g., voice or data), or signaling information, which is required rapidly.

Logical Channels

The time-slot structure above defines a physical channel. This allows data to be exchanged between transmitter and receiver. Allied to this are "logical channels"; a number of logical channels are allocated to one physical channel. For example, during a call, both speech and data information need to be exchanged. The speech information is allotted a *traffic channel* (TCH), and the data are allocated a *slow associated control channel* (SACCH). The time sharing of the two channels is shown in Figure 3.13(c). Note that each slot in the diagram corresponds to one mobile (i.e., the mobile is receiving, transmitting, and monitoring other base stations within the allocated slot—seven others offset in time can use the same carrier).

The sharing pattern is based on a 26-frame repeat pattern. The TCH uses 24 of the frames, the SACCH one and one frame is idle (i.e., no data). The reason for this can be seen when half-rate channels are used. In the future, speech coders are expected to become available and will produce acceptable quality at half the data rate of the one currently defined. Then, only half of the TCH time slots will be required, and two channels to two different mobiles can be fitted into the structure above by using 12 TCHs and one SACCH per 26-frame repeat. This then explains the idle frame on the full-rate channel.

The control channels use a 51-frame repeat pattern. Because of this, the 26-frame and 51-frame patterns "slide past" each other. This allows a mobile monitoring another base station during a call the opportunity to listen to all the types of control information as they "slide past" (see Figure 3.13(d)).

3.6.4 Digital Speech

From the above, we can see that the transmission of the signal occurs in bursts, whereas obviously a voice conversation is continuous. The discrepancy is handled by digitizing the speech. The digits can then be stored, passed across the radio channel in a burst, and "stretched" at the other end to reconstitute the speech.

Speech can be digitized by a number of methods. On the fixed telephone network, speech is routinely digitized, being sampled at 8000 samples per second and using eight bits (256 different levels) to represent each sample. This is known as PCM (*pulse code modulation*) and results in a data rate of 64 kb/s. Speech can be digitized at much lower rates (e.g., 1200 b/s), although there is a progressive loss of quality as the data rate is reduced. To achieve these low data rates, a speech coder is used, which, rather than taking a sample of the speech signal at one instant, evaluates the speech over a longer period of time and extracts longer term trends in the signal. The speech encoder chosen for the GSM system is the RELP (*residually excited linear predictive*) coder. This encodes the speech at 13 kb/s and produces "toll quality" output. Improvements in speech coder technology are expected to occur within the time scale of the GSM system, and so it is being designed to cope with speech coders that can operate at about 6.5 kb/s. This data rate will release every other time slot for a different conversation, and the provision of these "half-rate encoders" will double the capacity of the system.

3.6.5 Error Correction

Although the speech encoders can provide high quality speech on an error-free channel, they degrade rapidly when errors occur. On a normal radio channel, interference and propagation anomalies tend to make their presence felt as noise, with which the human ear can cope relatively well. On a digital radio channel, interference and propagation problems result in errors in the data bits. When errors are fed into a speech decoder, the errors result in speech-like sounds that are difficult for the human ear to differentiate from speech and are very disturbing. Error correction techniques are thus used to alleviate this problem. The redundancy introduced by error correction increases the data rate "over the air" to some 22 kb/s.

In addition to error correction, a technique known as *interleaving* is used. Errors have a tendency to occur in bursts, caused, for example, by fading of the signal. Error correcting codes are better at correcting randomly distributed errors and interleaving is used to modify the bursty distribution of errors to a more random one. The basic principle of interleaving is to transmit the data bits in an order different from that in which they are generated. The process of reordering the data bits at the receiver then changes the distribution of a burst error. A simple example of bit interleaving is shown in Figure 3.14. The burst of four consecutive errors spreads when the bits are reordered into their proper sequence.

Error correcting codes tend to produce a perfect output when the error rate is below the design limit and to fail catastrophically if above that limit. The speech output would therefore also follow this pattern, which would be very disturbing to the user. Because the error correcting circuits know how many errors are being corrected at any time, this number can be used as a measure of quality for the signal. When the quality degrades, this fact can be signaled to the base station, which can initiate a handover to a better base station before catastrophic failure occurs. Note that this quality factor will detect problems due to interference as well as those due to low signal strength.

```
Order of bits        1  2  3  4  5  6  7  8  9 10 11 12 13 14 15 16 ...
Order transmitted    2  5 10 15  1  7 11  3 12  4  8 14  6  9 13 16 ...
Burst of errors                  ^  ^  ^  ^
Re-ordered errors    ^     ^           ^           ^
```

Figure 3.14 Interleaving of a bitstream: The bits corrupted by the error burst are redistributed by interleaving and can then be corrected more easily by the error-correcting code.

3.6.6 Data Transmission in GSM

All information transmission on the GSM system is digital, so the transmission of data may seem to be relatively easy. However, on the air interface, the data are arranged in a fixed format and sent at a fixed data rate. Difficulties therefore occur in translating the external data format and data rate to be as required for GSM.

There are two basic modes of transmission: *transparent* and *nontransparent*. In transparent mode, the GSM system attempts to recreate the external data stream as faithfully as possible at the receiving end and with a constant time delay. No extra error correction precautions are taken, and the responsibility lies with the user for errors and action to be taken when the link fails. In nontransparent mode, extra protocols are used to detect errors in transmission, and the GSM system will re-transmit data found to be in error. This is a much more secure method of transferring data, and the overall probability of error is very low. Whereas the accuracy of the received data is virtually guaranteed, the time delay across the link may be variable. For example, if the link fails, the system will wait and only start again when the link is restored.

The different rates between the GSM link and any external data source are catered for by adding "dummy" data bits to the information transmitted over the air. These dummy bits are added in such a way that they can be identified and deleted before the data are presented to the user.

3.7 OTHER MOBILE RADIO AND TELEPHONE SYSTEMS

A good general introduction to the technology of mobile radio systems, as well as to the current situation in the UK, is given in Macario [1988].

3.7.1 Private Mobile Radio Systems—PMR

Private mobile radio systems have been used by public utilities and other organizations for many years prior to the introduction of cellular radio, and the utilities have had their own allocation of radio frequencies. More recently, in the UK, frequency bands vacated by the old 405-line VHF television services (the so-called "Band III") have been allocated to trunked PMR systems operated by commercial organizations. Unfortunately, the details of these systems are beyond the scope of this chapter. Further details will be found in Robb and Preston [1987] and Macario [1988].

3.7.2 Mobile Cordless Telephone Systems—CT2 and "Telepoint"

CT2 Cordless Phones

Cordless telephones were an early manifestation of mobile communication. The cordless telephone, as its name implies, replaces the cord to the handset with a radio link, giving the user a degree of mobility. The range to the fixed "base station" of the phone can be many tens of meters, dependent on siting. The earliest cordless telephones in the UK were restricted in that only eight channels were available, each phone being allocated one of these eight. There was consequently a possibility that your neighbor would have a cordless phone using the same channel. Simultaneous use of the phones would then create interference.

A proposal was put forward in the UK in 1986–87 that would overcome some of these deficiencies. The system has become known as CT2 (*cordless telephone generation 2*). This system uses a dynamic allocation of channels to overcome the interference problem. CT2 also uses digital speech, which allows encryption. The transmitters use very low power (10 mW) in the 860 MHz band, which gives a maximum range under good conditions of some 100 m. The same frequency is used for transmitting and receiving, which makes allocation of frequencies simpler and has some diversity advantages. Thus, the radio transmits for 1 ms and then receives for 1 ms. Each 2 ms of speech is stored and then transmitted rapidly across the link to give the illusion of a continuous link to the user. This process is very easy because the speech is digitized.

In operation, when a user wants to make a call, the radio searches around the 40 allocated channels and chooses the one which is most free of interference. The user then sends a "paging" signal on this channel. Meanwhile the base station is "listening" to each channel in turn, searching for the paging signal. Each handset–base-station pair is uniquely coded so that all can recognize each other. When the paging signal is correctly received, the dialing information can be transmitted and the call progresses as normal. If interference should disrupt the radio link during a call, another clearer channel will be sought and the call may be re-established on this channel with minimum disruption to the user.

Because of the limited power output, the range of the phone is relatively small and any interference to other users is localized. The dynamic channel allocation feature also ensures that the best use is made of the available spectrum at any time. Therefore, the density of subscribers can be relatively high, and the system has also been proposed as suitable for providing cordless communication in the office environment. Eventually, of course, as the number of users at any time increases, interference between users will increase because more than one phone will try to use the same frequency. The only solution then is to move closer to the base station.

Telepoint Systems

As a development of CT2 technology, the *phonepoint* or *telepoint* concept has been introduced. This is a modification of the basic concept in that, with a handset, a user can communicate with a public base station as well as the fixed station at home. The handset has a unique identity so that the provider of the public base station can bill the user—or bar access if an unauthorized user attempts to access the service. The base stations can be sited in convenient public places, for example, town centers, railway stations, and highway service areas. This concept is essentially a partial cellular radio system. The differences are that the mobile handset cannot be called because the network does not know where each handset is at any moment, and the user must be stationary because no "handover" facility is provided. Not providing these two features in the infrastructure results in a considerable saving in cost, and the cost to the user of this type of service is anticipated to be similar to normal calls from fixed phone booths. Several consortia have recently been licensed to operate telepoint services in the UK (see Table 3.9).

Table 3.9
The UK "Telepoint" Consortia

1. Ferranti
2. The "Phonepoint Consortium" (British Telecom, NYNEX (New York and New England Telephone Companies), France Telecom, STC
3. Shaye Communications, Motorola, Mercury Communications
4. Philips, Barclays Bank, Shell

Note that the licenses last for 12 years in the first instance.

3.8 FUTURE DEVELOPMENTS IN CELLULAR AND MOBILE TELEPHONY

3.8.1 Digital Short-Range Radio—DSRR

Hudson [1988] and Dettmer [1989] have described a new type of mobile communication service aimed at the business user, *digital short-range radio* (DSRR; originally known as the *private advanced radio service,* PARS). DSRR was announced by the UK's Department of Trade and Industry and Electronic Engineering Association in June 1987. DSRR is expected to be a low-cost, high-quality voice and data service, with no subscription costs apart from a low annual license fee and the service is designed to meet the need for local two-way communications (e.g., on a building site or locally within a town). DSRR should be less costly to the user than cellular or PMR, though probably more expensive than telepoint. A radio frequency band between 933 and 935 MHz (on the edge of the UK cellular radio band) has been allocated, containing 75 voice-data channels and 4 signaling channels.

DSRR has no base stations, *per se,* unlike cellular or other mobile radio systems. The system has been made possible because the complex signaling needed to set up calls and avoid interference can be done by low-cost microprocessors. The main features of DSRR are listed in Table 3.10. After a three-year collaborative research and development program, a consortium involving GEC-Marconi, Orbitel, Motorola, Multitone, Philips, and Nova Radiotelephones demonstrated a prototype

Table 3.10
Digital Short-Range (DSRR) (from: Hudson [1988])

- 2-way digital voice and data; voice is digitized using the GSM RELP codec.
- privacy via speech encryption is possible because speech is digitized.
- 75 voice or data channels and 4 signaling channels between 933 and 935 MHz (a CEPT recommendation, therefore pan-European).
- channel selection is automatic—the radio itself selects a free channel—so a low (possibly zero) annual license fee is possible.
- mobile-to-mobile or mobile-to-fixed station, but no base stations, and hence no infrastructure costs.
- group calling is available.
- short-range operation (0.5 to 10 km).
- efficient use of radio spectrum by frequency reuse, due to the short range; it is estimated that the 2 MHz band can accommodate 500,000 subscribers in the UK.
- efficient use of radio spectrum by trunking.

in March 1989. An installed base of one million units in Europe is predicted by the consortium [Utley, 1989].

3.8.2 The Universal Mobile Telephone System—UMTS

The telepoint systems will bring some of the benefits of mobile radiotelephony to people who cannot afford, or do not want, all the facilities of a cellular telephone. Judging by the incredible growth in the number of cellular subscribers (see Chapter 11), telepoint systems should also be popular. People apparently have an insatiable demand for mobile communication. However, telepoint base stations will be installed only at certain locations. Ideally, users should be able to make telephone calls from anywhere. The logical next step therefore is the so-called *universal mobile telephone system* (UMTS) [Thrower, 1986; McFarlane and Mohamed 1987; MacNamee *et al.*, 1987].

The UMTS will evolve from the CT2 and telepoint systems, probably via the CT3 or *digital European cordless telecommunications* (DECT) [Carpenter, 1988]. The handset will be small enough to carry in the pocket or handbag; it may fold up for greater compactness (Figure 3.15). As well as working in the home and at telepoints, UMTS will also work in the office via a cordless PBX, in airplanes, on ships (see Chapter 4), and indeed anywhere in urban areas via a network of "microcells" linked in the cellular telephone systems and the PSTN. Although such a device will be complicated, it is within the bounds of present-day technology, is less complex than a cellular telephone, and with the low transmitting power needed, batteries will last much longer, as Thrower [1986] indicates.

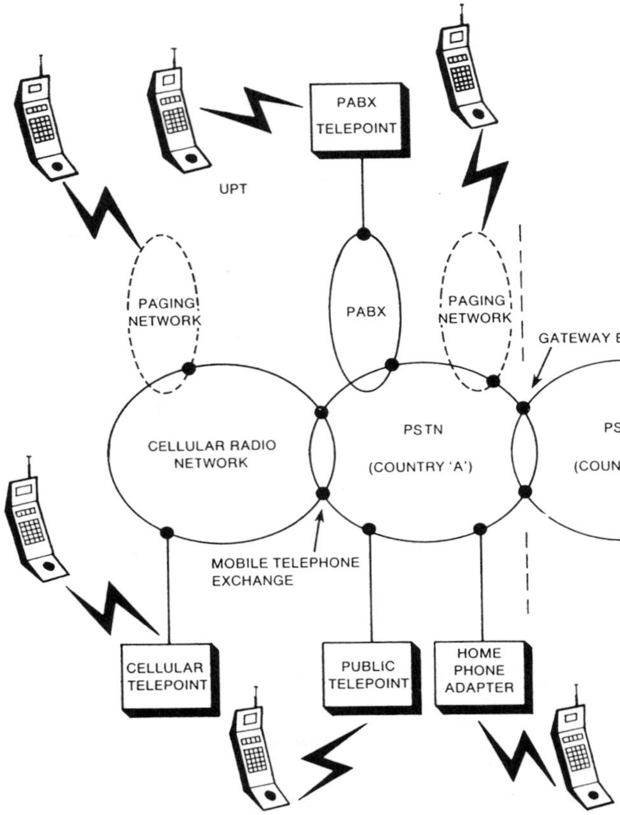

Figure 3.15 The Universal Portable Telephone (UPT): This is a pocket-sized, low-cost unit with low transmitting power requirements, capable of interfacing with various telecommunication networks.

A key word is "evolve." The UMTS will not appear overnight, but will develop from existing systems. UMTS will also need the allocation of new frequencies, the 1 GHz to 2 GHz band being particularly suitable. A RACE project (*research and development into advanced communications technologies for Europe*—a collaborative European research program; see Appendix B) has already made an initial study of the concept, and this work will continue. Experiments using microcells are underway, but a great deal of work remains to be done on definition of interfaces, propagation, addressing, security and authentication, signal processing, and digital techniques before the UMTS becomes a reality in the late 1990s [MacFarlane and Mohamed, 1987; MacNamee *et al.*, 1987].

3.8.3 Other Developments

Cellular telephones are currently seen as a business tool that is too expensive for the consumer; hence, we have the lower-cost telepoint systems. However, this perception is from the urban perspective of a developed country. The cost of cellular handsets and infrastructure (base stations, *et cetera*) will decrease with increasing production runs and advances in electronic technology. Cellular systems may therefore become more cost-effective to install in less developed and more sparsely populated countries, in preference to the landlines of the "normal" PSTN, probably in combination with satellite services (see Chapter 4).

On the business front, we are already seeing the internationalization of cellular radio, not only in the supply of subscriber equipment, but also in system operators. For example, France Telecom and the US phone company NYNEX are in UK telepoint consortia. This trend will continue.

Data communication has not, so far, surpassed cellular—not surprising because users would need to learn not only about cellular, but also the technology of data communication. However, this is bound to change in the future as the electronic office becomes more of a reality and users learn about the advantages of electronic mail, as well as word processing and spreadsheets. We will then see desk-top data communication becoming personal data communications, just as cellular radio and UMTS have given us mobile voice communication.

The use of cellular radio for traffic information broadcasting and route guidance, and cellular's relationship to other mobile information systems, is covered in Chapters 5, 6, 9, and 11.

REFERENCES

Appleby, M.S., and J. Garrett, "The Cellnet cellular radio system," *British Telecommunications Engineering,* Vol. 4, July 1985, pp. 62–69.

Balston, D.M., "Pan-European cellular radio: or 1991 and all that," *Electronics and Communications Engineering Journal,* January-February 1989, pp. 7–13.

Bear, D., "*Principles of Telecommunication-Traffic Engineering,*" Peter Peregrinus, Stevenage, UK, IEE Telecommunications Series, Vol. 2, 1980.

Bergqvist, J. T., "The colourful world of mobile communications," *Discovery* (*Nokia Telecommunications*), Vol. 16, February 1989, pp. 35–39.

Calhoun, G., *Digital Cellular Radio,* Artech House, Norwood, MA, 1988.

Carpenter, P., "From mobile to personal communications," *Third Nordic Seminar on Digital Land Mobile Radio Communication,* Copenhagen, 12–15, September 1988, paper 1.7.

Cole, R., C. Hall, M. Hassall, A. Pell and J. Walker, "Demonstrating the Mobile Office," *UK IT88 Conference,* IEE, London, 1988, pp. 597–600.

Crown, J. D., S. Hannigan, and C.A. Ward, "An Experimental Investigation of Intelligibility and Acceptability of Speech Communication Using a Hands-free Carphone," HUSAT Human Sciences and Advanced Technology Research Center, University of Loughborough, UK, July 1987.

Dettmer, R., "Digital short-range radio," *IEE Review,* July-August 1989, pp. 243–245.

DiPiazza, G.C., A. Plitkins, and G.I. Zysman, "AMPS: the cellular testbed," *Bell System Technical Journal,* Vol. 58, No. 1, January 1979, pp. 215–248.

Fluhr, Z.C., and P.T. Porter, "Advanced Mobile Phone Service: Control Architecture," *Bell System Technical Journal,* Vol. 58, No. 1, January 1979, pp. 43–69.

Fontana, G.M., "Transmission of data over the cellular telephone network," *37th IEEE Vehicular Technology Conference,* June 1987, pp. 528–531.

Frazer, E.F., I. Harris and P.J. Munday, "CDLC—a data transmission standard for cellular radio," *Journal of the Institution of Electronic and Radio Engineers,* Vol. 57, No. 3, May-June 1987, pp. 129–133.

Furuya, Y., H. Fukagawa, and H. Matsui, "High speed digital mobile facsimile with error protection," *37th IEEE Vehicular Technology Conference,* June 1987, pp. 32–37.

Huff, D.L., "AMPS: The developmental system," *Bell System Technical Journal,* Vol. 58, No. 1, January 1979, pp. 249–269.

Jakes, W.C., *Microwave Mobile Communications,* John Wiley and Sons, New York, 1974.

Jarvis, R., "CDLC—A new general purpose communications protocol for cellular telephone networks," *Conference on Cellular and Mobile Communications International,* London, November 1985 (Online, UK), pp. 159–175.

Jarvis, R., "Data services on cellular radio," *Mobile Communications Guide* (IBC Technical Services, London), pp. 56–60.

Lee, W.C.Y., *Mobile Communications Engineering,* McGraw-Hill, New York, 1987.

Macario, R.C.V., *Mobile Radio Telephones in the UK,* Glentop Press, London, 1988.

Macdonald, V.H., "The cellular concept," *Bell System Technical Journal,* Vol. 58, No. 1, January 1979, pp. 15–51.

MacNamee, R.J.G., S.K. Vadgama, and R.W. Gibson, "Universal Mobile Telephone System—a concept," *IEE Fourth International Conference on Land Mobile Radio,* Warwick, UK, December 1987 (IEE, London), pp. 19–26.

Makitalo, O., "Land mobile radiocommunication," *Telecommunication Journal,* Vol. 45, No. VII, 1978, pp. 389–394.

McFarlane, D.A. and S.A. Mohamed, "Personal communications—fact or fantasy," *IEE Fourth International Conference on Land Mobile Radio,* Warwick, UK, December 1987 (IEE, London), pp. 15–18.

Noble, J., "Simulation of an adaptive echo canceller for car phone hands-free units," *UK IT88 Conference,* Swansea, UK (IEE, London), 1988, pp. 456–459.

Okumura, Y., E. Ohmori, T. Kawano, and R. Fukuda, "Field strength and its variability in VHF and UHF land mobile service," *Review of the Electrical Communication Laboratory,* Vol. 16, No. 9–10, September-October 1968, pp. 825–873.

Robb, A.D. and R.J. Preston, "Private Mobile Radio Systems," *1987 European Conference on Cellular Radio and Mobile Communications,* (IBC Technical Services Ltd), pp. 1–16.

Siemens Aktiengesellschaft, "Telephone traffic theory tables and charts," Siemens, Munich, 1981.

Thrower, K.R., "Mobile Radio Possibilities," Presidential Address to the Institution of Electronic and Radio Engineers, London, UK, October 1986, pp. 1–37. (Also published in *Journal of the Institution of Electronic and Radio Engineers,* Vol. 57, No. 1, January-February 1987, pp. 1–11.

Wickham, R.L., "Cellular technology at home and abroad," *Cellular Business,* December 1988, pp. 30–38.

Young, W.R., "Advanced mobile phone service: introduction, background and objectives," *Bell System Technical Journal,* Vol. 58, No. 1, January 1979, pp. 1–14.

FURTHER READING

General

Calhoun, G., *Digital Cellular Radio,* Artech House, Norwood, MA, 1988.
Jakes, W.C., *Microwave Mobile Communications,* John Wiley and Sons, New York, 1974.
Lee, W.C.Y., *Mobile Communications Engineering,* McGraw-Hill, New York, 1982.
Macario, R.C.V., *Mobile Radio Telephones in the UK,* Glentop Press, London, 1988. (Despite its title, this book is an introduction to mobile radio technology in general.)

AMPS

"Advanced mobile phone service," *Bell System Technical Journal,* Vol. 58, No. 1, January 1979.

GSM

Alvernhe, M., "Services of the future pan-European mobile communication network," *International Conference on Digital Land Mobile Radio Communications,* Venice, June–July 1987, pp. 7–17.
Beddoes, E.W., "Roaming in the pan-European cellular system," *Telecommunications,* September 1988, pp. 38–46.
Beddoes, E.W. and R.I. Germer, "Operational requirements of the GSM system," *International Conference on Digital Land Mobile Radio Communications,* Venice, June–July 1987, pp. 476–485.
Mallinder, B.J.T., "The pan European cellular system: the GSM project," *European Conference on Cellular Radio and Mobile Communications,* February 1987 (IBC), pp. 1-11.
Mallinder, B.J.T., "An overview of the GSM system," *Digital Cellular Radio Conference,* Hagen, Federal Republic of Germany, October 1988, pp. 1a/1–1a/13.
Third Nordic Seminar on Digital Land Mobile Radio Communication, September 1988, Copenhagen, Denmark.

NMT

Anon. *Nordic Automatic Mobile Telephone System: Technical Outline*. (Available from the Nordic PTTs.)
Billstrom, O. and B. Troili, "A future automatic mobile telephone system," *Ericcson Review*, No. 1, 1980.
Hammer, T., T.M. Iversen, E.T. Mortensen and P. Aagard, "Automatic nordic Mobile telephone system," *Teleteknik,* No. 1, 1982, pp. 1–17.
Haug, T., "The Nordic mobile telephone system, an extension of the telephone network," *IEEE Global Telecommunications Conference,* 1983, pp. 1405–1409.

TACS

Hughes, C.J. and M.S. Appleby, "Definition of a cellular mobile radio system," *IEE Proceedings,* Vol. 132, Pt.F, No. 5, August 1985, pp. 416–424.

Chapter 4
Satellite Communication for Mobiles

P.D. BRITTEN and J.G. SHOENENBERGER*
RACAL RESEARCH LTD. RACAL AVIONICS

4.1 INTRODUCTION

The increased availability of cellular radio systems means that many individuals will expect to have communication facilities at their disposal whenever they are traveling, including on transoceanic flights and in parts of the world where cellular radio is not available. The frequencies generally used for terrestrial mobile communication restrict coverage to within the line of sight from the transmitter. Although HF can be used over long ranges, its poor reliability makes it undesirable for general public use. Satellites offer a distinct advantage in these situations because a single satellite can, for example, provide coverage over virtually the whole Atlantic Ocean. Another factor is that erecting base stations may not be economically viable in remote areas where vehicle traffic density is low. In these circumstances, satellite communication may offer a more cost-effective solution. Thus satellites can be seen as essential in providing a worldwide personal communication capability.

Satellites were initially used for fixed point-to-point communication. The restrictions on the size and weight of the mobile terminal mean that more efficient transmitting and receiving systems are required on the satellite for mobile communication. Although satellites with sufficient performance to make mobile communication technically viable have been available since the early 1970s, economic viability has dictated the introduction of services. Experiments in mobile satellite communication were conducted in the 1960s and early 1970s, but not until 1979 was the *International Maritime Satellite Organization* (INMARSAT) formed to provide the first mobile satellite service. Thus far, it has provided only a maritime service,

*Now with the PA Consulting Group.

but an aeronautical service is due to start soon, and a land mobile service may follow later.

Frequencies at L-band (1.5 and 1.6 GHz) have been allocated for mobile satellite communication. Figure 4.1 shows the spectrum allocations made at the 1987 *World Administrative Radio Conference* (WARC). Other frequency allocations for mobile satellite services have been made (117.975–137 MHz, 121.5 MHz, 243 MHz, 406 MHz, 235–322 MHz, 335.4–399.9 MHz, 806–890 MHz, 942–960 MHz, 2500–2535 MHz, 7250–7375 MHz, 7900–8025 MHz, and a number of others above 10 GHz). However, most of these are either employed by the military or carry various restrictions on their use, and so are of little current importance for civil mobile satellite applications.

At present, only satellites operating in the band allocated for maritime satellite services are in use, and this band is far from congested. Consequently, there is an appreciable amount of spectrum available for future mobile satellite services. The number of users that current satellites can support is limited by the power available on the satellite rather than by bandwidth. As a result, the modulation and coding schemes are optimized for power rather than bandwidth efficiency. This may change as higher powered satellites become available.

The number of users of the INMARSAT maritime system has grown steadily since its introduction, and there are now over 7000 installed terminals, so that traffic on the existing satellites is reaching saturation. The number of potential users of an aeronautical satellite service is several thousand, and for a land mobile service it could be hundreds of thousands if not millions. Future generations of satellites will have spot beam capabilities to permit efficient frequency reuse. Although mobile satellite communication is still at an early stage, problems of spectrum overcrowding appear unlikely before the year 2000.

4.2 BENEFITS OF MOBILE SATELLITE COMMUNICATION

Mobile satellite communication is particularly important over parts of the globe, such as oceans, which cannot be adequately served by terrestrial-based communication. Because these areas comprise the greater part of the earth's surface, ocean travelers who are reliant on terrestrial-based communication may be without adequate communication for hours or even days. As markets become increasingly global in scale, the time lost and money wasted because of the inability to communicate with individuals in transit will become increasingly important. There also are, of course, great benefits to individual safety by having reliable mobile communication worldwide.

There are also indirect benefits to the user from mobile satellite communication. A data link to the mobile could provide information on its position and operational condition virtually anywhere in the world, and would enable journey times to be shortened and the travelers' safety to be enhanced. For aircraft, for example,

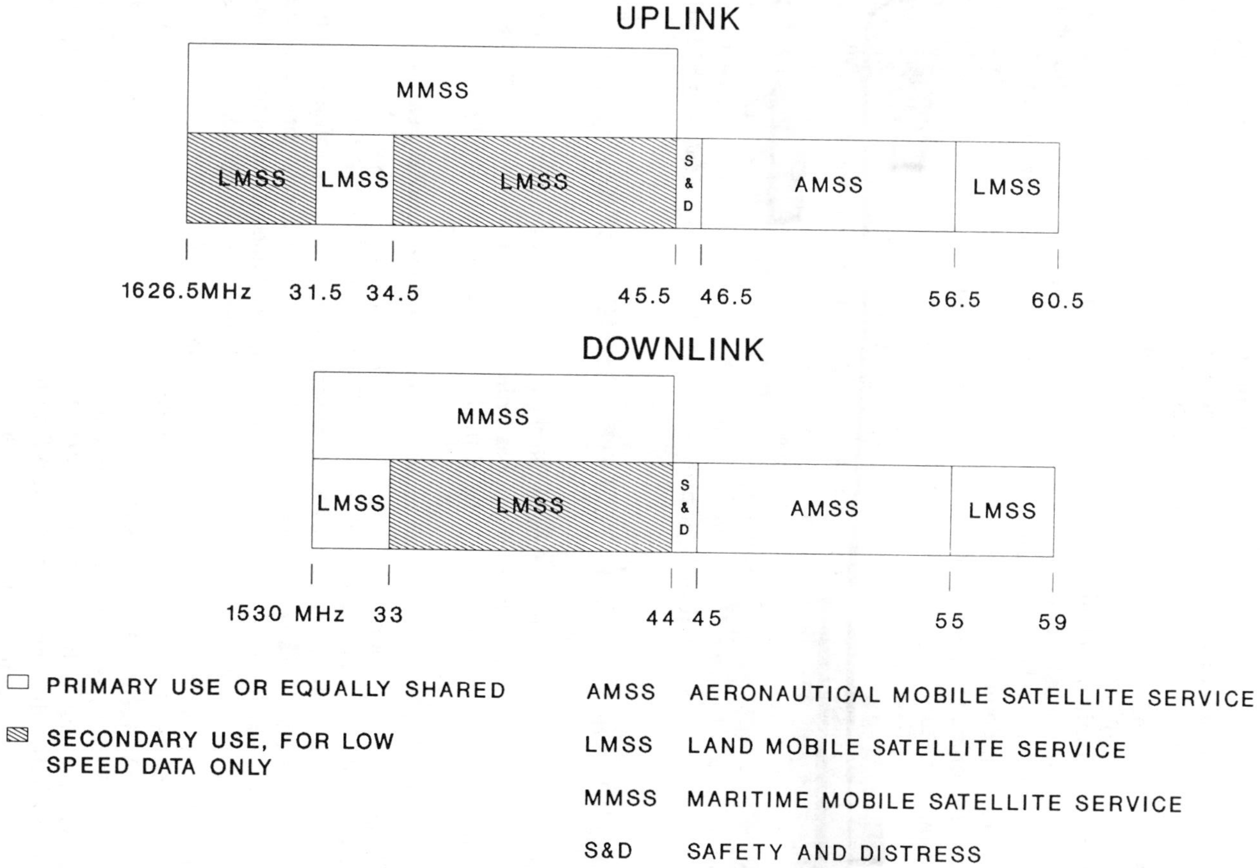

Figure 4.1 WARC 1987 allocation of L-band mobile satellite frequencies.

mobile links will permit greatly improved air traffic control for transoceanic flights. At present, only communication via HF is available for such flights. HF is unreliable and, hence, allowances must be made in planning the flight for loss of communication during parts of it. A reliable data link throughout the flight, together with a suitable navigation system, would permit improved aircraft separation on transoceanic flights. As a result, more aircraft could be accommodated, resulting in more frequent services. Similarly, changes in flight plans could be made during the flight to take account, for example, of changes in weather conditions close to the destination, thereby avoiding possible delays. These issues are discussed further in Section 4.4.1.

Because the communication terminals must be readily transportable, they could also be used for rapidly establishing communication facilities if the local communication infrastructure were destroyed by a natural disaster. Satellites are already in use as part of an international search and rescue facility known as COSPAS/SARSAT. When activated, the *Emergency Position Indicating Radio Beacons* (EPIRBs) transmit bursts at 406 MHz, which are received by one of the low-earth-orbiting COSPAS/SARSAT satellites. The position of the EPIRB then can be determined, and the search and rescue authorities are automatically alerted. EPIRBs operating at 1.6 GHz and using INMARSAT satellites will also be available in the future. Under the *International Maritime Organization* (IMO), the *Global Maritime Distress and Safety System* (GMDSS) will make it mandatory that ships carry a satellite EPIRB.

4.3 HISTORY OF MOBILE SATELLITE COMMUNICATION

The civil maritime user is in the first of the three categories of civil mobile user for whom a satellite communication service has been established. Fewer restrictions on the mobile terminal size and weight have made much easier the development of a maritime system.

The first satellites dedicated to provision of a maritime service were the three MARISAT satellites launched in 1976, located at 15° W, 176.5° E, and 72.5° E. They are owned by the US *Communication Satellite Corporation* (COMSAT) and are still in operation, although currently only in a standby role for civil maritime communication. INMARSAT was established in 1979 to provide an international maritime satellite communication service, but the system did not become fully operational until 1982. Initially, the MARISAT satellites were used, but then transponders on three INTELSAT V satellites were leased from the *International Telecommunication Satellite Organization* (INTELSAT), and two MARECS satellites were leased from the *European Space Agency* (ESA). Figure 4.2 shows the coverage provided by the current INMARSAT system. A second generation of satellites is being procured, the first of which is due to be launched in 1991. The growth in maritime satellite communication traffic through the INMARSAT system has been rapid despite the early years being at a time of depression in world trade (Figure 4.3).

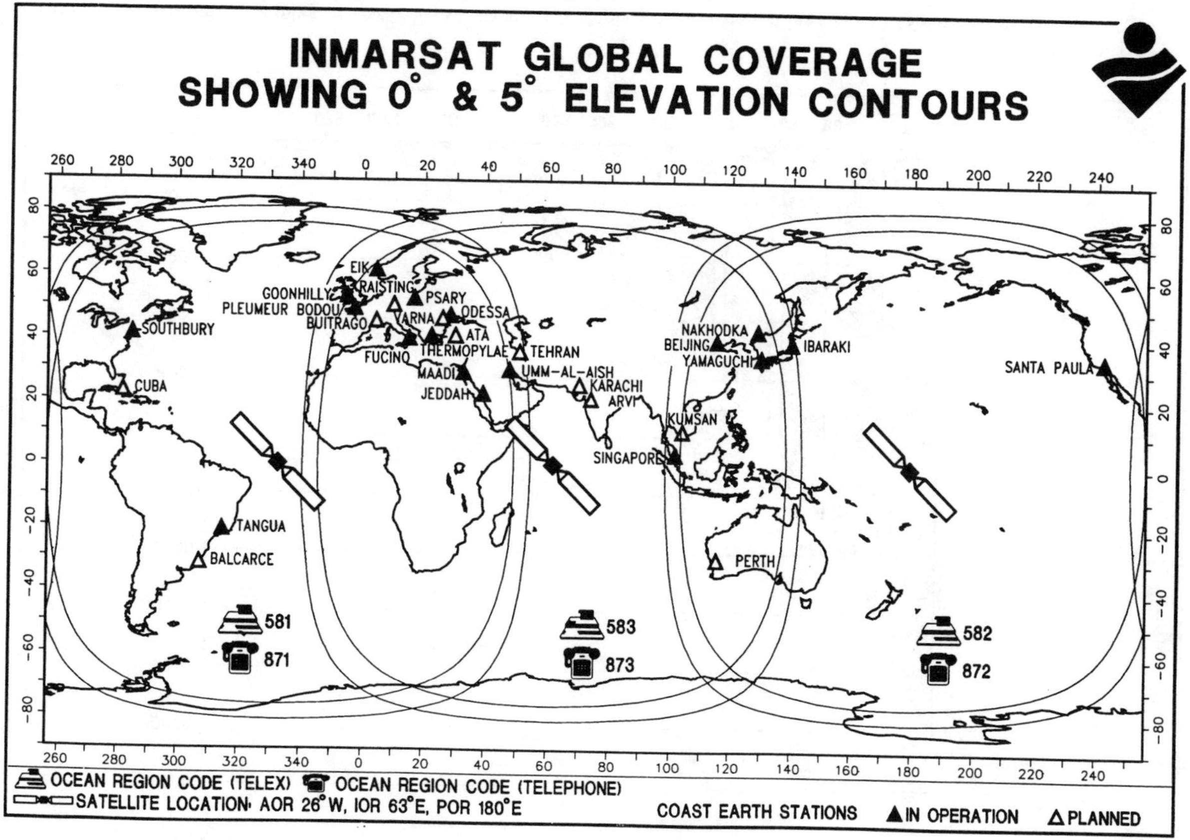

Figure 4.2 Map showing the coverage provided by INMARSAT satellites (*Source*: INMARSAT).

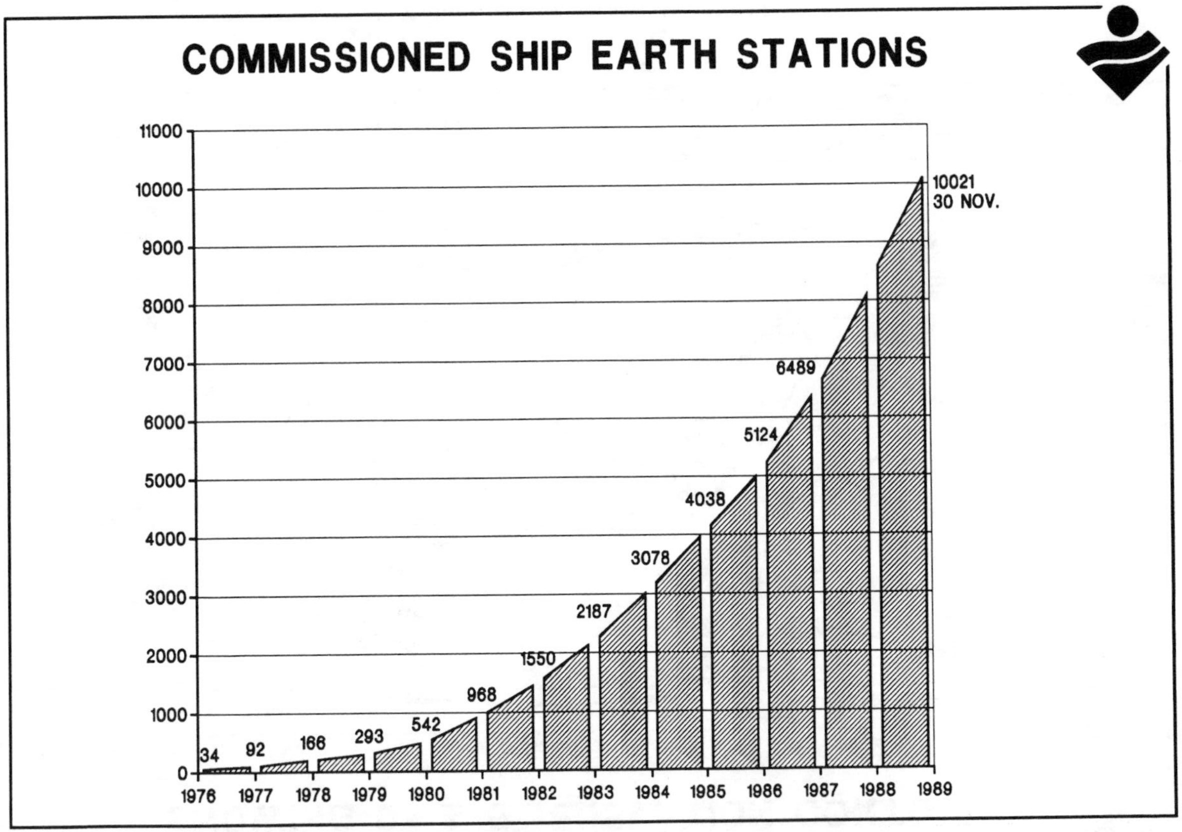

Figure 4.3 Graph of growth in the number of INMARSAT standard-A shipboard earth station installations (*Source*: INMARSAT).

The benefits of satellite communication for aeronautical use have been apparent since the earliest satellites were launched. However, the progress has been slow for a variety of reasons. The technical feasibility was proved during the mid-1960s in experiments using the ECHO satellite and in the early 1970s using the US National Aeronautics and Space Administration (NASA) ATS-6 satellite. The AEROSAT project was started jointly by the US Federal Aviation Administration (FAA) and the European Space Research Organization (ESRO) in the early 1970s with the aim of setting up a worldwide communication facility for aircraft via satellite. However, disagreements over the scale, form, and potential cost to the airlines resulted in the demise of the project. After this there was a break in experimental work for nearly ten years, although a number of studies and analyses were undertaken. Following introduction of the maritime service by INMARSAT in 1982, a study into the applicability of its system to aeronautical communication was commissioned. It [Racal Decca Advanced Development, 1984] concentrated on a low-rate (200–400 bits/s) data system, although voice communication was considered technically feasible but more difficult to implement. Subsequent experiments and trials demonstrated the feasibility of low-rate data and voice communication, with aircraft using the INMARSAT space segment.

In 1983, the International Civil Aviation Organization (ICAO) set up the committee on *Future Air Navigation Systems* (FANS) to study future potential air navigation systems and to make recommendations for their coordinated evolutionary development over the next twenty-five years. Air navigation systems in this context cover both communication and radio navigation. In addition, the Airlines Electronic Engineering Committee (AEEC), which specifies the characteristics of avionic equipment used by most airlines, set up a satellite systems subcommittee to define the characteristics of equipment for aeronautical satellite communication. As a result of the work of these two committees, an international standard has now been set for aeronautical satellite communication.

Land mobile satellite communication is far less developed than either aeronautical or maritime communication. This is a result of greater constraints on mobile equipment size and cost. Land mobile satellite communication also differs in that there is not the same need for global operation over land. As a result, its development has been far more fragmented. So far, a few experiments have been carried out and some proposals made for services, but as yet it is not clear when the first commercial service will be introduced.

In addition to work relating to specific mobile environments, there have been programs studying a unified approach to mobile satellite communication. The first of these was the ESA Prosat program, which started in 1982 and was an investigation of future mobile satellite communication techniques. The first phase involved a number of propagation experiments, the results of which were used to define experimental systems. The second phase involved the development and trials of the experimental systems [Jongejans, 1986]. The Prodat system, one of these experimental systems, was a low-rate data-only system, which provided satellite communication facilities for land, maritime, and aeronautical users via the MARECS Atlantic Ocean satellite. It was the first fully operational system to provide simultaneous satellite communication facilities for land, maritime, and aeronautical users.

In the United States, NASA has taken a slightly different approach with its MSAT-X program, started in 1983 [Rafferty, 1988]. This program has concentrated on the development of specific aspects of mobile satellite technology rather than the development of a complete mobile satellite system as with Prosat. Aspects that have been investigated are vehicle antennas, nearly toll-quality digital voice coding, modem design, access techniques, and channel characterization. Work has also been undertaken in Canada on development of mobile satellite technology.

4.4 CURRENT STATUS

4.4.1 Current Status of Aeronautical Satellite Communication

The first commercial aeronautical satellite communication service is now being implemented. This initial service will be enhanced in the future, with additional satellites and new types of service being offered.

4.4.1.1 The User Community

Services to be provided to the aeronautical community vary widely depending upon the group concerned. The general aviation community, defined primarily as light aircraft flying short sectors (usually under *visual flight rules* with minimum equipment), are not now considered to be potential users of satellite communication for economic reasons.

The business or corporate aviation community, however, desires worldwide communication, particularly voice, but also telex and facsimile. These users are accustomed to high-quality voice communication from their office, car, or ship (yacht). The lack of such a capability (except via VHF-UHF over land or via an unreliable HF link) results in a high level of demand for satellite communication. Data communication is also valid here, primarily for flight plan and worldwide weather reporting in a form similar to Global-Wulfsberg's VHF-UHF *Airborne Flight Information Service* (AFIS) system [Larson, 1985].

Offerings under development for airlines include data services for the aircraft cockpit and the cabin crew, cockpit voice communication, and passenger telephony. Requirements depend on the airline route structures involved. For example, some airlines servicing major transoceanic routes wish to offer passenger communication as a free service (as a marketing tool), whereas others want the services to generate additional revenue [Bean, 1987]. Services offered may include telephony and advance bookings for flights, hotels, and cars. Airlines operating nonoceanic routes are unlikely to offer passenger telephony via satellite for commercial reasons, but may use satellite communication for aircraft operations.

Airline data service requirements are divided into two main areas: *Airline Administrative and Airline Operational Control* (AAC-AOC) and air traffic services. Efficient operation of modern aircraft requires that automatic engine and airframe health monitoring is available wherever the aircraft is flying. In addition, worldwide access to remote data bases while in flight is required for optimum flight profile planning [Haapala, 1987]. The *ARINC* (Aeronautical Radio, Inc.) *Communication and Reporting System* (ACARS) and the SITA (*Societé Internationale de Télécommunications Aeronautique*) *AIRCOM* system currently provide these facilities via VHF, but there are plans to use satellite communication to make these services available worldwide.

4.4.1.2 The Service Providers

At present, INMARSAT is the only organization developing a worldwide aeronautical satellite communication system. The Aviation Satellite Corporation (AvSat), a subsidiary of ARINC in the US, has abandoned its plans to launch its own satellites. The AvSat system is described here for historical interest.

The AvSat system was formulated as an "airline industry owned and operated" service [Dement, 1987], to cater uniquely and totally to the needs of the world's airlines by launching a dedicated constellation of six or more satellites with both global and spot beam (concentrated on areas of major traffic flow) coverage.

AvSat used digital transmission at L-band using a *time division multiple access* (TDMA) scheme with QPSK modulation. TDMA was viewed favorably by the airline community because it avoided any need for linear high power amplifiers as required, for example, by the INMARSAT system when more than one channel is required. Up to eight voice or data channels, each at 8 kb/s, were planned for each aircraft, providing ample service growth capability within a single set of avionics equipment. AvSat is no longer attempting to start up a new service by launching its own satellites. ARINC is likely to start an international ACARS service via INMARSAT satellites and according to INMARSAT specifications in 1990.

The INMARSAT Aeronautical System has been widely misunderstood to be a development of the INMARSAT maritime standard-A system. The system design is, in fact, totally new, derived from many studies and tailored specifically to aeronautical requirements. The system is also totally digital, using *offset binary phase shift keying* (OBPSK) modulation at low bit rates and *offset* QPSK (OQPSK) at high bit rates, together with convolutional coding and interleaving. A *frequency division multiple access* (FDMA) scheme using a *single channel per carrier* (SCPC) is employed for the high rate data channels, and a TDMA scheme is used for the low rate channels [Smith, 1987; INMARSAT, 1987].

In its initial form, the system will use existing leased satellites shared with the maritime system, but will quickly use satellites being procured by INMARSAT (scheduled for launch beginning in 1991). Further dedicated satellites may be added as traffic grows.

As with INMARSAT's maritime system, the service is provided to users via the INMARSAT signatories, who are predominantly national PTTs. Two competing service provider agreements have been announced, one operated by U.K., Singapore, and Norway (the "Skyphone" service), and the other by Canada, France, and Australia in conjunction with SITA. This situation is viewed favorably by the aeronautical community, which is assured of competition in both voice and data services. The Skyphone service initially will be telephone only, but will quickly expand to include data services. The SITA consortium is expected to offer similar services, but in reverse order. Agreements may also be made with other service providers.

Like AvSat, the INMARSAT system makes use of a 12 dBic gain aircraft antenna, and avionic boxes of about 0.04 m^3 volume. The use of FDMA-SCPC necessitates linear power amplifiers for multichannel operation which, from an aircraft installer's point of view, is undesirable in terms of the size, weight, and cooling required. However, the use of offset QPSK, with its constant envelope characteristic, means that a class-C amplifier can be used where only single-channel operation is required. Each voice channel is coded at 9.6 kb/s (AvSat was 8 kb/s), although a variety of data rates (down to 300 b/s) can be supported.

4.4.1.3 Trial Programs

Many initial technical trial programs have taken place using the INMARSAT space segment; a number of preoperational service trials are in progress, and are described below.

A UK consortium of Racal Decca Advanced Development, British Telecom International, and British Airways conducted Skyphone service trials using two British Airways Boeing 747 aircraft during 1988 and 1989, employing the INTELSAT MCS B satellite, leased by INMARSAT from INTELSAT and located over the Atlantic Ocean at 18.5° W. Passengers were able to make direct dialed telephone calls into the public telephone network from the aircraft. Technical trials were carried out during 1988 using a Racal owned Jetstream aircraft [Schoenenberger, 1988].

Trials were made in 1987 of a telephony system using a Japan Airlines Boeing 747 flying on Pacific Ocean routes [Makita, 1988]. Trials have been done in Canada of a service that provides paramedics aboard an air ambulance with a high quality voice link to surgical staff at their destination; calls from the aircraft are linked into the Canadian PSTN via a Teleglobe, Inc. ground station [Butterworth, 1988]. The

INMARSAT Atlantic Ocean Region (AOR) satellite is used, and an operational service is due to start in 1990.

Trials of aeronautical data links via satellite have also been done [Britten, 1988; Anderson, 1985].

4.4.1.4 Aeronautical Satellite Communication Equipment Standards

To obtain freedom of choice in supply of avionic equipment, airlines for many years have supported the *Airlines Electronic Engineering Committee* (AEEC) under the chairmanship of ARINC. Although the deliberations of the AEEC are of little direct relevance to the nonairline community, most avionic equipment manufacturers follow the AEEC's recommendations in constructing systems.

The AEEC is producing *Characteristic 741* [Aeronautical Radio, 1988], which details the form, fit, and function of avionic equipment and antennas for airline satellite communication. First, the size of boxes is decided, then how to interconnect them, and finally what goes inside them. This process ensures that airframe installers can wire any aircraft in accordance with Characteristic 741, obtain boxes from any manufacturer, and expect them to work the first time.

A prime concern comprises the size, weight, and power consumption of the equipment. Characteristic 741 ensures that these parameters are given due consideration in the design process. The form and fit aspects of the Characteristic were ratified in late 1987, and the function aspect in late 1988.

4.4.2 Current Status of Land Mobile Satellite Communication

At present, there are no commercially operational land mobile satellite communication services. However, there have been some experiments in the provision of telex type of services [Britten, 1988] and paging services [Casewell, 1988].

One of these experiments is part of ESA's Prodat experiment, discussed in Section 4.3. Trucks used by a French long-distance hauling company have been equipped with Prodat terminals, enabling telex messages to be prepared by the driver (using a small personal computer) and transmitted to the company's base. Messages may also be transmitted to the driver, and displayed on a small printer installed in the cab. Experiments are also being conducted by INMARSAT into the use of their standard-C system for land mobile communication.

Some large countries are opting to introduce a mobile satellite service using spot beam satellites to provide telephony as well as data communication, although the full specifications are not yet final. These systems will provide facilities for land mobile traffic in the area and for local maritime and aeronautical traffic. However, as mentioned in Section 4.1, much of the aeronautical and maritime traffic operates on a global scale, so these systems will be of limited use unless they are coordinated with other national or international services.

In the United States, an eight-company consortium has been formed known as the *American Mobile Satellite Consortium* (AMSC) [Agnew, 1988]. Although its service is primarily intended for land mobile use, it intends to provide voice and data services for aeronautical and maritime users as well. Users will be permitted to have terminals with different antenna gains, with the lowest charges being for the terminal with the highest antenna gain. The detailed specifications are still being prepared, but the system will use three geosynchronous satellites, with the first expected to be launched in 1992.

Telesat of Canada is planning to introduce a mobile satellite system known as MSAT [Sward, 1988]. It is similar in concept and specification to the AMSC system, and in fact the satellite and ground system development, production, and operation will be closely coordinated with that of the AMSC. However, Telesat, in conjunction with the Canadian government, is developing interim services using INMARSAT satellites.

AUSSAT, the Australian satellite communication operator, will introduce a mobile satellite service [Nowland, 1988] in 1992, which will be compatible with AMSC and MSAT systems.

INMARSAT is proposing a low-cost mobile satellite telephony system, known as *standard-M,* which will be compatible with the services proposed by AMSC, AUSSAT, and MSAT.

Mobile terminal cost is a crucial factor in the development of land mobile satellite communication. A purchase price of a few thousand pounds or less will be necessary for the system to be successful, which will require terminals to be manufactured in large numbers, particularly for telephony. To serve a large number of users, the system must be designed to be bandwidth-efficient. In addition, to keep the cost of the mobile terminal low, its antenna gain must be low, which dictates the use of power-efficient techniques. There is much activity in developing techniques to give both bandwidth and power efficiency. Even with the use of power-efficient modulation techniques and spot beam satellites, a relatively high gain (>10 dBic) antenna will be required for telephony services. Adequate coverage therefore dictates a steered antenna; this represents a significant engineering challenge in the vehicular environment. Antennas are discussed further in Section 4.5.4.

One major problem with L-band land mobile satellite communication in urban areas concerns signal blockage due to bridges, tunnels, and tall buildings. A low-rate data system can be designed to overcome this problem by incorporating ARQ, but the inherent delays mean that it cannot be used for telephony. Therefore, a satellite system is unlikely to be used for urban mobile telephony. In addition, calls by satellite would be more expensive than those via a terrestrial-based urban system. However, in areas of North America and Australia, satellite communication will be more cost effective in nonurban areas. A land mobile satellite system therefore, will need to work with a terrestrial system. To date, efforts have concentrated on developing the basic systems, so this problem has received relatively little attention.

4.4.3 Current Status of Maritime Satellite Communication

The present INMARSAT maritime communication system, *standard-A*, provides telephone and telex services. Standard-A is an analog system using FM. A receiver G/T of -3 dB/K and a transmit *effective isotropic radiated power* (EIRP) of 26 dBW are required, which results in an antenna of about 1.2 m in diameter being needed. The size and cost of the terminals (about $35,000) means that they are generally only fitted to large ships; there are currently over 7000 terminals installed. Telephone calls can be dialed directly into the international public telephone network, and telexes sent from the ship. Calls may also be received on the ship. A major use is the transmission of voice-band data or facsimile. Each terminal provides one telephone and one telex channel. A high-speed data transmission facility is also available, which enables data to be transmitted at 56 kb/s from the ship.

A low-rate data service providing telex and broadcast message facilities, known as *standard-C*, is due to be introduced. A small terminal using an omnidirectional antenna and costing a few thousand pounds (or dollars) will be needed, which allows provision of satellite communication facilities on even the smallest ships. The terminal will operate in half-duplex mode. Telex messages can be sent and received. Facilities for broadcasting messages to groups of ships will also be available.

A new maritime telephony system, known as *standard-B*, has been proposed by INMARSAT. Standard-B is a digital system using 16 kb/s voice coding. The EIRP and G/T of the terminal are the same as for standard-A terminals. It will allow for additional services to be provided and connection into the *integrated services digital network* (ISDN) in the future. The date of introduction for standard-B has not yet been announced. The proposed standard-M system will also be suitable for maritime use.

4.5 SATELLITE COMMUNICATION TECHNOLOGY

4.5.1 System Technology and Design

The satellites now in use for civil mobile communication employ nonregenerative transponders in which signals are received by the satellite, amplified, translated in frequency, and retransmitted. This applies in both ground-to-mobile (forward link) and mobile-to-ground (return link) directions. The ground station transmits to the satellite at a C-band frequency around 6 GHz, and the received signal is translated to an L-band frequency (around 1.5 GHz) for transmission to the mobile. The mobile transmits to the satellite at an L-band frequency of around 1.6 GHz, and the received signal is translated to C-band at around 4 GHz for transmission to the ground station. Future spot beam satellites will employ on-board processing, which will require the signals to be demodulated, processed, and remodulated.

4.5.1.1 The Link Budget

A key aspect of the design of any satellite system is the *link budget*. An example of a simple downlink budget is

$$C_D/N_{0D} = 10\ \log P_T G_T - 20\ \log L_P + 10\ \log G/T + 10\ \log L_S - 10\ \log k$$

where

C_D/N_{0D}	=	the average wideband carrier power-to-noise density ratio at the receiver,
$10\ \log P_T G_T$	=	the satellite EIRP, where P_T is the power into the antenna and G_T is the antenna gain,
$20\ \log L_P$	=	the free-space loss,
$10\ \log G/T$	=	the mobile terminal gain-to-equivalent-noise temperature ratio,
$10\ \log L_S$	=	an additional loss in the link (e.g., due to a propagation effect such as ionospheric scintillation),
k	=	Boltzmann's constant.

A similar equation can be derived for the uplink. If the satellite uses a non-regenerative frequency-translating transponder, as is the case with all current mobile communication satellites, then the two link budgets are related as follows:

$$C/N_0 = \frac{1}{(N_{0U}/C_U) + (N_{0D}/C_D)}$$

where C_U is the uplink carrier power, and N_{0U} is the uplink noise density.

For a digital system, the bit energy-to-noise ratio is related to the carrier-to-noise density ratio as follows:

$$C/N_0 = (E_b/N_0) \times R$$

where R is the bit rate.

4.5.1.2 *Access Schemes*

A satellite will have a receiver figure of merit (G/T), and saturated transponder output power in both uplinks and downlinks. The satellite serves several mobiles, so a means must be found to service each user. In an SCPC system, as used by INMARSAT, each carrier is allocated a fraction of the available satellite EIRP on each link. The total power allocated to the carriers must be lower than the saturated transponder output power to keep intermodulation products at acceptable levels; therefore, the relation between numbers of channels and EIRP per channel is not quite linear. Nevertheless, the more channels there are, the lower is the satellite EIRP for each carrier. To increase the number of users is obviously desirable, but in most cases, the available EIRP for each carrier on the satellite-to-mobile link is the limiting factor, rather than the transponder bandwidth.

The method used to access the satellite is an important aspect of the system design. FDMA using a single channel per carrier was mentioned in Section 4.4.1, along with TDMA. *Code division multiple access* (CDMA) uses a spread-spectrum technique and relies on each channel employing a separate spreading code. There are also several variants of each access scheme. The design of access schemes is an extensive topic in itself, and the reader is referred to Feher [1983] for more information on TDMA and FDMA, and Dixon [1976] for spread-spectrum techniques. Most access schemes in use or proposed for mobile satellite communication are FDMA using SCPC, so, in the remainder of this section, we assume that such an access scheme is used.

4.5.1.3 *Other System Design Options*

Given the limitation discussed above, a worthwhile exercise is to examine other options that the system designer has available. If an existing satellite is being used, looking at the forward link budget apart from the satellite EIRP for each channel, only the receiver G/T and the received C/N_0 can be varied.

In a digital system, the required C/N_0 is determined by the acceptable bit error rate (BER), implementation losses, and the bit rate. Implementation losses below 1 dB can be achieved for most mobile applications with digital signal processing using low-cost LSI devices, and is not a significant factor in the overall system design. The acceptable bit error rate will depend on the application. However, as we can see from Figure 4.4, by using an efficient modulation and coding scheme, the gradient of the curve over usable bit error rates is sufficiently steep that the difference in E_b/N_0 is small. Most mobile applications are low-rate data (<2400 b/s) or voice communication. Voice coding must be employed if a digital system is being used. Looking at the example link budget in Table 4.1, which is for 9.6 kb/s voice coding with half-rate convolutional coding, an additional 8 dB (which would be required

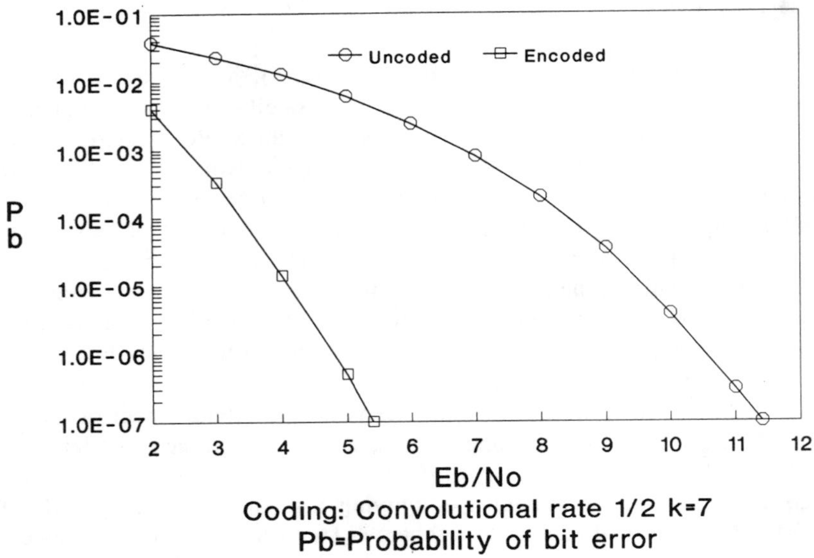

Figure 4.4 Bit error rate curve for BPSK modulation with convolutional encoding.

for 64 kb/s voice coding on the public telephone network) would be very difficult to find. Consequently, good quality voice coding techniques at rates as low as 9.6 kb/s are crucial to the commercial viability of digital mobile satellite voice communication. This topic is discussed in more detail in Section 4.5.3.

In an analog system, the voice quality is largely determined by the signal-to-noise ratio of the recovered signal. There is a minimum C/N_0 that represents the lowest acceptable voice quality, although the degradation is much more gradual than for digital voice coding as C/N_0 is reduced.

4.5.1.4 Receiver Figure of Merit

We will now examine the mobile terminal G/T and, in particular, the receiver noise figure. Small, low-cost, low-noise amplifiers (LNAs), operating at L-band, can now be produced with noise figures below 2 dB. The small size means that LNAs can be located very close to the antenna in virtually all mobile applications. Consequently, there can be little improvement in receiver total equivalent noise temperature. Hence, the antenna gain is the remaining variable. Table 4.1 shows that a G/T of -13 dB/K is required at the mobile terminal, which typically results in an antenna gain of 12 dBic. To achieve this gain and enable the system to be used with

Table 4.1
INMARSAT Aeronautical System Return Link Budget for 21,000 b/s Voice Channels at 5° Elevation Using a MARECS Satellite

Return Link		
	Link Requirement	
	Required Link C/N_0 (dBHz)	47.9
	Uplink (AES to Satellite)	
	Frequency (GHz)	1.64
	AES Elevation (°)	5
	Path Loss (including Atmospheric; dB)	189.4
	Satellite G/T (dB K)	−11.0
	AES EIRP (dBW)	25.5
	Uplink C/N_0 (dB Hz)	53.7
	Satellite	
	Satellite Gain (dB)	151.3
	Satellite C/IM_0 (dB Hz)	57.5
	Downlink (Satellite to GES)	
	Frequency (GHz)	4.2
	GES Elevation (°)	5
	Path Loss (including Atmospheric; dB)	197.6
	GES G/T (dB K)	32.0
	Satellite EIRP (dBW)	−12.6
	Downlink C/N_0 (dB Hz)	50.4
	Link Performance	
	Achieved C/N_0 (dB Hz)	48.2
	Margin (dB)	+0.3
Forward Link		
	Link Requirement	
	Required Link C/N_0 (dB Hz)	47.9
	Uplink (GES to Satellite)	
	Frequency (GHz)	6.42
	GES Elevation (°)	5
	Path Loss (including Atmospheric; dB)	201.3
	Satellite G/T (dB K)	−14.0
	GES EIRP (dBW)	62.0
	Uplink C/N_0 (dB Hz)	75.3
	Satellite	
	Satellite Gain (dB)	161.3
	Satellite C/IM_0 (dB Hz)	69.0

Table 4.1 (cont'd)

Downlink (Satellite to AES)	
Frequency (GHz)	1.54
AES Elevation (°)	5
Path Loss (including Atmospheric; dB)	188.9
AES G/T (dB/K)	−13
Satellite EIRP (dBW)	22
Downlink C/N_0 (dB Hz)	48.7
Link Performance	
Achieved C/N_0 (dB Hz)	48.7
Margin (dB)	+0.8

the mobile in all normal orientations requires the use of some form of steered antenna. To support low-rate data services, a G/T of about −26 dB/K is required, which can be achieved with a 0 dBic antenna; consequently, good coverage can be obtained without the need for steering. Antennas are discussed in more detail in Section 4.5.4.

4.5.1.5 Propagation

There are marked differences in the propagation characteristics of each of the three mobile environments. Surface scattering, ionospheric effects (and to some extent tropospheric effects) are the main propagation phenomena affecting the L-band mobile satellite link. Tropospheric and ionospheric effects depend on geographical location, and will affect all mobile types in the same way, although aircraft flying above the turbulent tropospheric layer will be unaffected by tropospheric effects, which have the least influence, accounting for only fractions of a dB loss. Ionospheric scintillation can be severe at L-band in equatorial regions, but margins of 1 to 2 dB will provide a system reliability of 99% in most areas. Surface scattering is the dominant effect in the maritime and aeronautical environments, the differences being a result of the different link geometries and mobile velocities. The maritime fading rate is lower, with typical fading rates of 0.2 to 5 Hz, but the fade depth is greater, and so a greater margin must be allowed. Signal-to-interference (S/I) ratios of 7 dB due to *multipath* (signals reaching the receiver after traversing different paths, resulting in destructive interference and signal cancellation; see Chapters 2 and 3) are typically allowed for in the maritime environment. Aeronautical fading rates of 50 to 250 Hz are typically encountered; signal-to-interference ratios due to multipath are generally better than 15 dB, although link margins down to 7 dB, for example, are included in the link budget shown in Table 4.1. The land mobile environment has multipath effects and signal blockage from buildings, bridges, hills, trees, the extent of which depends on where the vehicle is traveling. Work on modeling the land mobile environment is covered in Lutz [1988] and Barts [1988].

Thus, we can see that design of a mobile satellite system, particularly for voice communication, requires a careful trade-off between the various conflicting requirements.

4.5.2 Signal Processing

Although there are differences in the packaging of terminals for the different mobile environments, generalized block diagrams can be produced (see Figure 4.5).

A single antenna is the most common technique because it simplifies installation, although the antenna design is complicated because of the wide bandwidth to be covered. A diplexer is required if a duplex link is used, but a switch will suffice if only half-duplex operation is employed. (A diplexer prevents the strong transmitted signals from getting into and overloading the sensitive receiver circuits.) A low-noise amplifier is located close to the antenna to maintain a low receiver noise figure. In many installations on ships and aircraft, there are long cable runs to the remainder of the equipment. In marine installations, colocating a down-converter with the low-noise amplifier is wise to avoid excessive cable losses or the use of expensive cable.

A *high power amplifier* (HPA) is required and, for many installations, to install it separately and closer to the antenna will avoid losses in long cable runs. The type of HPA and the amount of power to be generated depends on the system and the type of service. For multicarrier operation, a linear amplifier is required. The INMARSAT aeronautical system, for example, needs a 40 W class-A amplifier when operating in multicarrier mode. However, for single-carrier operation, use of class-C is possible, as the offset QPSK modulation technique employed results in very little spectral spreading.

The down-converters and up-converters are normally housed with the baseband processing and control electronics, the exception being the maritime installations discussed above, where the first stage of down-conversion and the final stage of up-conversion may be remotely located. SCPC operation requires channel frequency selection. A synthesizer controlled from the receiver processor is normally used to generate one of the local oscillator (LO) frequencies. In some systems, transmitting and receiving frequencies are paired, which enables a single synthesizer to be used for transmitting and receiving channel selection. However, in other systems, two separate synthesizers must be employed.

Aircraft velocities result in doppler shifts of up to 2 kHz. An aircraft traveling at 365 m/s in the direction of the satellite will experience a shift of 1.89 kHz on a 1.55 GHz carrier. The aircraft terminal must correct its transmitting frequency by an amount that is opposite but adjusted for the difference between transmitting and receiving frequencies. Land and maritime mobile velocities are sufficiently small that

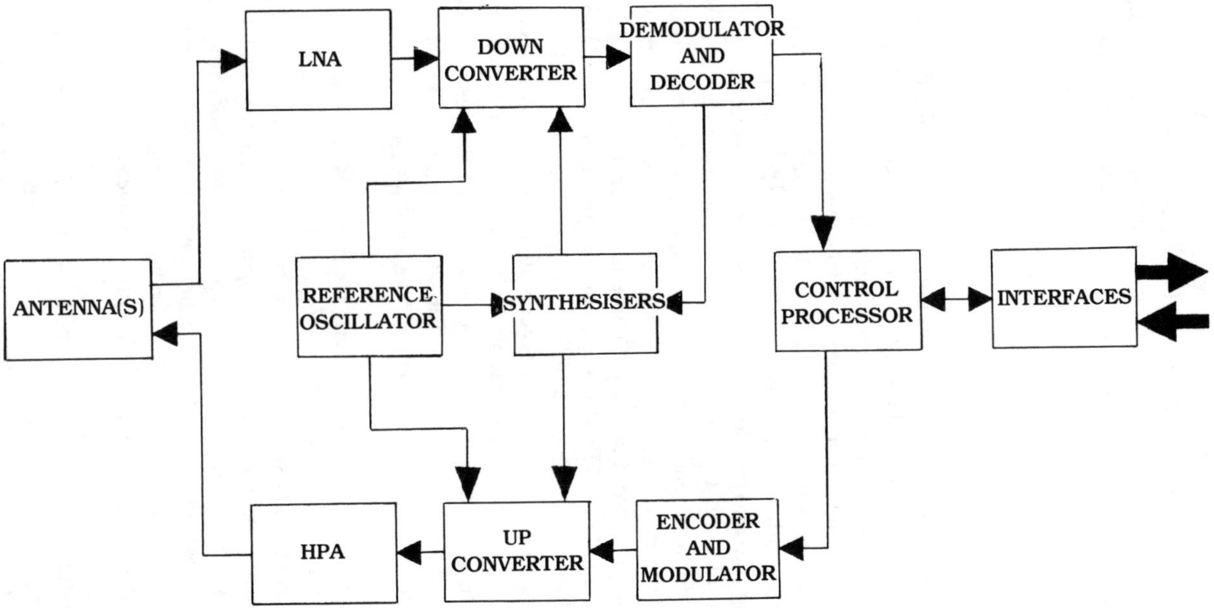

Figure 4.5 Generalized block diagram of a mobile satellite communication terminal.

correction is not required. However, land and maritime terminals, as well as aeronautical terminals, must accurately set their transmitting frequencies. To minimize the size of guard bands between channels, and hence make most efficient use of the bandwidth available, tolerances of ±300 Hz, for example, are specified for the INMARSAT aeronautical system, which represents 5 in 10^6 at 1.6 GHz. Consequently, to avoid frequent recalibration, an ovened crystal oscillator is necessary to provide the terminal reference frequency. However, some systems have proposed the use of active frequency control via the satellite, which avoids this requirement.

Low-cost digital signal processors are commonly used to provide the demodulation function [Oppenheim, 1975]. Figure 4.6 shows a typical signal processor for an aeronautical terminal. Analog-to-digital conversion of the complex baseband signal is shown in the example, although IF conversion can be employed, but more processing may be required. Following the data detection filters, the signal in the two arms is passed to the carrier recovery and symbol timing recovery units. A Costas loop is typically employed for carrier recovery [Lindsey, 1973]. The early-late gate and in-phase midphase synchronizers are commonly used for symbol timing recovery. *Automatic gain control* (AGC) is usually applied prior to the analog-to-digital (A/D) converter to ensure that a constant signal plus noise power is applied to the converter. This control may be locally generated, or a signal-plus-noise power measurement in the digital signal processor may be used to control amplifier gain

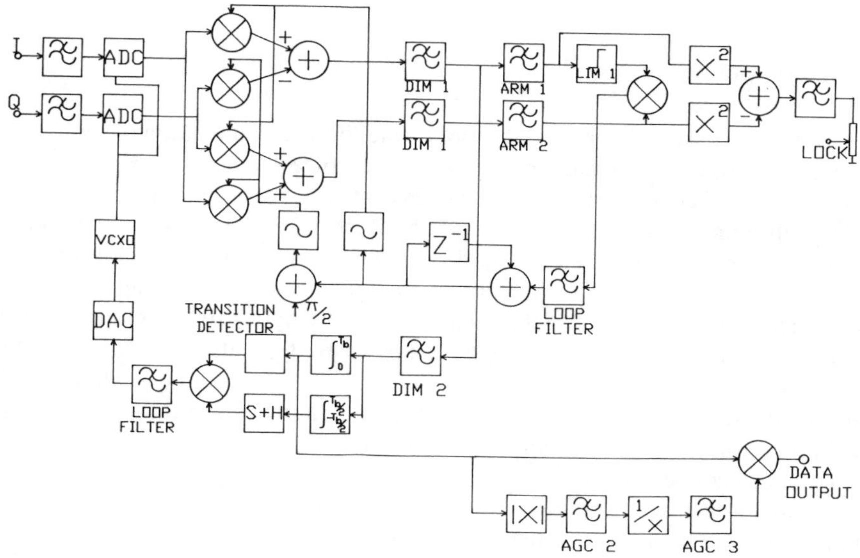

Figure 4.6 Block diagram of a BPSK demodulator.

prior to the converters. The carrier recovery loop can also be used to measure any doppler offset.

Our discussion thus far has concentrated on examining just one method of implementing the demodulator function for one type of system. Problems of signal acquisition and reacquisition, data filtering, and other factors in the design of this subsystem have not been discussed; the reader is referred to [Spilker, 1977] for more information. Similarly, we have looked at a BPSK demodulator, but other digital modulation techniques such as *trellis coded modulation* and *minimal shift keying* (MSK) have been proposed, as well as analog techniques such as *companded* (compressed-expanded) *FM* and *amplitude companded single sideband* (ACSSB). Alternative demodulation techniques are required for these, but digital signal processors typically can be used.

As most mobile satellite systems are power limited, to employ FEC techniques is advantageous with the digital systems; Figure 4.4 shows that an advantage of 5 dB at an error rate of 1 in 10^4 is obtained. This is gained at the expense of an increased bandwidth, but can easily be accommodated because the satellite is not bandwidth-limited. Block codes and convolutional codes are both used; see [Lin, 1983] for more information on coding theory. Block codes can be decoded using standard microprocessors, but to use a specially designed LSI device is often better. Convolutional decoders, such as the Viterbi decoder, are best implemented using LSI devices for rates above a few thousand bits per second.

4.5.3 Voice Coding

Voice coding is required for digital mobile satellite communication systems.

There is inevitably a trade-off between the voice quality and coding rate, and the distribution of errors occurring on the link must be taken into account. All three mobile environments are subject to multipath effects of varying degrees, the numbers of errors and interval between bursts depending on the mobile's environment and velocity. The effect of burst errors can be minimized by interleaving, but only to a limited degree because of the delays introduced. To optimize the system capacity, it must operate with the highest acceptable error rate; typically 1 in 10^3 is used with links employing a 9.6 kb/s voice coding rate. Some voice coding techniques deteriorate rapidly when the error rate falls below a certain threshold, and this must be taken into account when a technique is selected.

To define an acceptable quality level is difficult for digitally encoded speech. A professional user accustomed to VHF might find a particular voice coding technique acceptable, while a member of the public may not, because it is not to the same quality as the PSTN. Another factor is that, for public use, speaker recognition is often regarded as being the minimum requirement, whereas professional users will prefer accurate message delivery.

Until recently, 64 kb/s PCM was the standard PSTN voice coding technique; a 32 kb/s standard has now been introduced. Nevertheless, much lower voice coding rates are needed for mobile satellite communication systems; good quality is now being achieved at 9.6 kb/s, and some satisfactory systems operating at 4.8 kb/s have been demonstrated.

Finally, a voice coding and decoding device should be tone-transparent to enable multitone dialing, dial tone, and other PSTN tone signals to permeate the system without distortion, especially if personal computer modems are to be used (a desirable feature for many users).

Security can be provided on analog systems only by the use of complex techniques. In a digital system, security can be introduced by appropriate manipulation of the bit streams between any encoder-decoder pair. This is often important to business users, although sometimes requiring special consideration within the PSTN.

4.5.3.1 Voice Coding Algorithms

All voice encoding and decoding devices are based on algorithms that aim to minimize distortion and retain maximum tonal quality while minimizing the number of bits used in the process [Stansfield, 1987]. Every algorithm makes use of the redundancy inherent in the spectral composition of human speech, attempting to code the tonal and temporal information into the minimum number of bits. This process, if poorly performed, can result in lack of depth and clarity.

Language becomes a key element in algorithm design because, although the tonal and temporal contents of all human tongues are similar on first inspection, the peculiar subtleties of tone, often related to nasal tract usage, can present major difficulties in encoding and reconstruction. This is particularly true of the French language, whereby a voice coding algorithm that performs well for English can be totally unacceptable to native French speakers.

The algorithm must also be readily capable of implementation in a compact form, and may be based on custom chip sets or standard digital signal processing devices; the choice depends on commercial, rather than technical, considerations.

4.5.3.2 Performance of Predictive Coding

Adaptive predictive coding (APC) and *residual excited linear predictive* (RELP) coding produce good speech quality at 16 kb/s, with somewhat poorer performance at 9.6 and 8 kb/s. Both are prone to degradation if errors are present in the transmission path, and the sensitivity to errors increases as the rate is reduced. APC is generally more tolerant of errors than is RELP. Error protection, judiciously applied to the most sensitive bits in the transmitted data, can be particularly effective. Commercial implementations predominate here, with the subjective quality being highly dependent on the error and noise environments, as well as the skills of the algorithm and codec designers in overcoming them.

Assessment of perceived quality is particularly difficult. CCITT procedures and various national and international standards for assessing subjective quality have been devised, using standard listening panels to assess nonrepetitive recordings of diagnostic rhyme tests and phonetically balanced phrases to provide unbiased opinions of quality. Male and female speakers and listeners are used, ideally from differing cultural (and possibly native) origins, at a variety of speech and background noise levels. The quality assessment methods were originally devised for the assessment of voice coding techniques such as PCM, which rely on the waveform being sampled at the encoder and reconstructed at the decoder. Techniques such as APC and RELP rely on modeling the vocal tract and synthesizing the speech. There is some concern that methods developed for assessing waveform techniques may not be as effective for assessing the synthesis techniques because of the different way in which distortions and imperfections manifest themselves.

Another factor that is particularly important for mobile communication is the surrounding acoustic environment; some mobiles can impose a severe acoustic environment, which degrades the perceived quality.

The assessment procedures are clearly time-consuming and expensive, although they are essential because the perceived voice quality is the most important factor by which users will judge a mobile satellite voice communication service.

4.5.4 Antennas

The antenna is the most significant variable in the system. Nevertheless, there are some severe limitations on the available options. There are many antenna parameters that must be specified to ensure successful system operation, all of which affect the fundamental antenna design. Some of these parameters are discussed below.

4.5.4.1 Antenna Gain, EIRP, and G/T

Antenna gain cannot be considered in isolation..The primary performance parameters of the overall communication system depend on the mobile terminal EIRP and the receiver figure of merit, G/T. The EIRP is governed by power amplifier output level, cable losses, and antenna power gain (which is given by antenna directive gain less antenna losses). The receiver G/T is governed by antenna gain, receiver noise figure, and other noise-generating components such as antenna losses, sky and ground noise, and the effects of transmitting while receiving in a duplex system.

Antenna gains above 0 dBic (omnidirectional) imply the formation of beams which must be pointed in the direction of the satellite by some means, which also implies the formation of sidelobes (i.e., radiation and reception in undesired directions).

Discrimination

Some radiation or reception of signals in undesired directions will always be present if the antenna has gain in a wanted direction. Antenna design is concerned with minimization of these unwanted sidelobes without compromising main-lobe performance.

Fourier theory implies that sidelobe level can be reduced only by degrading (widening) the main lobe, thereby decreasing directive gain. Therefore, the gain of the basic antenna aperture may far exceed the stated specification requirements so as to avoid antenna sidelobes occurring in the direction of other satellites, as well as to optimize multipath performance.

Polarization

Antenna polarization must be matched to the signal in space for optimum power transfer. Polarizations can be *linear, elliptical,* or *circular*. The direction of polarization is also important: *vertical* or *horizontal* for linear, and *left-hand* or *right-hand* for elliptical and circular. Right-hand circular polarization is commonly used on the L-band mobile link.

Cross-polar performance may also be of importance. In theory, polarization diversity can double the capacity of the available spectrum. In practice, this is usually limited by the cross-polar performance of the antennas used, particularly for phased arrays scanned well off-axis, and so polarization diversity is not employed.

Coverage

The coverage provided by any mobile satellite communication antenna affects the geographical locations and vehicle attitudes available to the mobile. If coverage is limited to 10° and upward in elevation, for example, most geographical locations are covered, but vehicle attitude will become increasingly important at extreme latitudes.

Coverage is conventionally stated as that zone toward zenith where minimum system EIRP and G/T are maintained, ignoring all installation effects. The structure on which the antenna is mounted can cause ray blockage and interference, so careful siting is always needed.

Beam Steering

For optimum performance, the antenna beam must be pointed directly at the satellite. Two methods exist for maintaining the beam pointing angle: *open-loop* and *closed-loop* beam steering.

In the former, the mobile's geographical location and attitude are used to calculate the relative azimuth and elevation angles of the satellite. These angles are fed to the beam steering electronics, which points the beam either mechanically or electronically (see Section 4.5.4.2). Pointing angle errors may be minimized by using this method.

In closed-loop steering, a measure of the received signal level drives a feedback loop to the antenna steering electronics, which attempts to maintain a "constant" position. In practice, the production of an error signal to drive the feedback loop requires the antenna to be off its ideal position for a certain percentage of time, resulting in a reduction in effective available gain of typically 0.5 to 1 dB.

4.5.4.2 Antenna Technologies

Antenna technologies for mobile satellite communication come under two categories if high gain and beam steering are required: *mechanically scanned* and *electronically scanned*. In either case, the base radiating elements can be derived from monopoles, dipoles, microstrip patches, helices, *et cetera*.

Mechanically Scanned Antennas

Usually comprising a radiating aperture or reflector, mechanically scanned antennas are steered in three dimensions by motors with appropriate pointing information. A rotating joint or wrap-around cable is used to provide RF continuity between connector and radiating element. Coverage is usually excellent, being steerable by 360° in azimuth and from +90° to below the horizon in elevation. This type of antenna often provides optimum electrical performance.

The drawback of mechanical scanning is the size of the resulting installation and the maintenance problems associated with the mechanism exposed to a relatively harsh environment.

Electronically Scanned Antennas

Electronic scanning uses phased array technology to steer the antenna beam; different elements within the array are fed with different phases, building up a phase gradient to steer the beam.

Scan angles are generally limited to ±60° from the array normal. Hence, providing coverage over 360° in azimuth and greater than 90° in elevation needs several arrays, mounted appropriately on the host vehicle. For limited elevation angles (i.e., from 25° or 30° upward), a single, horizontally mounted array may be suitable. Scanning beyond ±60° is not generally possible because of the lack of gain of the individual elements, and the occurrence of very high sidelobe levels.

The main advantages of electronically scanned antennas are their flexibility in providing rapid scanning over large angles and their reliability. Electronic scanning can imply smaller physical size, an advantage on trucks, cars, and aircraft, where physically large structures are undesirable from either aesthetic or aerodynamic viewpoints.

4.5.4.3 Antenna Implementations and Installations

Low-gain, omnidirectional (0 dBic) antennas pose few problems in either implementation or installation. The antennas are usually monopoles, dipoles, helices, or patches in radomes suitable for their host vehicle and environment. High-gain antennas (10 dBic gain or greater), can pose considerable problems unique to the class of mobile being considered, as discussed below.

Maritime Mobiles

INMARSAT standard-A service antennas are of 20–23 dBic gain, requiring a dish of 0.9 or 1.2 m diameter. Such antennas are almost invariably mechanically scanned, using a three-axis mount that responds rapidly to the ship's motion. Scanning to angles some 30° below the horizon is necessary to accommodate the pitch and roll of the vessel.

Antenna installation is ideally on the ship's superstructure or a dedicated mast to avoid blockage effects. (See Figure 4.7.)

Land Mobiles

There is currently no standard for land mobile terminals. Antennas with gains of around 10 dBic are being proposed for the North American MSAT system. Mechanically scanned antennas may be suitable for mounting on truck cabs, but care is needed in overcoming blockage by the trailer. For cars, coaches, and some trucks, electronically steered phased arrays are preferred because they can be constructed in planar form and mounted conformally with the top surface of the vehicle. (See Figure 4.8.)

Aeromobiles

Coverage for aircraft attitudes of 360° azimuth and −20° to +90° in elevation relative to the satellite is ideally required, with a minimal increase in aerodynamic drag and

Figure 4.7 INMARSAT standard-A shipboard earth station (*Source*: INMARSAT).

Figure 4.8 Land mobile satellite communication antenna. (Photograph courtesy of Ball Communications Systems Division, Broomfield, Colorado.)

no induced acoustic noise. Both AvSat and INMARSAT aeronautical satellite communication systems require the use of 12 dBic gain at L-band.

Electronic scanning is generally preferred to permit the use of conformal antennas on each side of the aircraft, or multiple-array blades mounted on the top surface. (See Figure 4.9.)

For safety services, reliable communication is desirable for all aircraft attitudes, resulting in difficult mounting problems because there will always be some attitudes at which a part of the airframe blocks the direct satellite path.

4.6 RELATED APPLICATIONS AND OTHER SYSTEMS

There are a number of systems that are relevant to mobile satellite communication, but are not addressed elsewhere in the book.

The first of these is an air-to-ground telephony system currently operating in the US, by GTE Airfone [Dennis, 1985]. The system operates at 900 MHz using ACSSB modulation. A ground station network of about 50 sites provides coverage over most of the continental US. Each aircraft can be equipped to handle up to four simultaneous two-way telephone conversations. Direct dialing into the PSTN is offered, but calls cannot be initiated from the ground.

Each ground station radiates a channel indicating the number of channels currently available. When a passenger wants to make a telephone call, the aircraft terminal searches through the ground stations for the one with the largest positive doppler offset (indicating the aircraft is heading toward it), plus a low but increasing signal level (making it likely that the aircraft has some distance to go). The aircraft terminal then requests a channel, the call is set up, and the same ground station is used throughout the call. If the aircraft goes out of range, the call is terminated, but

Figure 4.9 12 dBic aeronautical satellite communication antenna (*Source*: Racal Antennas).

under most circumstances the call can last for about 30 minutes before this happens. Two types of cabin equipment are offered. One uses a cordless telephone, and the other uses handsets built into the seat backs.

A Japanese telephony system uses six ground stations to provide coverage over most of Japan, and passengers can make directly dialed telephone calls into the PSTN.

A service for business jet users in the US is operated by Global-Wulfsberg and uses VHF. This system also uses a network of ground stations, and provides passengers with the capability to make PSTN calls.

A related system is the *radiodetermination satellite service* (RDSS) [Rothblatt, 1987], which also has a message transmission capability, and finds application in the control of vehicle fleets where an operations center needs to know the location of its vehicles and exchange short text messages with the drivers. The control center transmits an interrogation signal to a number (≥ 3) of satellites in turn. Each satellite translates the signal in frequency and retransmits. User terminals receive the signals and transmit a response, which is received by all the satellites, translated in frequency, and retransmitted to the control center. The user's position in three dimensions can be calculated by using the delay between interrogation and response for each satellite. The position can then be inserted in the next interrogation message and sent to an individual user by addressing his or her identity code. Short messages may also be sent by a similar means in the user terminal's response. One such system being set up in the US is known as Geostar; a European version known as Locstar is also being established.

4.7 FUTURE DEVELOPMENTS

The introduction of more satellites with L-band mobile communication transponders and higher powers will provide additional capacity. Spot beams will enable terminals with lower EIRP and higher G/T to be used, allowing terminal costs to be reduced.

For all users, especially at L-band, the integration of communication and navigation functions is certain to occur. Whether the navigation function is provided by dedicated systems, such as the US Navstar *Global Positioning System* (GPS) or the USSR Glonass system or by ranging from communication satellites, the similarity of receiver requirements makes the integration straightforward. From a commercial viewpoint, one of the most desirable applications of satellite communication is position reporting, for which an integrated package is by far the best solution.

The drive toward future global systems, the technologies used, and the facilities offered depend on whether one considers maritime, land, or aeronautical systems. Clearly desirable is that the benefits of economy of scale be applied, if possible, by using common system elements. In practice, the packaging and environmental aspects of each mobile class may reduce this economy, and the differences in doppler behavior and fading characteristics, for example, may well result in different design compromises being required.

Maritime use of satellite communication will continue to grow. Services such as INMARSAT's standard-C will open satellite communication to a new and larger category of user. The existing standard-A service will continue for some time, but gradually will be replaced by a digital system (standard-B). The developments occurring in land mobile communication will reduce the size and cost of maritime terminals.

Land mobile communication is the major growth area for mobile satellite communication. For North and South America, Australia, Africa, and parts of Asia, satellite communication offers the most effective means of providing a communication facility for long-distance road traffic, although the problems of integration with cellular radio systems are yet to be addressed. How growth will occur in densely populated areas such as Europe, where availability of cellular radio coverage will be high, remains to be seen. (See also Chapter 3.)

Aeronautical use of satellite communication is likely to grow steadily for many years. Initial users will be the business or corporate aviation community and airlines requiring simple data communication and passenger telephony services. As worldwide pressure grows for more efficient and safer use of airspace, automatic position reporting and air traffic control will make more use of satellite services. New service areas are also likely, including uplinking of news and financial services for display to passengers via cabin video screens or seat-back terminals.

Owners of personal portable telephones will also demand that they operate with satellite systems available on aircraft, boats, or coaches. Although only starting to receive attention, this method of providing the passenger interface has a number of advantages. First, use of portables avoids taking up space with what are effectively public telephones. Second, portables provide a convenient means for interworking between terrestrial and satellite systems. Finally, for aeronautical use, portables are indeed desirable to prevent the use of cellular telephones from aircraft because of the high speed at which aircraft move between cells and the large number of cells illuminated. Therefore, aircraft are likely to require some form of cellular telephone interface, whether it is fitted with satellite communication or not.

Nongeostationary orbits have some advantages for users at higher latitudes, particularly land mobile users, because nongeostationary satellites reduce the blockage effect of buildings and trees. In addition, a combined geostationary and nongeostationary satellite constellation could also help to reduce the range of angles over which aeronautical and maritime antennas would be steered. Such a constellation could also form the basis for a positioning system having true global coverage; studies are already well advanced [Ashton, 1988; Rouffet, 1988; Norbury, 1988; Aghvami, 1988; Jayasuriya, 1988].

As the use of mobile satellite systems increases, there will be more pressure for greater spectrum availability. The frequency spectrum is already under considerable pressure, and in the long term much higher frequencies will be used if, for example, a personal communication service using satellites is to be considered. Already, experiments using satellites operating at 51/40 GHz have been conducted [Shimada, 1988].

Currently designed systems use satellites that are mostly power-limited. When large area spot-beam systems are introduced, bandwidth will also be limited. Bandwidth-efficient modulation and coding schemes will become important. This importance is already seen in land mobile systems where coverage has to be provided by multiple spot beams to permit use of small and cheap mobile terminals. Trellis coded modulation and ACSSB are being considered in place of PSK schemes, and proposed for aeronautical use.

In RF technology, advances in amplifier design may improve the efficiency and cost of the power amplifier stages of the mobile terminal. Advances in low-noise amplifier technology will not result in significant improvements in performance, although further reductions in cost are likely. As faster digital signal processors become available, to implement more parts of the IF chain in the digital signal processor will become attractive.

The antenna's physical size is unlikely to change because it is determined by the required frequency, gain, and bandwidth. However, progress is likely in the introduction of truly active antennas, where the current phased array technology is extended to include distributed power and low-noise amplification for each element within the actual antenna structure. Such advances require the commercial availability of *monolithic microwave integrated circuits* (MMICs) coupled with the appropriate technologies to integrate them economically and reliably into the antenna structure.

Mobile satellite communication thus will grow in several ways: technical advances will make cheaper mobile terminals possible, and commercial pressures will ensure the development of hitherto unexploited market areas.

REFERENCES

Aeronautical Radio, "ARINC Characteristic 741," 1988.

Aghvami, D.H., A. Clarke, B.G. Evans, P.G. Farrell, J.G. Gardiner, J.R. Norbury, and E. Vilar, "Land mobile satellites using highly elliptic orbits—the UK T-SAT mobile payload," *Proc. 4th Int. Conf. Satellite Systems for Mobile Communications and Navigation,* London, October 1988.

Anderson, S.A., *et al.,* "Aeronautical satellite data link development and evaluation of an experimental system," *30th Annual Air Traffic Control Association Fall Conf.,* September 1985.

Ashton, C.J., "Archimedes—land mobile communications from highly inclined satellite orbits," *Proc. 4th Int. Conf. Satellite Systems for Mobile Communications and Navigation,* London, October 1988.

Barts, R.M., and W.L. Stutzman, "Propagation modelling for land mobile satellite systems," *Proc. Mobile Satellite Conf.,* Pasadena, CA, May 1988.

Bean, C., "Trends in passenger telephone services," *Proc. 2nd Annual AvSat Conf., Aeronautical Radio,* September 1987.

Britten, P.D., J.T. Ryan, D.T. Taylor, P. Jupille, and F. Meuleman, "Terminals for ESA's PRODAT programme," *Proc. 4th Int. Conf. Satellite Systems for Mobile Communications and Navigation,* London, October 1988.

Butterworth, J.S., "Satellite communication experiment for the Ontario air ambulance service," *Proc. Mobile Satellite Conf.,* Pasadena, CA, May 1988.

Casewell, I.E., I.C. Ferebee, and M. Tomlinson, "A satellite paging system for land mobile users," *Proc. 4th Int. Conf. Satellite Systems for Mobile Communications and Navigation,* London, October 1988.

Dement, D.K., "AvSat: an aeronautical terminal for satellite communication," *Proc. ICC'87,* 1987.

Dennis, T.L., "On the phone at 30,000 feet," *Aerospace America,* June 1985.

Dixon, R.C., *Spread Spectrum Systems,* John Wiley and Sons, New York, 1976.

Feher, K., *Digital Communications,* Prentice-Hall, Englewood Cliffs, NJ, 1983.

Haapala, D., "Flight Operation Prospectus," *Proc. 2nd Annual AvSat Conf., Aeronautical Radio,* September 1987.

INMARSAT, "INMARSAT Aeronautical System Definition Manual," International Maritime Satellite Organization, London, 1987.

Jayasuriya, D.A.R., and S.B. Lynch, "Comparison of frequency sharing aspects of satellites in elliptical orbits and the geostationary orbit," *Proc. 4th Int. Conf. Satellite Systems for Mobile Communications and Navigation,* London, October 1988.

Jongejans, A., A. Dissanayake, N. Hart, H. Haugli, C. Loisy, and R. Rogard, "Prosat Phase 1 Report," ESA STR-216, European Space Agency, Paris, May 1986.

Larson, G.C., "Global's AFIS plugs into the GNS-1000," *Business and Commercial Aviation,* September 1985.

Lin, S., and D.J. Costello, *Error Control Coding: Fundamentals and Applications,* Prentice-Hall, Englewood Cliffs, NJ, 1983.

Lindsey, W.C., and M.K. Simon, *Telecommunication Systems Engineering,* Prentice-Hall, Englewood Cliffs, NJ, 1973.

Lutz, E., "Land mobile satellite channel—recording and modelling," *Proc. 4th Int. Conf. Satellite Systems for Mobile Communications and Navigation,* London, October 1988.

Makita, F., H. Nakamura, S. Kashiwabara, H. Saitoh, K. Kosaka, and K. Maekita, "Field trials of aeronautical satellite communication system," *Proc. 4th Int. Conf. Satellite Systems for Mobile Communications and Navigation,* London, October 1988.

Norbury, J.R., H. Smith, V.S.M. Renduchintala, and J.G. Gardiner, "Land mobile satellite service provision from the Molniya orbit—channel characterization," *Proc. 4th Int. Conf. Satellite Systems for Mobile Communications and Navigation,* London, October 1988.

Oppenheim, A.V., and R.W. Schaefer, *Digital Signal Processing,* Prentice-Hall, Englewood Cliffs, NJ, 1975.

Racal Decca Advanced Development, Study of the Applicability of the INMARSAT System to the Aeronautical Mobile Satellite Service, Contract Number INM/83-054, March 1984.

Rafferty, W., K. Dessouky, and M. Sue, "NASA's Mobile Satellite Development Programme," *Proc. Mobile Satellite Conf.,* Pasadena, CA, May 1988.

Rothblatt, M.A., *Radiodetermination Satellite Services and Standards,* Artech House, Norwood, MA, 1987.

Rouffet, D., J.F. Dulck, R. Larregola, and G. Mariet, "Sycomores: a new concept for land mobile satellite communications," *Proc. 4th Int. Conf. Satellite Systems for Mobile Communications and Navigation,* London, October 1988.

Schoenenberger, J.G., and R.A. McKinlay, "An airline passenger telephone system—design, development and early trials," *Proc. 4th Int. Conf. Satellite Systems for Mobile Communications and Navigation,* London, October 1988.

Shimada, M., Y. Suzuki, Y. Arimoto, and T. Shiomi, "A millimeter-wave mobile satellite communication system for personal use," *Proc. 4th Int. Conf. on Satellite Systems for Mobile Communications and Navigation,* London, October 1988.

Smith, G.K., and P. Branch, "INMARSAT Plans for Early Introduction of Aeronautical Satellite Communication," *Proc. ICC'87,* 1987.

Spilker, J.J., *Digital Communication by Satellite,* Prentice-Hall, Englewood Cliffs, NJ, 1977.

Stansfield, E.V., D.P. Martin, and M. Newman, "Adaptive filters in speech coding," *Proc. IEE,* Pt. F, Vol. 134, No. 3, June 1987.

PART III
TRAFFIC INFORMATION, GUIDANCE, AND CONTROL

Chapter 5
Traffic Information Broadcasting and RDS

S.R. Ely
British Broadcasting Corporation
Research Department

and

D.J. Jeffery
Transportation and Road Research Laboratory
Department of Transport

5.1 INTRODUCTION

On occasion, all motorists have been delayed in a long traffic jam, possibly caused by an accident or road work, which they could have avoided if given advance warning. Broadcast traffic information services could give such warning and, as a consequence, prevent delay and a number of secondary accidents, particularly those due to fog. Even where drivers could not divert to avoid the problem, knowledge of its cause and likely duration would lessen frustration in dense traffic conditions. Valuable information on holiday routes, weather conditions, and car parking availability could also be given.

A system that enabled the police to give advice and instructions quickly and reliably to all drivers would provide an incentive to improve traffic control strategies, which would yield better road safety, reduced travel times, and lower fuel costs. From these strategies, significant benefits would accrue not only to the individual motorist, but also the community as a whole. The sum of all such benefits (which include social and environmental factors as well as economic ones) seems likely to be large, but is difficult to quantify. However, the potential value of all such savings was recently [1987] estimated by the UK Transport and Road Research Laboratory (TRRL) to be worth around $45 million per annum in the UK.

5.2 NEED FOR IMPROVED TRAFFIC INFORMATION BROADCASTING SYSTEMS

For many years, broadcasters have interspersed traffic news with other radio programs. This valuable service provides motorists on all road classes with information about traffic and road conditions, information which they are unable to receive by any other means.

In practice, however, there are two main factors that limit the effectiveness of conventional radio broadcasts as a traffic management tool. First, there is the editorial problem that the traffic news bulletins must be fitted in among other programs in a schedule intended to satisfy all kinds of listeners, many of whom have no interest in traffic information. These editorial considerations limit the number of traffic messages and updates which can be broadcast. Second, if (as is the case in many countries) one particular station is designated to carry most of the traffic messages, drivers can hear the messages only when they tune to that station. If drivers tune to another station, listen to a cassette tape, or drive in silence, they will miss the traffic messages. Consequently, many drivers do not hear the traffic messages that are broadcast.

Moreover, in some countries, the service areas of broadcast transmitters are tailored to cover the main centers of population, rather than to achieve universal coverage along all classes of road. Hence, the broadcasts may not be capable of reliable reception on a significant proportion of road networks.

These limitations mean that the police and traffic authorities do not perceive conventional radio broadcasts as a means of quickly conveying information to a large proportion of drivers. The potential for this is large, however, and there is increasing interest in developing specially designed and more effective traffic information broadcasting systems. The UK and other Western European countries have taken the lead in this area, so that, while an account is given of relevant US and Japanese activities, this chapter is for the most part devoted to describing European developments.

5.3 BACKGROUND

In 1976, a seminar on the use of broadcasting for traffic information was held at TRRL. The seminar was attended by representatives of the two UK broadcasting authorities, police, motoring organizations, radio manufacturers, universities, and government departments. The seminar's attendance confirmed the interest in developing improved systems, but highlighted the need for a better understanding of the part that broadcasting can play and the benefits, costs, and organizational structure of a complete system. A working group under TRRL chairmanship, with representatives of the police, broadcasting authorities, the Home Office, and the Department of Transport, came to the following conclusions [TRRL, 1979]:

(1) The essential characteristics of an ideal traffic information broadcasting service are:
- a rapid response time. The time between an incident and a broadcast should be short (15 minutes or less);
- regular updates and repeats of messages;
- clear, concise, and accurate messages;

- drivers should hear only broadcasts that concentrate on incidents in the area where they are driving;
- positive information; where a diversion avoiding a long line is available, it should be clearly described;
- low-cost, in-car equipment;
- no interference with the driver's choice of entertainment.

(2) A comprehensive and well-coordinated traffic information broadcasting system could achieve benefits, mainly from reduced congestion and delays, worth around $45 million per year (in 1987 traffic conditions), and more as traffic volumes increase.

(3) To achieve such benefits, adoption of improved procedures would be necessary for the collection and processing of information. Speed would be essential, and advice would need to be more comprehensive than at present.

(4) A dedicated network of transmitters broadcasting an appropriate mixture of local and national information is desirable.

(5) The vehicle's receiver should be simple and convenient to use.

(6) The requirements for an individual message informing about a traffic incident are that it should describe:
- the location;
- the cause;
- the effect;
- the duration; and
- any advice that may be needed.

Current work is investigating the vocabulary, grammar, and syntax needed to convey traffic messages accurately and concisely in different languages and the data compression techniques that will be needed if a full range of messages is to be adequately conveyed in channels of very limited capacity (see Section 5.6).

5.4 TRAFFIC INFORMATION BROADCASTING SYSTEMS

Special arrangements for broadcasting traffic information via radio exist in most countries. The systems are used mainly to advise drivers of traffic incidents and hazards, and to encourage them to divert or postpone their journeys to alleviate congestion and delays. The organization of such services, and a number of systems that have been developed in different countries, are considered in the next section. The RDS, which is becoming available across most of Western Europe, is a significant advance, and is described in detail in Section 5.5.

5.4.1 Organization and Facilities in the UK

In the UK, radio broadcasting is shared between the public service BBC (British Broadcasting Corporation) and a number of independent commercial companies, which use transmitter equipment provided by the Independent Broadcasting Authority (IBA).

The BBC provides four major channels, which are available throughout the UK, using a large number of high-power FM transmitters. AM transmitters broadcasting at medium (MF) and low (LF) frequencies also provide duplicate coverage of most of these "national" services. Three of the four BBC national radio services carry traffic information at some time during the day. However, only incidents affecting the main national trunk routes, or very urgent information, usually can be broadcast in the very limited time available within the schedules of the main channels.

The BBC also operates about 40 local radio stations. Each of these BBC local radio stations typically uses two medium-power FM and one AM transmitter to serve an area of a few hundred square miles, usually being a major city or an English "county." The Independent Local Radio (ILR) stations are similar in coverage to the BBC local radio stations, and number about 50.

Many of the programs carried by BBC local radio and the ILR stations cover items of local interest, and many kinds of traffic information fall naturally into this category. To give much more detailed local traffic information via these local radio transmitters is hence feasible, and this is done by both the BBC and ILR stations, especially at peak hours.

As in the US, and most of Western Europe, more than 85% of cars in the UK are fitted with radios. Most of these can receive medium and long wave transmissions. About 50% are capable of receiving on VHF, and this proportion is rising steadily.

For London and the southeast of England, the Automobile Association's (AA) Roadwatch Unit is one major supplier of road condition and traffic information. The unit gathers information from a number of sources, including AA members and patrols, the police, and weather forecasts, and provides daily reports for the press, radio, and television. Roadwatch has its own radio studio facilities, and provides some 900 live traffic information inserts to 11 local radio stations each week.

In addition, all police forces provide input to the BBC Travel Centre, which arranges for traffic incident and roadwork information to be broadcast over the BBC national and local radio services.

The Travel Centre is manned 24 hours a day, and is responsible for editing the information supplied from police forces and other sources. The edited information is used to compile traffic messages, which are broadcast as special inserts to BBC national or local programs. Guidelines for editing and presenting travel information have been agreed with police. The importance of any supplied information is based on whether the road is a principal road or a highway; if so, whether it will be blocked for more than an hour is assessed, as is the reason for the blockage. A further requirement is that any diversion instructions must be simple. These criteria, together with the urgency of the information and availability of air time within the program schedule, are the principal factors used to decide whether a particular message is broadcast.

Information on "predictable events" such as road work is also received by the Travel Centre from local authorities, public utilities, and motoring organizations, but the information is checked for accuracy with the appropriate police force before being broadcast. Predictable information is usually broadcast at regular times on three of the four BBC national stations, but information on emergency incidents is usually transmitted as a "flash" interruption to a program. Motoring broadcasts initiated by the BBC Travel Centre averaged about 1600 per month in 1987, and peaked at over 2000 during October of that year, when severe gales were experienced in southern England. Typically, one in three broadcasts are urgent "flashes," mainly concerning accidents and their traffic effect.

Drivers in the UK need no special equipment to receive these broadcasts.

5.4.2 The US Shadow Traffic Network

Shadow Traffic Network is the name of a commercial news service that specializes in providing traffic information and news. Shadow, which operates in the New York, Chicago, and Philadelphia areas, employs a fleet of vehicles on the ground and 12 spotter planes in the air to keep watch on traffic conditions. Other services, such as *Metro Traffic Control* serving Boston and elsewhere, operate in a similar manner.

The drivers and pilots in Shadow's fleet pass their information to a traffic operations center using mobile radio. The cars are also equipped with citizen's band (CB) radios (so that the driver can speak with other motorists), cellular radiotelephones, and other equipment that monitors police and fire service radio.

At the traffic operations center, a computer is used: first, to sort incoming messages as they are received from the drivers, pilots, public safety officials, and others; and then to produce scripted messages for announcers. Users of the information currently include 175 radio and television stations, which may receive scripted messages by telex, or may engage in live on-air discussions with Shadow's traffic-watch announcers.

Drivers in the US need no special equipment to receive these broadcasts.

5.4.3 The German ARI System

Broadly similar procedures to those used in the UK are found in most Western European countries and the US although parts of West Germany, Austria, Switzerland, and Luxembourg have adopted the Blaupunkt ARI (Autofahrer Rundfunk Information) system, which has some special features.

The main function of the ARI system is to assist in tuning to the broadcasting stations that provide traffic information, and to alert the driver when a traffic broadcast is imminent. In West Germany, some 65% of cars are fitted with FM receivers, and traffic information is broadcast over a network of about 40 stations, each transmitting on its own frequency in the FM band. Because of the characteristics of radiation in this band, the range of each station is usually limited to several tens of miles. The country is divided into about 12 traffic information zones, each served by several of these stations. With such a system using a large number of short-range stations, each radiating on its own frequency, to provide some tuning aid is essential to select the frequency of the station covering the area in which the vehicle happens to be traveling.

Each transmitter equipped with ARI radiates a special subcarrier at 57 kHz. This subcarrier is modulated with one of several low-frequency tones, which identifies the stations serving the zone wherein the vehicle is traveling, and is modulated by a further tone whenever a traffic announcement is being made. To utilize these special signals, the vehicle must be fitted with a decoder associated with its VHF radio, which, in its simplest form, detects the subcarrier and lights a pilot lamp when the driver tunes to an ARI station. A slightly more complex decoder also allows the driver to mute his or her radio (thus avoiding the necessity of listening to the rest of the program broadcast by the ARI station), and restores the volume to a preset level upon detecting the special modulation associated with a traffic message. A more complex decoder can be fitted, which illuminates the pilot light only when the driver tunes to an ARI station serving the zone where he or she is driving (the driver must previously select a code corresponding to the zone of travel, which he or she obtains from a map or roadside sign). Another, much more expensive version incorporates an additional radio frequency section, which automatically tunes to the strongest ARI station serving the vehicle's zone and interrupts the driver's chosen program whenever a traffic announcement is made. About 80% of the car radios in West Germany are estimated to have an ARI decoder, although most are of the simplest kind.

Several manufacturers produce car radios that are capable of decoding ARI signals. This additional feature adds between $15 and $30 to the basic price of the radio receiver.

5.4.4 The American HAR System

HAR (Highway Advisory Radio) is now widely used in the US [Turnage, 1980]. Two frequencies, 530 and 1610 kHz, are allocated for this system. The respective frequencies fall just below and just above the standard AM broadcast band, and can therefore be received by most car radio receivers in the US. Local transmitters using one of these frequencies and with a power such that reception is confined to a radius of up to about one mile, are sited at the approaches to hazardous areas, where they broadcast prerecorded hazard warning and diversion messages from a continuous-loop tape. Because of the local nature of the information, a fixed sign must be used to tell drivers to tune to the appropriate frequency so that they can hear the broadcasts. The second frequency can be used at the far end of the hazard, so drivers approaching from the opposite direction can be given a separate set of messages.

Subsequent development of an AHAR (Advanced HAR) system [Mammano, 1984] attempts to avoid the need for manual tuning of the car radio by using a subsidiary FM receiver, which automatically mutes the car radio, tunes to the AHAR frequency (45.80 MHz), and returns the radio to its initial state after the message has been repeated twice. Tests on this system used the transmission of actual voice messages, and the possibility of transmitting digitally encoded speech for synthetic reproduction in the vehicle was also examined.

5.4.5 The Japanese HAIR System

HAIR (Highway Advisory Information Radio) is virtually identical to the American HAR systems. Examples are found on the Tokyo Metropolitan Expressway and on the Tomei Expressway. Many systems are now operational in Japan, mainly at permanent sites on highways, where they are used to advise drivers of conditions on the road ahead. The Japanese have standardized on a single frequency of 1620 kHz and, depending on the location, may use a leaky coaxial cable, a parallel pair of cables, or a pole as an antenna. A variable message sign is used upstream of the antenna to warn drivers when the system is operational, and to indicate that they should tune their car radios to 1620 kHz to receive a traffic message.

The single frequency of 1620 kHz is within the tuning range of most home and car radios in Japan. Drivers who have radios equipped with preset tuning buttons can therefore preselect the HAIR frequency, and need only press the button to receive the broadcasts.

5.4.6 The BBC CARFAX System

CARFAX, a broadcasting system proposed in 1977 by the BBC, would have used a special network of transmitters dedicated to broadcasting traffic information. The proposed system was specially developed to meet the requirements of an "ideal" traffic information broadcasting service, as discussed previously in Section 5.3. CARFAX has not been implemented, but it is described here because of its strong influence on the development of newer systems, described later in this chapter.

CARFAX would consist of a network of low-power MF transmitting stations operating in TDM on a single channel and sited 30 to 40 miles apart to cover all or selected parts of the country (see Figure 5.1). Whenever a traffic broadcast was required in a particular area, the appropriate transmitter would be energized for the duration of the broadcast, and simultaneous transmission by other transmitters within mutual interfering range would be prohibited. A special technique known as the *ring system* was developed (see below) for accurately defining local coverage from each

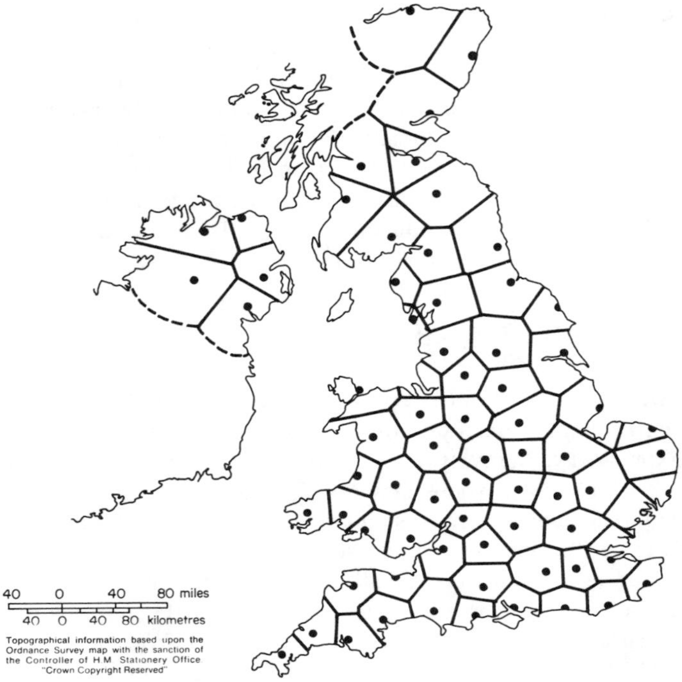

Figure 5.1 Cellular network of MF transmitters which was proposed for CARFAX.

transmitter within a nationwide cellular network operating on a single MF frequency. The system was to be controlled by a computer at a broadcasting center, which would allocate messages to be broadcast by the transmitter serving the appropriate area, and could give precedence to messages allocated a high priority by the traffic control authorities.

The motorist would receive the broadcast for his or her particular area by means of a fixed-tuned radio receiver costing between $15 and $30. This unit could operate in conjunction with an existing car or home radio, be installed as an independent unit, or form an integral part of a normal radio. In any such system based on local broadcasts, the motorist should not be forced to listen to messages that serve other areas, and should also be able to listen to entertainment stations or a cassette player while awaiting a relevant traffic message. These requirements are met in the CARFAX system by preceding each transmission with a "start of message" code, which is detected in the special receiver. The receipt of the signal is necessary to prepare the receiving equipment for the traffic broadcast, and it can mute other programs being received until the receipt of a "finish" code or the elapse of a preset time interval.

Although the system primarily uses amplitude modulation, the "start of message" code and the "finish" code are broadcast by using frequency modulation, and the short burst of FM is accompanied by simultaneous FM transmissions from the ring of adjacent transmitters (see Figure 5.2); these simultaneous transmissions, however, do not include the "start" code. The ring system procedure is designed to take advantage of the property of FM receivers that they are "captured" by the strongest of two or more FM transmissions radiating on the same frequency and, unlike AM transmissions, can demodulate the signal from the strongest station with little interference from any others. Thus, a receiver in the zone for which a broadcast is imminent would be captured by the FM transmission containing the "start" code, and would be activated to receive the traffic message. A receiver in the adjacent zone would be captured by the stronger simultaneous burst of FM from its own local transmitter, but, because this does not include the "start" code, the receiver would not be activated, and the message could not be received. The shape, size, and degree of overlap of these zones to some extent could be tailored to meet the requirements of the traffic authorities by adjustment of the transmitters' output powers.

The CARFAX system is fully described in an article by Sandell and Edwardson [1977], and in a BBC Research Department report [Sandell, 1984]. Field-trials of

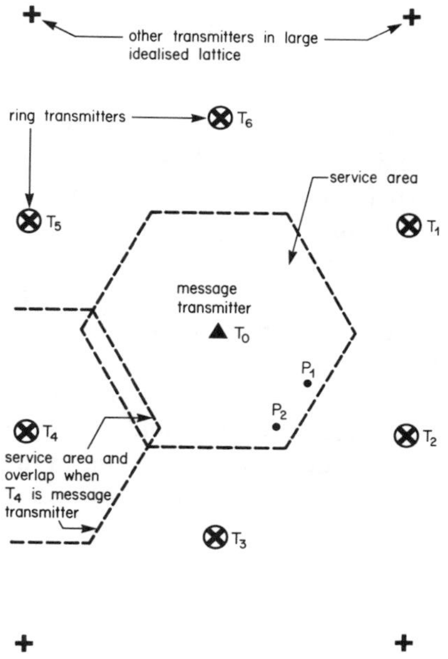

Figure 5.2 The CARFAX ring system.

the system were technically successful, but it was not implemented because of problems in allocating a suitable frequency in the MF band and the capital cost of the dedicated transmitter network.

5.4.7 Cellular Radio

The cellular structure of the ring system used in CARFAX is similar to the infrastructure required by a cellular radio system.

The cellular radio system adopted in the UK is typical of many such systems used throughout the world. UK cellular radio networks are based on the US AMPS system. They provide portable, mobile, and fixed station radiotelephone services using a "cellular" radio coverage plan. The systems operate in the frequency bands between 890 to 915 MHz and 935 to 960 MHz, and the coverage area is divided into cells of radii that vary from about ten miles in rural areas to about one mile in busy urban areas (see Figures 3.2 and 3.5). The system is controlled by digitally encoded control codes, but the speech messages themselves are analog. The system is described in detail in Chapter 3.

A particularly interesting development is the pan-European digital cellular radio system. Unlike other cellular systems that are essentially "point-to-point," the pan-European system has a "broadcast short message" service. This allows the same message to be sent to a group of subscribers, and therefore can be used to transmit traffic information. More details are given in Chapter 3.

In essence, cellular radio enables the PSTN to be extended into vehicles, so anything that can be achieved on that network, in principle, can be achieved on the cellular system. Cellular radio therefore offers potential not only for broadcasting traffic information, but also for interactive inquiry systems, which provide added value with regard to route planning. A cellular radio could be used in combination with a portable microcomputer and a modem, thus enabling a driver to interrogate a remote computer directly from his or her vehicle. The driver could thus have immediate access to computer-based route planning services such as ROUTE-TEL [TRRL, 1983], or to a database providing updated versions of the electronic maps required by the CARIN, EVA, and NAVIGATOR types of navigational aids (described in Chapter 6). However, developments such as these would be conditional upon sufficient capacity being available in the cellular radio system. Often, cellular radio networks in parts of the UK are already saturated at peak times of the day.

5.5 THE RADIO DATA SYSTEM—RDS

5.5.1 Introduction

The Radio Data System comprises an inaudible data-modulated subcarrier, which is added to the stereo multiplex signal at conventional FM broadcast transmitters. (See

Figure 5.3.) The prime purpose of RDS is to facilitate automated tuning by conveying data labels, which identify the broadcasts. However, RDS also presents possibilities for conveying traffic information to motorists.

The specification of the RDS system was developed under the auspices of the European Broadcasting Union [EBU, 1984] by various European broadcasters, notably Swedish Telecom Radio (STR) and the BBC. RDS is also the subject of CCIR Recommendation 643 [Dubrovnik, 1986]. The European electrotechnical standards organization, CENELEC, is developing a standard for RDS receivers.

Most national (public service) radio broadcasters throughout Europe have implemented RDS or have plans to do so. RDS could also be used in other parts of the world, although to date there are few known implementations outside Europe. Car radios that implement RDS first became available in 1987, but, because of high cost (around $1000 for the first model) and limited availability, only a few thousand had been sold at the beginning of 1989. Many models, including those in the $400 to $500 price range, are expected to incorporate RDS, and the market for them is expected to develop rapidly. Figure 5.4 shows a block schematic diagram of a typical RDS receiver.

5.5.2 Origins of RDS

The concept of using a subcarrier to convey additional information along with a monophonic or pilot-tone stereo (Zenith-GE) VHF-FM broadcast dates back many

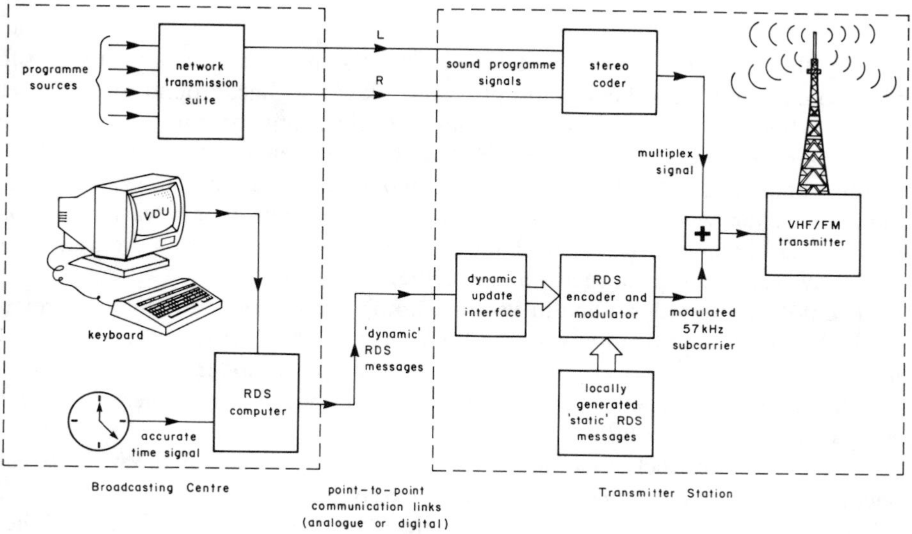

Figure 5.3 RDS transmission with dynamically updated information.

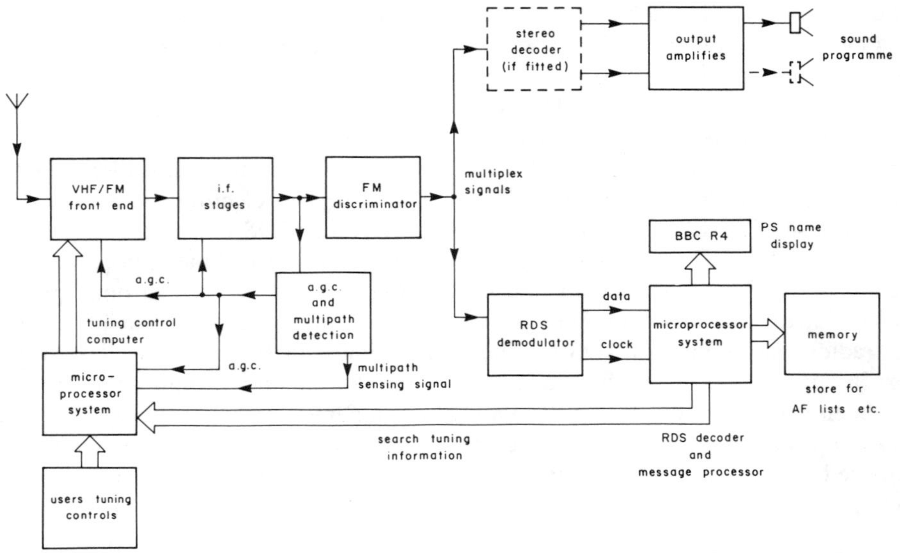

Figure 5.4 VHF-FM receiver with RDS assisted automatic tuning.

years. By the mid-1960s, many FM broadcast stations in the US were using subcarriers to convey a subsidiary audio program, played as background music in restaurants or shops, which gave rise to the name "storecasting" for this kind of system. Such subcarriers were the subject of a Federal Communications Commission (FCC) regulation known as Subsidiary Communications Authorization (SCA). Such SCA systems, however, were not found suitable for use by European broadcasters because there was unacceptable crosstalk from the SCA program into the main program when received by many European receivers [Whythe, 1977].

The ARI system (described in Section 5.4.3) was developed in West Germany as a simple traffic identification system in the early 1970s. ARI is a relatively simple tone-signaling system, and requires only a very simple decoder. However, by the mid-1970s, a data-modulated subcarrier clearly could provide much greater information-bearing capacity and flexibility.

One of the first such data-modulated subcarrier systems was the MBS paging service, which was developed in Sweden by STR. (See Chapter 2 and Appendix 2.B.) MBS provided the basis of the system developed as the EBU's RDS, and there are many similarities in the basic data transport system.

5.5.3 Development of the EBU RDS System

Although some initial work was done unilaterally by individual broadcasters and manufacturers, for a radio data system to be successful, an internationally standardized system obviously would be needed; the market in any one European country would be too small to support the development costs of the receivers. Thus, from about 1976, the work on the development of the system was coordinated by a group of technical experts working under the auspices of the EBU. The specification was unanimously agreed by all EBU members in 1983, and the specification was published as EBU Technical Document 3244 in 1984. A summary of the RDS specification is given in Appendix 5.A.

5.5.4 Functional Requirements of RDS

The first step in the development of the system was to establish a set of functional requirements:

(a) The radio data signals must be compatible; they must not cause interference to the reception of sound program signals on existing receivers or to the operation of receivers which use the ARI system.

(b) The data signals should be capable of being reliably received within a coverage area at least as great as that of the main program signal.

(c) The usable data rate provided by the channel should support the basic requirements of station and program identification, and provide scope for future developments.

(d) The message format should be flexible to allow the message content to be tailored to meet the needs of individual broadcasters at any given time.

(e) The system should be capable of being reliably received on low-cost receivers.

5.5.5 Choice of Subcarrier Level, Frequency, and Modulation Method

As was evident, these requirements, in part at least, conflict. For example, the system obviously should convey the data using a subcarrier added to the conventional stereo multiplex signal, but the level of the subcarrier relative to the program signal was a compromise between compatibility and ruggedness. If the level of the subcarrier were too large, it might cause interference to reception of the sound program signal. If it were too small, it would be impossible to receive the data reliably under poor reception conditions.

The EBU RDS specification allows the broadcaster to choose a level of subcarrier which corresponds to between 1 kHz and 7.5 kHz deviation of the main FM carrier, with 2 kHz deviation recommended for most circumstances. This is rather low compared with the deviation allowed for SCA subcarriers, for example, but has been found to give the best compromise between compatibility and ruggedness. In West Germany, 1.2 kHz deviation is used throughout to help ensure compatibility with the ARI system; in France, some stations use 4 kHz deviation to provide extra ruggedness for a paging application of RDS.

The choice of frequency for the subcarrier was far from clear. Inspection of the spectrum of the stereo multiplex signal (see Figure 5.5) suggests two regions where a subcarrier may be added: (a) in the region around the 19 kHz pilot tone; or (b) in the region above the stereo difference signal (i.e., 53 kHz to 76 kHz).

The upper limit is given by CCIR Rec. 451, and by the need to avoid problems of *adjacent channel interference* (ACI). (*Note:* in Europe, the channels occupied by stations in the FM band are planned on the basis of 100 kHz channels instead of the 200 kHz channel spacing used in the US.)

The use of a 57 kHz subcarrier is favored because it is a harmonic of the 19 kHz pilot tone. Therefore, the frequency, and hence the audibility, of any unwanted intermodulation products in a stereo receiver would be minimized. However, the ARI system, which was already established in some European countries, used a 57 kHz subcarrier, and at first this seemed to rule out 57 kHz as a choice of subcarrier frequency for radio data.

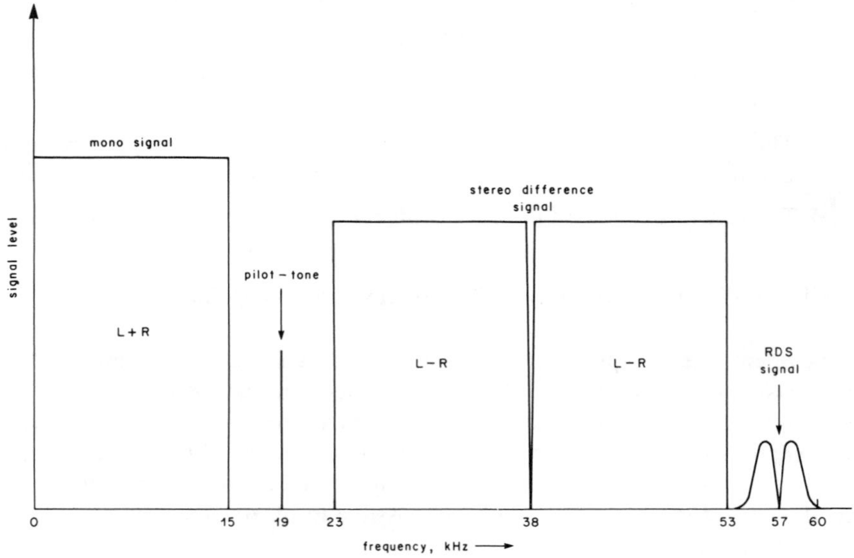

Figure 5.5 Spectrum of a pilot-tone stereo multiplex signal with RDS.

However, in Sweden the MBS paging service had been developed using a double-sideband amplitude-modulated suppressed 57 kHz subcarrier and biphase (sometimes also called "Manchester") coding of the data. The effect of such coding and modulation is to produce a notched spectrum, as shown in Figure 5.6. Comparison with the spectrum of the ARI signal indicates that the two systems could coexist, even when radiated simultaneously from the same transmitter, provided that the radio data receiver has a narrow notch centered on 57 kHz and the ARI receivers are narrowband. The latter condition was met by existing ARI receivers, and the former condition can be met at reasonable cost in radio data receivers.

The use of biphase coded data also helps compatibility with the audio program signal because coherent components at and around 57 kHz were found to introduce data-modulated crosstalk in receivers that used a phase-locked loop (PLL) stereo decoder.

The bit rate of the RDS data stream is 1187.5 b/s (1187.5 = 57,000/48), which, with biphase coding and the specified 100% cosine roll-off filtering, gives an overall bandwidth for the data signal of approximately 4.8 kHz.

The choice of a 57 kHz subcarrier modulated by biphase-coded data was confirmed in 1980 by international field tests for the four radio data systems proposed. These EBU tests were done in the mountains of Switzerland because, as had already been established, one of the most difficult, yet potentially most rewarding, applications of a radio data system was mobile reception in vehicles. In particular, multipath caused major disturbances to the reception of subcarrier data signals.

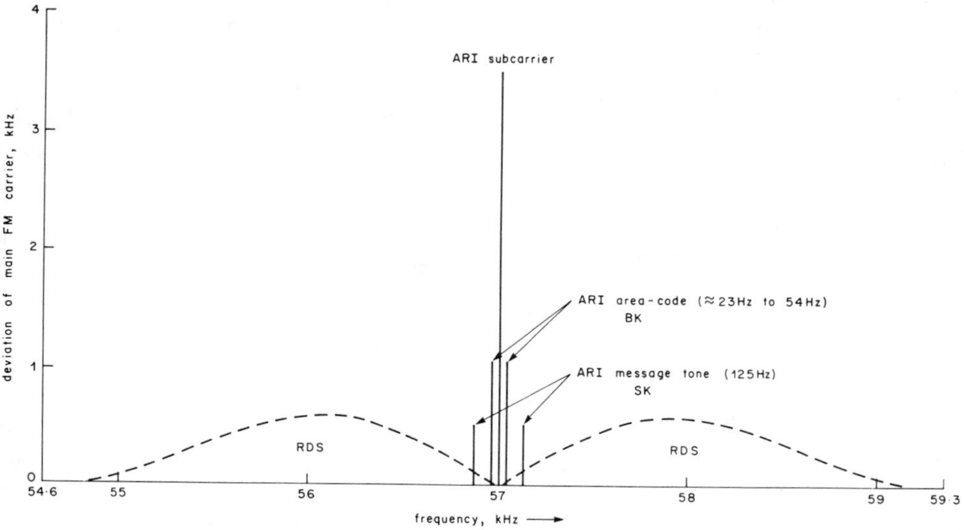

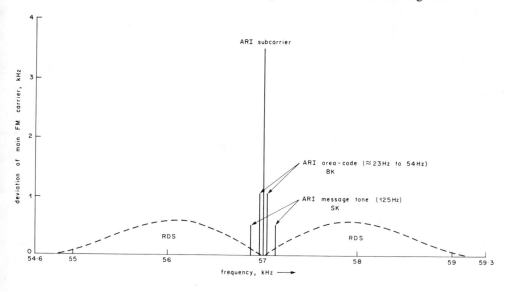

Figure 5.6 Spectra of RDS and ARI signals.

Multipath in an FM system produces distortion in the demodulated signal. The distortion components resulting from the relatively large amplitude sound program signal components can easily swamp the data signal. When a vehicle moves along a road characterized by multipath interference, the quality of the received FM signal varies rapidly; at some moments, the demodulated program is undistorted; at others, it is completely broken up. The very important lesson learned from these 1980 field tests was that, for reliable mobile reception, the radio data message stream must be divided into small independent entities, each of which can be received, decoded, and applied independently of other parts of the data stream. This factor was crucial to the basic design of the system and must be clearly understood when proposing further developments of the radio data system.

5.5.6 Choice of Baseband Coding

As with most serial data transmission systems, radio data partitions the data stream into blocks. To interpret the received data correctly, the decoder must identify the beginnings and ends of the received blocks; thus, a system for block synchronization is needed. However, in a radio data system, the need to be able to extract useful information from isolated fragments of the data stream posed special problems. These were solved in the EBU RDS system by the development of a novel self-synchronizing code in which the checkwords of a burst-error-correcting block code also serve to indicate correct block synchronization and address the contents of the block.

Figure 5.7 illustrates the baseband coding structure of RDS. The largest comprises four 26-bit blocks. Each block comprises a 16-bit information word and a 10-bit checkword formed with a cyclic code shortened by burst error correction.

The 10-bit checkwords are modified (by modulo-2 addition of various "offset" words) to form a self-synchronizing code, with which the decoder can reliably ac-

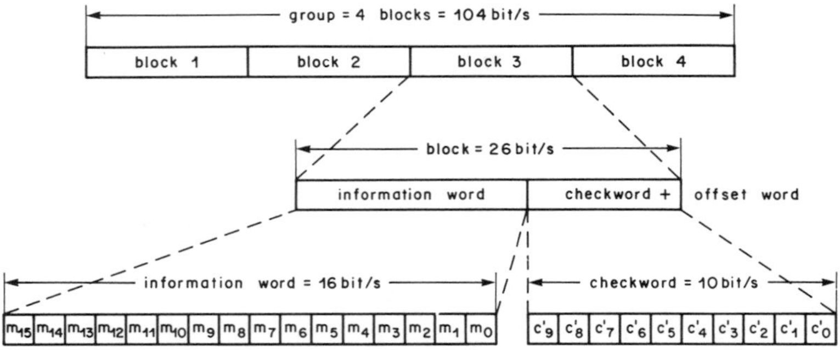

Figure 5.7 Structure of the baseband coding of RDS.

quire synchronization within the period of one block. By using different offset words for the four blocks comprising each group, the decoder obtains both block and group synchronization.

In every group, some data are the same and some are variable. The applications of the variable elements are identified by the 4-bit "group type" code contained in the second block of the group. Of the 16 group types available, 8 were defined in the RDS specification, leaving the remainder for future applications.

Subject to meeting the minimum repetition rates needed to support the basic tuning-assistance functions of RDS, a mix of different RDS groups can be transmitted in any order. The broadcaster can vary the mix and order of group types from one minute to the next to meet changing needs at different times of the day. Subject to capacity being available, new group types can be introduced into the transmitted data stream without affecting the operation of existing RDS receivers.

5.5.7 Applications of RDS

In parallel with the development of the data-transport layer of RDS, also essential was to develop the applications of RDS and to devise suitable ways for coding and presenting the messages. The main applications of RDS are summarized in Table 5.1.

We note that the baseband coding of RDS is designed such that the codes for PI, TP, and PTY features occupy fixed locations in all RDS group types. Therefore, they can be decoded from the information contained within just one block of 26 bits. All other information requires successful decoding of the group type address information contained in the second block of each group. The purpose of this mixture of fixed and variable format is to give rapid access to the information needed for search tuning and yet retain flexibility to allocate capacity to various applications to meet the different needs of individual broadcasters at any given time.

Some RDS information is essentially "static" (i.e., the broadcast message does not vary with time). Examples of this can include the PI code, AF list, and PS name. Note, however, that these codes are not necessarily static for all transmitters: many transmitters broadcast one program at one time of the day and another service, with a different PI code at other times of the day. Such "static" RDS information can be stored in permanent (ROM) memory in the encoder at the transmitter, as indicated in Figure 5.3.

Other RDS information is essentially dynamic in that the data change from moment to moment. All messages which relate to specific radio programs are dynamic, as are applications such as Radiotext. In this dynamic category we include PTY, PIN, DI, TA, M/S. To operate these features, the broadcaster must provide data circuits from the studio center to the transmitters, and will incur other revenue

Table 5.1
Some Important RDS Applications

Tuning Functions		
PI	Program Identification	16-bit machine code giving a unique (within any one country) serial number to a program service.
PS	Program Service Name	8 ASCII characters for display on the receiver (e.g., "BBC R1," "Capital").
AF	Alternative Frequency	List of the frequencies on which a given program service can be found in adjacent service areas.
EON	Enhanced Other Network Information	PI, AF of other program services (usually those operated by the same broadcasting organization).
PTY	Program Type Code	Indicates the type of program material broadcast in a particular service (e.g., drama or classical music).
Traffic Station-Announcement Identification		
TP	Traffic Program Flag	One-bit code which indicates if a station carries traffic messages as part of its sound program.
TA	Traffic Announcement Flag	One-bit code which indicates that a traffic announcement is currently on-air from the tuned station.
Other Information		
CT	Clock-Time and Date	Transmitted as Universal Time Code and Modified Julian date.
RT	Radio Text	Text for display (or reproduction via a voice synthesizer).
PIN	Program Item Number	Scheduled start time and date for an individual program.
TDC	Transparent Data Channel	Unformatted text or data.
M/S	Music-Speech Code	One-bit flag to indicate music or speech.
DI	Decoder Identification	Four-bit code to indicate, for example, mono, stereo, *et cetera*.
IH	In-House Applications	Data which are used by the broadcaster for his own applications: not to be decoded by ordinary RDS receivers.
Radio Paging		
RP	Radio Paging Service	Paging service using FM broadcasts as the transport mechanism.

Table 5.1 (cont'd)

Traffic Message Channel		
TMC	Traffic Message Channel	A means of conveying traffic messages separately from the sound program signal by using densely coded predetermined messages which are broadcast within the RDS multiplex. (It is still under development.)

costs in entering and editing this program-related information. For this reason, these dynamic features are generally being introduced more slowly than the static features.

5.5.8 Allocation of RDS Capacity

The calculation of how the RDS channel capacity is divided between different features is somewhat complicated. This is due to two factors:

(1) The mixture of groups broadcast can, and preferably should, be varied dynamically. For example, before a travel announcement, it is wise to increase the frequency with which the TA flag (contained in Group Types 0A, 0B, 15B, as described in Appendix 1) is sent by increasing the proportion of capacity allocated to these group types.
(2) The proportion of capacity allocated to a given group type does not directly yield the proportion of capacity allocated to the associated application. That is because some applications, such as PI and PTY, are contained in all group types.

Calculations of RDS capacity must, however, be done by the broadcaster to determine a budget with appropriate priorities to ensure that essential features are properly serviced at all times. A simple example of such a capacity budget is given in Table 5.2. Examples of capacity allocations are shown in Figure 5.8.

Figure 5.8(a) shows the capacity allocation if only the repetition rates recommended by the EBU for the basic tuning features are used. Figure 5.8(b) shows the same allocation, but with the highly desirable *enhanced other network* (EON) feature implemented.

Figures 5.8(c) and (d) show examples of how two broadcasters, BBC (UK) and ARD (West Germany) have chosen to allocate channel capacity in 1988. Note that in the BBC case, the allocation of capacity is varied dynamically to meet the

Table 5.2
RDS Channel Capacity Budget Calculations

Overall RDS Channel Capacity = 1187.5 b/s

Subtract Overhead:

Error protection 40 bits in every 104-bit group:

$$= \frac{40}{104} \times 1187.5 \text{ b/s} \quad = 456.7 \text{ b/s}$$

Addressing 5 bits in every 104-bit group:

$$= \frac{5}{104} \times 1187.5 \text{ b/s} \quad = 57.1 \text{ b/s}$$

Total Overhead = 513.8 b/s

Hence, *Net Channel Capacity* available to users:

= 1187.5 − 513.8 = 673.7 b/s

This net capacity is divided between those applications which occupy fixed locations in every group, and those which are in some groups only.

The fixed capacity allocated to PI, PTY, and TP is

16 + 5 + 1 bits in every group $= \frac{22}{104} \times 1187.5$ = 251.2 b/s

which leaves 673.7 − 251.2 = 422.5 b/s for variable applications.

Within this allocatable variable capacity, it is recommended that 40% of groups transmitted contain basic tuning and switching information (i.e., AF, TA, DI, PS contained in groups type 0A, 0B, 15B).

Thus, capacity for these items $= \frac{0.4 \times 37 \times 1187.5}{104}$ = 169.0 b/s

which leaves 1187.5 − (513.8 + 251.2 + 169.0) = 253.5 b/s

for all optional applications.

The capacity allocated to each optional application is most easily calculated by remembering that within any 104-bit group, the maximum capacity available after deducting overheads and fixed codes is 37 bits.

Thus, if a particular application uses *n* Type GX groups per second, the capacity available thus used is

$$n \times 37 \text{ b/s}$$

For example, the EON feature uses 2 Type 14A or 14B groups per second which uses

$$2 \times 37 = 74 \text{ b/s.}$$

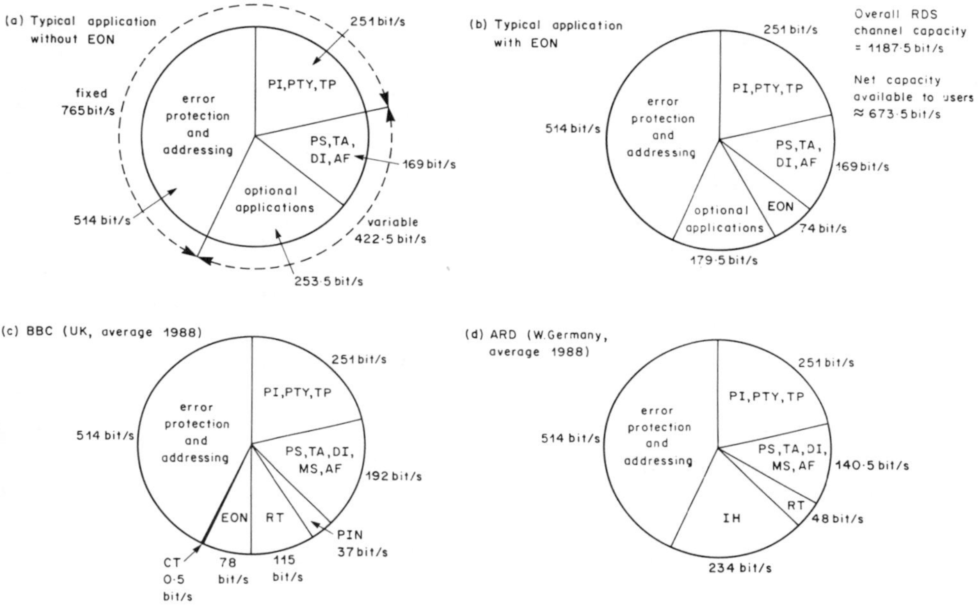

Figure 5.8 Allocation of RDS capacity.

priorities of different applications at different times of the day. Thus, the allocation shown in Figure 5.8(c), a long-term average, is subject to change.

The most important thing to note from these examples, which are typical of many European broadcasters' implementations of RDS, is that from the outset all the RDS channel capacity has been used. New applications can be accommodated only by reducing the repetition rate for some features, or by abandoning a feature on a given network.

5.5.9 RDS Reception Reliability

Curve (b) in Figure 5.9 shows the bit error rate of a prototype RDS demodulator measured as a function of the EMF applied in the aerial input of a VHF-FM receiver (noise figure 5.5 dB). For purposes of comparison, the theoretical bit error rate is given as curve (a). We see that the measured performance of this experimental RDS decoder is within about one decibel of that expected from theory.

Note that for these measurements, the deviation due to the RDS signal was set at 2 kHz. For other deviations, the results may be linearly scaled. For example, for

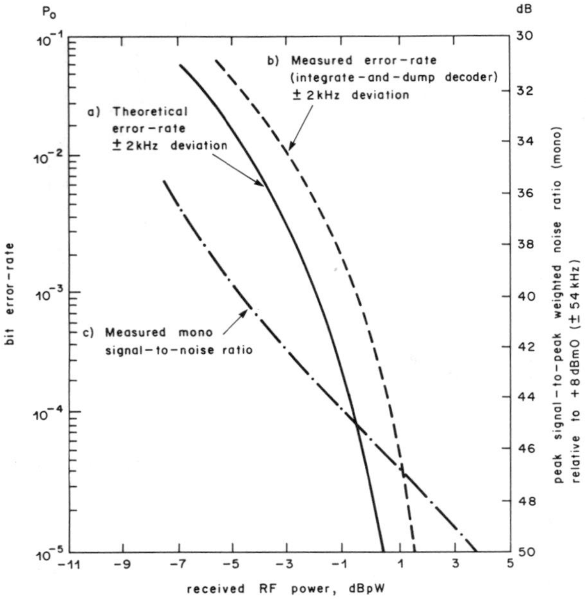

Figure 5.9 RDS bit error rate and mono signal-to-noise ratio as a function of the RF power at the receiver aerial input.

1.2 kHz, the deviation and RF input power needed to attain a given bit error rate would be about 4.4 dB greater.

When signals of the ARI system are broadcast simultaneously with RDS, most first-generation RDS demodulators show several decibels degradation of noise performance due to interference from the ARI signals. This is not fundamental, and later RDS demodulators may be significantly improved in this respect.

Also shown in Figure 5.9 is the monaural peak-signal to peak-weighted-noise ratio (measured according to CCIR Recommendation 468) obtained in the sound program channel of the same FM receiver used in the error-rate measurements of curve (a). Stereo reception would need about 20 dB more RF power to achieve the same signal-to-noise ratio as that shown for monaural. Thus, when the only impairment to reception is random noise due to low-field strength, we find that the RDS system operates satisfactorily at field strengths below the level needed for satisfactory reception of the stereo signal. At the field strength corresponding to the failure point of the RDS system, monaural reception is noisy but still intelligible, and remains so for aerial input levels down to about −10 dB pW.

In mobile reception, multipath propagation due to reflections from hills or buildings is the dominant factor in rendering the program signal unusable. To quantify the performance of RDS is difficult under these conditions, because the results are critically dependent upon the precise delay and relative amplitude of the echoes, and these obviously vary rapidly as the vehicles moves. Laboratory simulations using a single echo of fixed delay have not proved successful in realistically simulating mobile reception, and therefore mobile field tests must be made in appropriate areas. Even then, consistent results are difficult to obtain from day to day, because the level of interference on a given stretch of road varies considerably over a period of hours. The precise reasons for this variation are not fully understood although co-channel interference has been implicated [Edwardson, 1986].

Detailed studies of RDS reception reliability for moving vehicles were undertaken by broadcasters throughout its development [Mielke and Schwaiger, 1986; Lyner, 1987]. More recently, receiver manufacturers have made their own tests in the course of evaluating prototypes of their RDS receivers. One of the most useful methods of presenting the results of such mobile tests is based on that developed by Lyner [1987]. The results are presented as the "waiting time" between successful acquisitions of a particular RDS message. This indicates directly the proportion of time the receiver would have to wait to identify a given station or switch to a traffic program.

Figure 5.10 presents the result of such a waiting-time analysis based on Lyner's results. Figure 5.10(a) shows the waiting time for acquisition of the PI code, and Figure 5.10(b) the time for the PS code. Thus, for example, with 2.25 kHz deviation, in 97.5% of cases a correct PI code was received on the first attempt with no intervening erroneous PI codes. For 1 kHz deviation, the corresponding figure is 91.5%.

Similarly, for PS codes, in 79% of cases with 2.25 kHz deviation, a correct code was acquired on the first attempt, reducing to 57% with 1 kHz deviation. (*Note:* The BBC now uses 2 kHz deviation on all its services except BBC Radio 3, on which 1.2 kHz deviation is used.)

These results clearly show the improved ruggedness which results from the fixed format of the PI codes as compared with the variable addressed format of the PS codes. Nevertheless, both are reliable for their intended purpose, and this satisfactory result is generally being endorsed as experience is gained with the system.

5.5.10 RDS as an Aid to Tuning

The FM band in Europe is becoming increasingly crowded as the number of radio stations grows rapidly. In contrast to the US, sound radio broadcasting in Europe has traditionally been dominated by public service broadcasters such as the BBC, who broadcast the same programs throughout one country. Such national program services (the BBC operates four: Radio 1, 2, 3, and 4) need many transmitters to achieve coverage throughout the country. For example, the BBC uses stations at

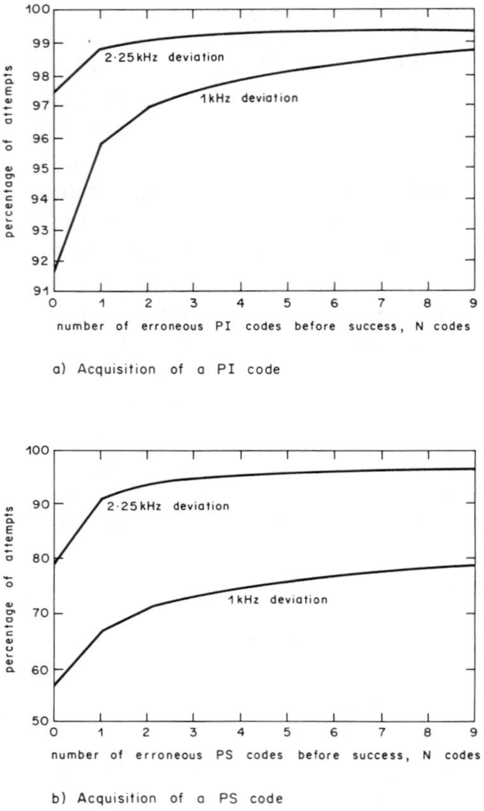

Figure 5.10 Percentage of the attempts when there were *N* or fewer erroneous receptions before successful acquisition of a correct code.

over 100 different sites (operating on 22 different frequencies) to broadcast each of their national radio services.

In these circumstances, to listen to the same program service throughout a long journey becomes feasible and desirable. Without RDS, however, to retune every 40 miles or so is both difficult and dangerous as the car moves from the service area of one transmitter into that of the next. Even finding the wanted program service can be difficult. RDS helps to solve both of these problems.

How does the listener find the program he wants? With an RDS car radio, all a user normally does is press the appropriate tuning button (i.e., button 2 for BBC Radio

2). Of course, push-button tuning is not new or even unusual in car radios, and some models already offer some forms of "auto-store" tuning. However, without RDS, these rely on the user preprogramming them with the appropriate frequencies, or the random "auto store" of the six strongest signals, regardless of what programs they carry.

To confirm the station which has been selected, the RDS *program service* name (PS) can be displayed on the front of the receiver instead of the tuned frequency.

How does the radio find the best frequency for the selected program in a given area? For program services which are broadcast on more than one frequency, RDS broadcasts a list of the *alternative frequencies* (AFs) (which can include AM stations) on which the same program is broadcast in adjacent areas. The receiver can then continually monitor the field strength on each of these frequencies, and retune if the signal on one alternative frequency looks consistently stronger than that of the tuned frequency. This signal strength checking takes only a few milliseconds, so it can be done almost imperceptibly by the same receiver used to derive the audio program signal. More sophisticated receivers check the level of distortion due to multipath as well as the signal strength.

When the receiver switches to an AF, it then checks the *program identification* (PI) of the broadcast on the AF. The PI code is uniquely assigned within any one country and between adjacent countries, so two broadcasts can carry the same PI if and only if the audio program signal is the same. Thus, an RDS receiver should never switch to another wrong program service. If the new transmitter carries a different AF list, then this will replace the previous one.

In some cases, the selected program service is available in a regional area only, and if the car moves out of the area, it is no longer possible to listen to the same audio program signal. In such cases, it is desirable to offer another regional variant of the same program. For example, the listener who normally listens to the BBC local radio station Greater London Radio (GLR) may wish to listen to other BBC local stations such as Radio Bedfordshire, or BBC Essex as the vehicle passes through service areas of these stations.

In such cases, the PI code of the relevant stations differs only in the penultimate 4 bits of the 16-bit PI code, and the RDS receiver can automatically offer this "generically linked" regional variant. In such cases, because the audio program signal of the generically linked stations is not the same, the receiver will usually switch only when triggered by the listener (e.g., by again pressing the preset button).

What happens if the listener changes stations? From the above description, it is easy to understand how the receiver keeps track of the best frequency for any one tuned program. Yet, what happens if, after traveling 200 miles listening to one station, the listener decides to retune to another service? Broadcasters can implement the RDS feature called EON information to update the AF lists associated with services other than the tuned station. So when the corresponding button is pushed, the receiver always finds an appropriate frequency stored in the AF list associated with that button.

Of course, broadcasters will usually carry EON information only for their own services. If the listener wants to tune between two services not cross-referenced by EON, then the receiver must do a PI search. In this, the receiver scans the band in 100 kHz steps, checking the PI code on each broadcast as it goes until it finds a broadcast with a PI code which matches that of the desired station. Such a PI search can take several seconds.

5.5.11 Conveying Traffic Information via RDS

There are three levels of sophistication of an RDS traffic information system:

(1) emulation of the ARI system;
(2) emulation of the ARI system with cross-referencing via EON information;
(3) digital *traffic message channel* (TMC).

The first phase, simple emulation of the ARI system, was a design objective for RDS from the outset. Table 5.3 indicates how the ARI features are accommodated in RDS.

The high repetition rate of the TP flag (and area code information) in RDS means that we can search in real time for stations with the TP flag set (i.e., those which, at some time, carry traffic messages as part of their normal audio program signal). Most RDS receivers have a "TP," "INFO," or "ARI" button, which modifies the search tuning operation so that the receiver will only search for stations having the TP flag set to 1.

In this basic emulation, the TA flag can switch the receiver from the cassette or increase the audio volume only if the receiver is already tuned to the station which gives the traffic announcement: it is not possible to listen to one program service and then automatically switch to another when a traffic announcement is imminent. To achieve that, a dual receiver would be needed (one tuned to the traffic program, the other to the wanted audio program), which is very expensive, or else the broadcast and receiver must implement EON information, as described below.

The second phase of the RDS Traffic Service enables listeners to tune to one program service which carries classical music and no traffic information, and still hear any traffic messages which are broadcast on other program services which are cross-referenced via the EON feature of RDS. Listeners are still given the choice, of course, of whether to listen to the travel information: usually an "INFO" or "TA" button is used to select whether the traffic messages are wanted.

The way this second phase of RDS traffic information operates will now be described in detail based on an example which is illustrated in Figure 5.11. This is based on the pilot-scheme of this kind of service which the BBC started to operate from five of its local radio stations in the summer of 1989.

Table 5.3
ARI Features in Radio Data Systems

ARI Feature	*RDS Emulation*	*Location in RDS*	*Repetition Rate per Second*
Traffic Program Identification (presence of 57 kHz subcarrier modulated by any area code)	TP Flag Bit	Block 2 of all RDS groups	11.42
Area Identification Code	Area Code	Second 4-bit nibble in PI code	>11.42
Message Announcement Code	TA Flag Bit	Block 2 of Type 0 Groups Blocks 2 and 4 of Type 15B Groups	Typically 4

The sequence starts when the police radiotelephone details of the incident to the police control room. The control room has a specially adapted personal computer which is used to input details of the incident using a menu driven system developed by the BBC. The duty officer enters details like road number (e.g., M20), direction of travel (e.g., west), cause (e.g., heavy truck has lost its load), effect (e.g., single-lane traffic, serious delays can be expected), advice (e.g., take an alternative route if possible), and duration (e.g., four hours).

After having entered all the details of the incident, the officer sends the information via dial-up modem to a computer system in the BBC Travel Centre at Broadcasting House in central London. There the message is automatically entered into the computer database, and an on-screen message alerts one of the operators. The operator then edits the message and passes it on to one of the announcers.

The Travel Centre then contacts the local radio station (e.g., Radio Kent) to let them know that it has a travel announcement for their area, and agrees on a cue for it to be broadcast. At the next convenient moment (e.g., at the end of a record) Radio Kent prepares to mix in the voice of the announcer in London and preface the announcement with a "jingle." The London announcer, listening to Radio Kent output via "clean feed," takes the jingle as his or her cue. On hearing the jingle, the announcer presses his travel button, which signals to the RDS Central Computer that Radio Kent is about to broadcast a travel announcement.

The Central RDS Computer, receiving the signal from the Travel Centre, sends out update commands to all RDS encoders for Radio 1, 2, 3, and 4 as an "other

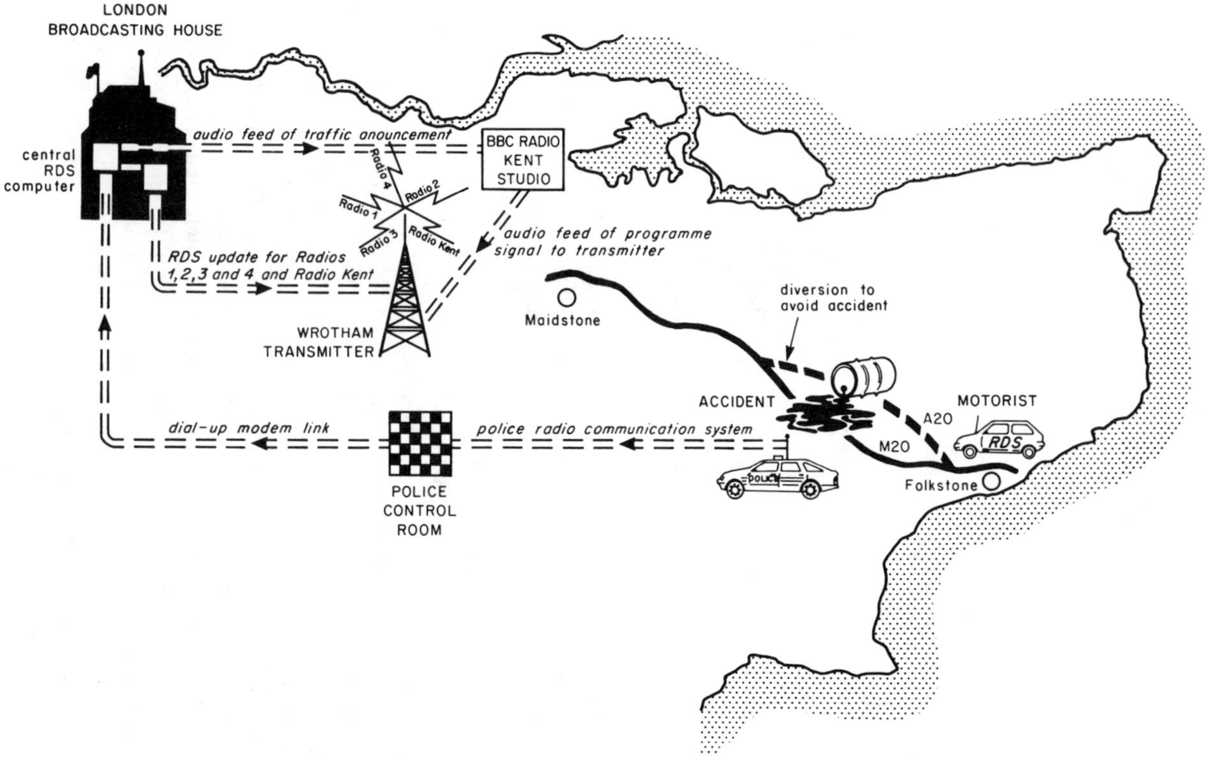

Figure 5.11 RDS traffic information using the EON feature (BBC pilot scheme, 1989).

network" to set the TA flag in the Radio Kent EON groups. Now, an RDS receiver (with the EON feature) tuned to BBC Radio 1, 2, 3, or 4 receives, via the EON data, the information that Radio Kent is signaling an imminent Travel Flash. If the listener has pressed the "TA" button on the receiver, it will respond by retuning to Radio Kent where the TA flag in the Type OA and Type 15B groups is already set to "1". That switching typically takes less than two seconds to achieve. The receiver is now tuned to Radio Kent for the duration of the Travel Flash.

By taking his or her cue at the end of the jingle, the London announcer reads the travel message, which is broadcast as the normal audio program signal on Radio Kent. This voice message is heard by both the Radio Kent listeners, and by listeners with RDS-EON sets who have pressed their "TA" buttons.

When the announcement is finished, Radio Kent fades out the London announcer and resumes its output with another jingle before returning to the interrupted program. The Travel Centre removes its signal to the RDS Central Computer, which in turn, resets the TA flag to "O" in Radio Kent's Type OA and Type 15B groups. The RDS-EON receivers, which had been captured for the Travel Flash, retune to their original program service.

As with the simple ARI emulation, the listener still has the option to listen to the cassette player or turn the volume down and still get the Travel Flashes. The Radio 1, 2, 3, or 4 listeners who do not depress the TA button or have ordinary non-RDS radios, will not hear the travel messages. However, all listeners to Radio Kent, even those with ordinary non-RDS radios, will hear each of the Travel Flashes because they are broadcast as part of the normal audio program signal.

This experimental BBC pilot scheme "Travel Service," which started from five BBC local radio stations in southern and central England in the summer of 1989, is the first example of this second-generation RDS travel information system. RDS receivers with the necessary EON traffic feature are expected to become available in mid-1990.

Even this second-generation RDS Travel Service has a number of drawbacks, which the third-generation digital system described in the next section is designed to overcome.

5.6 THE FUTURE

5.6.1 RDS-TMC

The TMC proposal uses the RDS system to provide a separate channel for traffic messages independent of the main audio program signal. This overcomes the drawback of the two systems outlined above; the broadcaster must somehow fit the traffic messages into his program schedule, which can lead to conflict between the need to provide timely information and the desire to maintain continuity and balance for listeners not interested in traffic messages.

As with the CARFAX system (see Section 5.4.6), the ability to broadcast traffic messages on a dedicated channel avoids the conflict outlined above, and has the potential to provide a much more detailed and comprehensive service of traffic information.

The use of RDS as the transport mechanism for this traffic information means that its capacity is very restricted; at most, some 200 b/s are available for this (assuming that the vital tuning assistance functions of RDS are maintained). However, traffic messages do lend themselves to dense coding, because they can easily be reduced to a number of preset phrases. Because the messages are reconstructed within the receiver, they can be reproduced (via a voice synthesizer) in the driver's choice of language, regardless of the language of the associated program. This feature is very important for a pan-European system. All that is needed to change the reproduced language is a different message decoding table, which can be supplied either as solid-state memory or as a CD-ROM.

The RDS-TMC system is now being developed by a consortium of broadcasters, road traffic experts, and receiver manufacturers as part of the Commission of the European Communities' DRIVE (*Dedicated Road Infrastructure for Vehicle Safety in Europe*) program (see Chapter 9 for details of DRIVE). This project is expected to be completed to the stage of a draft European standard on RDS-TMC by the beginning of 1991. A special receiver, which is likely to be independent of the normal car radio, will be needed to decode the RDS-TMC messages.

Because the RDS-TMC messages are digitally encoded, to integrate them with a route guidance aid carried in the vehicle will be possible. The possibilities here are considered further in Chapter 6. By using the route guidance system, the receiver could be instructed to decode only those messages which would be relevant to the driver's journey. Moreover, the messages can be reconstructed in the vehicle, in the language of the driver's choice.

5.6.2 Automated Data Gathering and Collation

Improved methods for gathering and dealing with traffic incident information are being investigated in many countries. These include methods for remotely and automatically detecting traffic incidents, as well as for collating and prioritizing the data and generating broadcast messages. These systems will provide more timely, more accurate, and hence more credible information than has been previously available with less need for human intervention.

Examples include the HAIR system on the Hanshin Expressway in Japan. This system is almost completely automatic. Queue detectors are used to automatically control variable message signs and warn drivers of conditions on the highway ahead. The control system monitors the queue detectors, sets the messages on the signs, and composes suitable messages for broadcasting. These messages are then broadcast over the HAIR system as synthetic speech.

In the Federal Republic of Germany, a system called ARIAM is being tested [Giesa and Everts, 1987]. ARIAM, which (loosely translated) stands for "ARI on the basis of actual measurements," involves the use of vehicle detectors distributed throughout a trial area of the highway network in Hessen. A control system continuously monitors the status of these detectors to decide when a traffic incident has occurred and to compose a suitable message to be broadcast over the ARI system, and then disseminated using other media and methods (e.g., teletext, videotex, and variable message signs).

An alternative approach under investigation in the UK [Allport, 1988a,b,c] is to use the techniques of *artificial intelligence* (AI) to interrogate police reports directly and in real time. As messages are relayed from police patrols to their control center, records of these messages, together with any actions taken, are typed into the control center computer. The TICC (*Traffic Information Collator-Condenser*) system under development will interrogate these records as they are put into the police control computer. Thus, (1) to decide if the text concerns a traffic incident, and if it does; (2) to deduce details such as type and location; (3) to deduce whether the incident is serious enough to warrant a broadcast; (4) to compose a message suitable for broadcasting; and (5) to decide when and where the message should be broadcast.

REFERENCES

Allport, D., (1988a), "Interpreting incident reports," *Proc. of IEE Colloquium on Applications of Expert Systems in Road Transportation,* IEE Digest No. 1988/9, London.

Allport (1988b), "The TICC: Passing interesting text," *Proc. 2nd Conf. on Applied Natural Language Processing,* ACL, Austin, Texas, 1988.

Allport (1988c), "Understanding RTA's," *IT '88 Conference,* Swansea, UK, July 1988, pp. 323–326.

European Broadcasting Union, "Specification of the radio data system RDS for VHF/FM sound broadcasting," Document Tech. 3244, EBU, Brussels, 1984.

Giesa, S., and K. Everts, "ARIAM: Car-Driver-Radio-Information on the basis of automatic incident detection," *Traffic Engineering and Control,* June 1987, pp. 344–348.

Lyner, A.G., "Experimental Radio-Data System (RDS): A survey of reception reliability in the UK," BBC Research Department Report No. 1987/17.

Mammano, F.J., "Speech synthesis for a motorist information system," *Proc. OECD Seminar on Microelectronics for Road and Traffic Management,* Tokyo, November 1984, pp. 182–191.

Mielke, J., and K.H. Schwaiger, "Progress with the RDS system and experimental results," *EBU Technical Review,* No. 217, June 1986.

Sandell, R.S., and S.M. Edwardson, "A proposed road traffic information service," *European Broadcasting Union* (EBU) *Technical Review,* December 1977, Brussels.

Sandell, R.S., "The 'Carfax' Road Traffic Information System," British Broadcasting Corporation Research Report RD 1984/1, February 1984.

TRRL, Report of the working group on the broadcasting of traffic information, Department of Transport, Transport and Road Research Laboratory Supplementary Report 506, Crowthorne, UK, 1979.

TRRL, ROUTE-TEL: A Viewdata route planning service for drivers, Department of Transport, Transport and Road Research Laboratory Leaflet LF952, 1983.

Turnage, H.C., "Highway advisory radio," *IEEE Transactions on Vehicular Technology,* Vol. 29, pp. 183–191, 1980.

BIBLIOGRAPHY

Duckeck, H.G., "ARI—Automatic Radio Information," *SAE Technical Paper series 840091, reprinted from P-142 Audio Systems, International Congress and Exposition,* Detroit, Michigan, February 27th March, 1984.

EBU, "Specifications of the radio data system RDS for VHF/FM sound broadcasting," EBU Doc. Tech. 3244 (and supplements 1 to 5), Brussels, 1984 and 1989.

EBU, "Guidelines for the implementation of the RDS system," Doc. Tech. 3260, Brussels-Geneva, 1989.

Edwardson, S.M., *Radio Receivers* (W. Gosling, ed.) IEE Telecommunications Series 15, Peter Peregrinus, London, 1986, Chapter 14. (The Broadcast Receiver (1) Sound.)

Ely, S.R., "VHF radio-data: Experimental BBC transmissions," BBC Research Department Report No. 1981/4.

Ely, S.R. (1982a), "The impact of radio-data on broadcast receivers," *The Radio and Electronic Engineer,* 1982, 52, 5, pp. 291–296.

Ely, S.R. (1982b), "VHF Radio-Data," *International Broadcasting Convention 1982,* IEE Conference Publication No. 220, pp. 282–287.

Ely, S.R., and D. Kopitz, "Design principles for VHF-FM radio receivers using the EBU radio-data system RDS," *EBU Technical Review,* No. 204, April 1984, pp. 50–58.

Ely, S.R., D.T. Wright, and C.C. Goodyear, "High-speed decoding technique for slip detection in data transmission systems using modified cyclic block codes," *Electronics Letters,* No. 3, February 1983, pp. 109–110.

Kamalski, T., "The Radio Data System from the Receiver Manufacturer's Perspective," *IEE Colloquium on "The RDS System—Its Implementation and Use,"* IEE Digest No. 1988/128, pp. 4/1–4/4.

Kopitz, D., "Development of VHF-FM Radio-Data Transmission from the European point of view," *International Broadcasting Convention 1982,* IEE Conference Publication No. 220, pp. 273–275.

Kopitz, D., "Future technical developments concerning broadcasting to motorists," *EBU Review* (Programmes, Administration and Law), No. 2, March 1982, pp. 37–40.

Kopitz, D., "Development of the European VHF/FM Radio-Data System 'RDS' for VHF/FM broadcasting and car reception," *SAE Technical Paper series 840090,* reprinted from P-142 Audio Systems, International Congress and Exposition, Detroit, Michigan, March, 1984.

Kopitz, D., "Radio-Data system permits receiver adjustments and special signalling by the FM broadcaster," *1986 NAB Engineering Conference Proceedings,* pp. 130–133.

Kopitz, D., "The development of the RDS system from a European point of view and the possibilities for distributing traffic messages with RDS," *IEE Colloquium on "The RDS System—It's Implementation and Use,"* IEE Digest No. 1988/128.

Lewis, A.R., S.J. Parnall, and J. Robinson, "Radio Data System (RDS): The Technical Realisation of a New Broadcast Service," *International Broadcasting Convention 1988,* IEE Conference Publication No. 293, pp. 311–315.

Livey, J., "Radio-Data services in independent local radio," *International Broadcasting Convention 1988,* IEE Conference Publication No. 293, pp. 316–317.

Lyner, A.G., "Experimental Radio-Data System (RDS): A survey of reception reliability in the UK," BBC Research Department Report No. 1987/17.

MacEwan, D., "Listeners' Options in the 1980s: Potential developments of radio sets," Document Sixteen of Second Symposium on Radio in the 80s, Ottawa, April 1976.

MacEwan, D., "Radio in the '80s: Broadcasting and the ideal sound receiver of the future," *Electronics and Wireless World,* May 1977, pp. 36–40.

Makitalo, O., "Utilisation of the FM broadcasting network for transmission of supplementary information," *International Broadcasting Convention 1978,* IEE Conference Publication No. 166, pp. 327–329.

Marks, B., (1988a), "Radio-Data System (RDS)—The planning and implementation of a new broadcast service (using high reliability system)," *International Broadcasting Convention 1988,* IEE Conference Publication No. 293, pp. 308–310.

Marks, B., (1988b), "What is happening to RDS?" *Electronics and Wireless World,* November 1988, pp. 1096–1100.

Marks, B., (1988c), "Radio-Data System (RDS)—The planning and implementation of a new broadcast service (using high reliability systems)," *IEE Colloquium on "The RDS System—Its Implementation and Use,"* IEE Digest No. 1988/128, pp. 2/1–2/5.

Marks, B., (1988d), "RDS—the way ahead for radio," *IEE Review,* April 1989, pp. 127–129.

Mayr, J., "FM data coder for the radio data system," *International Broadcasting Convention 1986,* IEE Conference Publication No. 268, pp. 89–93.

Mielke, J., and K.H. Schwaiger, "Progress with the RDS system and experimental results," *EBU Technical Review,* No. 217, June 1986.

Odmalm, C., "The development of the EBU VHF-FM radio-data system (RDS)," *EBU Technical Review,* No. 200, August 1983, pp. 186–192.

Ogawa, R., *et al.,* "Development of Radio-Data System decoder," *IC's, I.E.E.E. Trans. on Consumer Electronics,* CE-33, No. 3, August 1987, pp. 383–394.

Parnall, S., "Decoding RDS," *Electronics and Wireless World,* February 1989, pp. 148–152 (Part 1) and March 1989, pp. 284–287 (Part 2).

Peek, J.B.H., and J.M. Schmidt, "A 'Station Programme Identification' (S.P.I.) for FM sound broadcasting," *International Broadcasting Convention 1978,* IEE Conference Publication No. 166, pp. 321–323.

Riley, J.L., and J.D. Newland, "RDS receivers of the future," *IEE Colloquium on "The RDS System—Its Implementation and Use,"* IEE Digest No. 1988/128, pp. 5/1–5/4.

Shute, S.A., "RDS: The EBU Radio Data System and its implementation by BBC Radio," *International Broadcast Engineer* 18, no. 216, May 1987, pp. 65–67 and no. 217, July 1987, pp. 16–20.

Shute, S.A., "Towards the intelligent radio," *Electronics and Wireless World,* 93, October 1987, pp. 1023–1026.

Taura, K., and R. Tomotiro, "Automatic tuning car radio based on the radio data system," *I.E.E.E. Trans. on Consumer Electronics,* CE-23, No. 3, August 1987, pp. 319–326.

Wingham, J.K., and P.R. Kemble, "The implementation of RDS in the United Kingdom's local radio network," *IEE Colloquium on "the RDS System—Its Implementation and Use,"* IEE Digest No. 1988/128, pp. 3/1–3/9.

Wright, D.T., and S.M. Edwardson, "Review of broadcast radio-data systems," *International Broadcasting Convention 1986,* IEE Conference Publication No. 268, pp. 85–88.

Whythe, D.J., "The transmission of two programmes from Band-II FM transmitters: an assessment of 'storecasting'," *EBU Technical Review,* No. 161, February 1977, pp. 20–30.

Whythe, D.J., and S.R. Ely, "Data and identification signalling for future radio receivers," *International Broadcasting Convention* 1978, IEE Conference Publication No. 166, pp. 324–326.

APPENDIX 5.A
ABBREVIATED SPECIFICATION OF THE RADIO DATA SYSTEM (RDS)

5A.1 MODULATION OF THE DATA CHANNEL

1.1 Subcarrier frequency: 57 kHz, locked in-phase or in-quadrature to the third harmonic of the pilot tone 19 kHz (±2 Hz) in the case of stereophony. (Frequency tolerance: ±6 Hz.)

If used simultaneously with the ARI traffic broadcast identification system, the ARI subcarrier will have a phase difference of 90 degrees ±10 degrees, and the recommended nominal deviation of the main carrier will be ±1.2 kHz due to the RDS signal and ±3.5 kHz due to the unmodulated ARI subcarrier.

1.2 Subcarrier level: the recommended nominal deviation of the main FM carrier due to the modulated subcarrier is ±2 kHz. The decoder should, however, be designed to work with subcarrier levels corresponding to between ±1 kHz and ±7.5 kHz deviation.

1.3 Method of modulation: the subcarrier is amplitude-modulated by the shaped and biphase-coded data signal. The subcarrier is suppressed (see Figures 5.5 and 5.6).

1.4. Clock frequency and data rate: the basic clock frequency is obtained by dividing the transmitted subcarrier frequency by 48. Consequently, the basic data rate is 1187.5 bit/s ±0.125 bit/s.

1.5 Differential coding: when the input data level from the coder at the transmitter is zero, the output remains unchanged from the previous output bit, and when an input of one occurs, the new output bit is the complement of the previous output bit.

5A.2 BASEBAND CODING

2.1 Coding structure: the largest element in the structure is called a "group" of 104 bits. Each group comprises four blocks of 26 bits. Each block comprises an information word and a checkword, of 16 and 10 bits, respectively.

2.2 Order of bit transmission: all information words, checkwords, and addresses have their most significant bit transmitted first.

2.3 Error protection: the 10-bit cyclic redundancy checkword, to which a 10-bit offset word is added for synchronization purposes, is intended to enable the receiver-decoder to detect and correct errors which occur in reception.

2.4 Synchronization of blocks and groups: the data transmission is fully synchronous, and there are no gaps between the groups of blocks. The beginnings and ends of the data blocks may be recognized in the decoder by using the fact that the error-checking decoder will, with a high level of confidence, detect block synchronization slip. The blocks within each group are identified by different offset words added to the respective 10-bit checkwords.

2.5 Message format: the first five bits of the second block of every group are allocated to a 5-bit code, which specifies the application of the group and its version. The group types specified are given below. There is also space left to add applications yet to be defined.

Table 5A.1
Group Type Codes

Decimal Value	*Group Type Binary code A3 A2 A1 A0*	*Applications*
0	0 0 0 0 X	Basic tuning and switching information
1	0 0 0 1 X	Program item number
2	0 0 1 0 X	Radio text
3	0 0 1 1 X	Other network information (superseded by enhanced version in Type 14 groups)
4	0 1 0 0 0	Clock time and date
5	0 1 0 1 X	Transparent channels for text or other graphics (32 channels)
6	0 1 1 0 X	In-house applications
7	0 1 1 1 0	Radio paging
8	1 0 0 0 X	Reserved for traffic message channel
9–13		Applications not yet defined.
14	1 1 1 0 X	Enhanced other network information
15	1 1 1 1 1	Fast basic tuning and switching information

X indicates that value may be "0" (Version A) or "1" (Version B).

Chapter 6
Route Guidance and Vehicle Location Systems

D.J. Jeffery
Transport and Road Research Laboratory,
Department of Transport, UK

6.1 INTRODUCTION

The high economic, social and environmental costs of providing increased road capacity and the persistent growth in road traffic combine to place increasing emphasis on the need to use existing roads as efficiently as possible.

Advances in information technology can help. Recent examples include the widespread adoption in the early 1970s of fixed-time urban traffic control (UTC) systems, and in the 1980s of traffic responsive UTC systems such as SCOOT, which is now in use in many cities in the UK [Robertson, 1988]. Outside of urban areas, there are dynamic motorway control systems such as the MTCSS system used on the A13 motorway in the Netherlands [Jenezon, Klijnhout, and Langelaar, 1987] and New York State's IMIS (Integrated Motorist's Information System) on the New York Northern Long Island Corridor (see also Chapter 7).

For the 1990s, however, increased attention is being paid to the possibilities of route guidance systems. Unlike most conventional systems which control traffic in time alone, route guidance systems offer the potential for controlling traffic both in time and in space, by guiding individual vehicles over less congested paths, and hence by distributing traffic more evenly through the road network.

Fleet managers and drivers can see advantages, and industry has responded with a range of innovative solutions. Governments are also interested not only because of the potential benefits such systems could achieve for the community as a whole, but also because of concern about potential environmental impacts of rerouting traffic.

This chapter reviews the current state-of-the-art in the development of route guidance systems. Results of work from Europe, the US and Japan are considered to show, first, the range of possible techniques; second, the range of functions that these techniques can provide; and, third, the benefits that might be achieved. A number of developments that have progressed to a trials stage are then considered in more detail.

6.2 VEHICLE LOCATION

Various electronic aids have been developed to help drivers plan and follow routes. Some are self-contained, and can be wholly carried in a vehicle, while others depend on an infrastructure of equipment at the roadside. But all rely on a navigation aid to enable the vehicle to keep track of its position in the road network.

There are basically three techniques for navigation: 1. *dead reckoning;* 2. *trilateration;* and 3. *beacons*—any of which, either singly or in combination, may be used to update a vehicle's position.

6.2.1 Dead Reckoning

Navigational aids that use a dead reckoning technique rely on distance and heading sensors fitted to the vehicle (see also Chapters 8 and 10) so that progress from a known starting position can be continually monitored and location, in terms of grid reference, can be updated. Dead reckoning devices have been manufactured for military use for many years, but only recently with the advent of cheap microcomputers has it become economically worthwhile to adapt the technology for use in private vehicles.

The distance sensor may be a mechanical "take-off" from the odometer, or more often, some mechanism for sensing wheel revolutions (e.g., a series of magnets equally spaced around the wheel rim or drive shaft and sensed by a magnetically activated switch fixed to the body of the vehicle). A pulsed or digital output is clearly preferred for detection by a microcomputer. Accuracies of about two percent of distance traveled can be obtained. Higher accuracy is difficult because of variations in wheel diameter which occur as the tire wears, or as inflation pressures vary, but it can be achieved by regular calibration.

The heading may be deduced by measuring steering wheel rotation, or by monitoring the differential rotation of the rear wheels of the vehicle. Generally, however, a special purpose sensor is used, such as a gyro compass, or an "electronic" compass, (e.g., a flux gate sensor). These are usually an energized pair of coils wound at right angles. The resulting field, which is affected by the earth's magnetic field, is then measured to determine the orientation of the gate (and hence the vehicle) with respect to the earth's field. Careful positioning of magnetic heading sensors is required, because they are generally affected by the electric circuits and the mass of metal in the vehicle itself. As a result, they must usually be calibrated for each individual installation. However, they are also affected by metal masses (e.g., bridge structures, other vehicles) outside the vehicle, so that high accuracies are precluded.

Dead reckoning systems thus accumulate error both in distance and orientation. As a result, they typically need to be reinitialized after a few hours of driving, or after a few tens of kilometers have been driven.

6.2.2 Trilateration

Navigational aids that use a trilateration technique generally rely on the detection of radio transmissions from three or more fixed points. Decca and Loran-C are well known examples for use in the air and at sea. On land, however, the signals are confused by multipath reflections from tall buildings and hills, so that position accuracies of better than about 200 m are difficult to obtain.

Satellite navigation systems such as SATNAV and GPS are also based on trilateration, using transmissions from satellites in well behaved (mathematically speaking) orbits. Again, these systems are appropriate for sea and air use, but are confused on land, and particularly in urban areas, where tall buildings may prevent enough satellites from being within line of sight to enable a fix to be made.

GPS appears to be the most promising system: in good conditions, it offers the potential to determine the position of a moving vehicle to within a few meters. However, the topography, the number of satellites within sight, and their elevation above the horizon, all play a part in determining if and when a "good" position fix can be made. As a result, most investigators agree that in urban areas, a dead reckoning system will also be needed so that some interpolation of position can be made between good fixes.

These problems and the accuracies that can be achieved are discussed further by Held and Kricke [1985], who have tested GPS in Munich, West Germany; and by Mooney [1986], who has evaluated GPS in rural, suburban, and urban environments in the US.

6.2.3 Beacons

In these systems, beacons (which may be passive or active) are positioned at known points in the road network. A passing vehicle is then able to update its position by interrogating the beacons. In practice, these systems are rarely used in isolation because it is not usually economic to deploy sufficient beacons to provide very high location accuracy. Therefore, a dead reckoning technique is frequently used to interpolate between beacons. Such systems are commonly used by buses which follow fixed routes; in these cases, only distance information is required for interpolation.

The next three sections show how the three basic techniques of navigation may be incorporated into electronic aids for route planning and guidance.

6.3 AUTONOMOUS NAVIGATION AIDS

In this section, we consider the autonomous (i.e., wholly vehicle contained) systems. There are three main types: 1. simple directional aids; 2. map displays; and 3. route guidance aids.

6.3.1 Simple Directional Aids

These comprise a heading and distance sensor, a microcomputer, a numeric keypad, and a display unit. The driver enters the grid reference for his or her current position and destination on the key pad. The microcomputer then computes the vector connecting these positions and displays it in terms of a "crow flight" distance and a heading on the display unit. Heading is usually displayed as an arrow which physically identifies for the driver the direction he or she must take to reach his or her destination (see Figure 6.1).

As the vehicle heads off on its journey, the microcomputer continually monitors the heading and distance-traveled sensors to recompute the vehicle's current position, and to update the remaining "crow flight" distance and heading displays. Thus, as the driver approaches each junction on the journey, he or she has available a measure of how near he or she is to the destination and the direction he or she ought to take to reach it.

Examples of such systems are the NAVICOM from Toyota, and DRIVEGUIDE from Nissan. A similar device, CITYPILOT from VDO of West Germany, uses a specially prepared set of maps together with a lightpen connected to the ve-

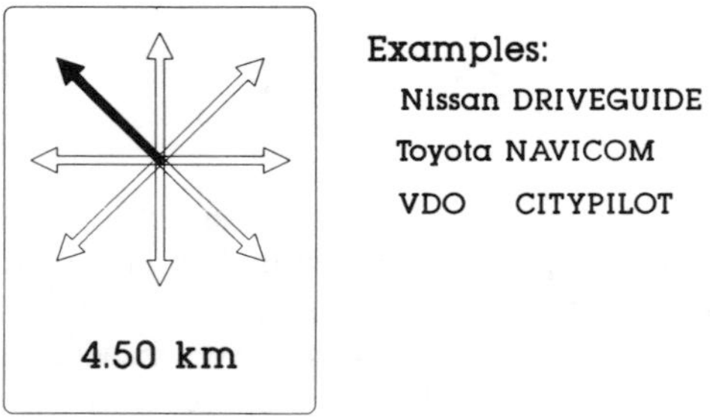

Figure 6.1 Simple directional aids.

hicle unit to facilitate the driver entering the grid references for origin and destination. These devices can be expected to work well in regular (i.e., matrix) networks, but may fail in irregular networks, such as in the UK, where roads do not always continue in the same direction as that from which they leave a junction. Nor can they help to identify an alternative route which may be faster when this requires the driver to head off in the wrong direction at some point in order to join, for example, a highway.

6.3.2 Map Displays

An alternative, and more usable, method of presenting navigation information is to superimpose the vehicle's position on a map of the surrounding road network (see Figure 6.2). Such techniques are not new; they have been used by the military for many years. However, only with the advent of inexpensive microcomputers has it become possible to adapt the technology for use in private vehicles.

Probably the first successful attempt at using map displays was made by Honda with its Electro-Gyrocator. This device relies on a gas rate gyro to sense heading and on a magnet type of distance sensor for dead reckoning the position of the vehicle at successive increments in time, in the same way as described above. The output, however, is a moving spot of light on a CRT. A suitable map foil is then placed over the screen, and the position of the spot of light is manually adjusted to correspond with the vehicle position. Other knobs and switches allow the driver to calibrate the system to correspond with the scale of the map foil. The driver then marks the destination on the map foil and sets off on his or her journey. After traveling a kilometer or so, the driver (or passenger) must align the map to correspond with the trace of the spot of light. Thereafter, the route thus far and the vehicle's present

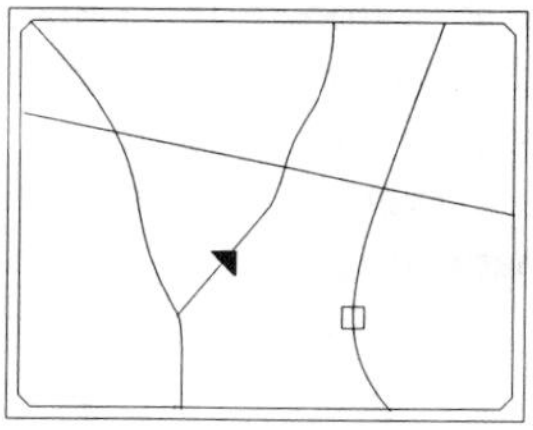

Examples: Philips CARIN
Honda ELECTRO GYROCATOR
Toyota ELECTRO MULTIVISION
ETAK

Figure 6.2 Map displays.

position on the map are clearly shown. A subsidiary spot of light shows the orientation of the vehicle on the map. Drivers can thus see where they are in relation to their destination, and which direction they must take in order to reach it.

Later developments such as ETAK from the US, ELECTRO-MULTIVISION from Toyota in Japan and CARIN from Philips in Holland (see Chapter 8) have replaced the map foil with a computer-generated map, which appears on the CRT together with the spot of light that shows the vehicle's position. In these systems, extra computer memory is required in which to store the map information in digitized form. ETAK uses cassettes, while CARIN uses a CD (compact disc) ROM, which is capable of storing around 600 Mbytes of data (several Mbytes are required for a digitized map covering a large town).

For the future, Philips proposes that CARIN use GPS. In situations where an accurate position fix can be made, this will save the driver from having to initialize the system. It should also improve the accuracy of position updates when used in conjunction with the dead reckoning system, but it is unlikely that it can substitute for that system. A further advantage of carrying the map data aboard the vehicle is the potential for map-matching. This is a technique which compares the vehicle's movements with the map data, and is used in some systems to correct the errors which accumulate in the dead reckoning system.

6.3.3 Route Guidance Aids

The simple directional aids and the map display devices described above tell the driver about his or her position relative to the destination, but cannot offer advice on which is the "best" route to follow. To do this, a more comprehensive description of the network must be stored in the vehicle, together with an algorithm which can operate on these data to compute a minimum path through the network.

In some devices, such as the Daimler Benz ROUTEN-RECHNER, Blaupunkt's EVA, and MICROMAP from Wootton Jeffreys and Partners in the UK, a description of the road network is stored on a suitable memory device in terms of the intersections (nodes) and the impedance of the sections of road (links) which connect them. A suitable algorithm (e.g., see [van Vliet, 1977]) can then be used to compute the minimum impedance path between any two intersections. Distance is most commonly used as a measure of impedance because it is the easiest to establish. But drivers generally prefer time, or cost (i.e., a trade-off between time and distance) as a criterion for route selection, and would sometimes like "most scenic."

In other devices (e.g., the TRRL's experimental NAVIGATOR), the network description is precompiled to provide a "Signpost" for each junction, with the resolution needed to give guidance to every other junction in the same network. The precompilation process can employ distance, time, or cost criteria, depending on the user's preference. In the CARGUIDE system from Carnegie-Mellon University, a hybrid system is used; the street network is divided up into "zones." Routes between zones are then precompiled, but routes within zones are computed on demand.

In all cases, the driver must initialize the system by keying in codes for the start position and the intersection which most nearly corresponds with his or her destination. The device then computes the "best" route through the network and gives the driver his or her first instruction. This may be via alphanumeric displays, speech synthesis units, or graphical display units, or a combination of these (see Figure 6.3).

MICROMAP, the first system demonstrated in a vehicle, was based on an Apple II microcomputer with 48K of memory and a 5.25-inch floppy disk drive. Special programs enabled the user to specify his own network of up to about 300 nodes and 450 links, and to specify both the lengths and impedances of the links. Origin and destination codes were entered on the Apple keyboard, and a speech recognition package was tried. The system used a speech synthesis unit to speak (i.e., "turn left") instructions to the driver as he or she progressed on the journey. It was assumed that the driver would not deviate from the planned route—a measure of distance traveled was then sufficient to "locate" the vehicle and to indicate when successive instructions should be given. The device was not developed beyond the demonstration phase.

ROUTEN-RECHNER from Daimler-Benz in Germany is similar. The system gives guidance on the West German highway network by using a minimum distance criterion for route selection, and instructions are presented to the driver on a small alphanumeric display unit. Again, we assume that the driver follows the instructions implicitly, so a measurement of distance alone tells the system when to display successive instructions.

The TRRL NAVIGATOR is also similar but uses a visual display comprising directional arrows and alphanumerics, as well as a speech unit to instruct drivers, for example, to "turn left on the A420." The journey criterion can be distance, time, or any combination of the two (i.e., cost), and the network is based on a 10,000

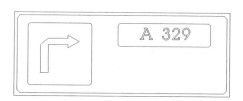

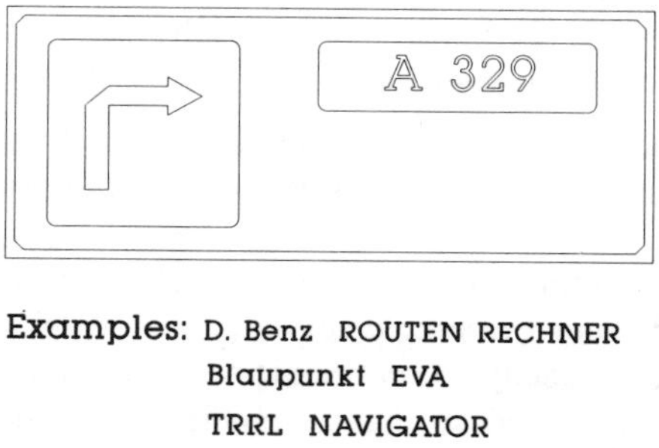

Figure 6.3 Route guidance aids.

node and 15,000 link network of roads in England maintained by the UK Department of Transport. Again, we assumed that the driver does not deviate from the planned route so that location can be deduced from a measurement of distance alone.

EVA from Blaupunkt in Germany employs both distance and heading sensors in order to detect when the driver has deviated from the planned route, to automatically reinitialize the system, and to work out a best route from the "new" position. EVA also employs a speech synthesis unit to give "turn left" instructions, and a liquid crystal display which shows the junction layout with the required route through the junction superimposed.

CARGUIDE, a system proposed by Carnegie-Mellon University, is based on a map display. The driver enters origin and destination coordinates in terms of intersections (using two street names to identify each), then receives instructions via synthesized speech: "straight, left, or right" and the name of the street to take. A visual display shows the surrounding road network with the current and next links flashing. The driver must press a button each time he or she obeys an instruction in order to go to the next. The CARIN system described in the previous section has also been developed to provide similar route guidance facilities, but without the need to press a button. (CARIN is described in detail in Chapter 8.)

The costs of vehicle-based devices can be expected to range from about $150 in the case of the simple directional aids, to a few hundred dollars in the case of NAVIGATOR and ROUTEN-RECHNER, and to $1,000 or more in the case of ETAK and CARIN, which incorporate sophisticated navigation systems and graphical displays. Mass production and the use of VLSI techniques may reduce these costs in the future.

6.4 REQUIREMENTS FOR A ROAD-VEHICLE COMMUNICATION LINK

The systems described above cannot take into account changes in the road network caused, for example, by new road building or changing traffic conditions. They could, however, be made to if the necessary information updates were made available over a suitable communication link. The essential requirements of such a communication link have been identified for use in Europe by COST 30 (1985) as:

(1) it should provide a means for the Traffic Control Authorities to communicate with drivers in their vehicles in order to facilitate more efficient use of the road network;
(2) it should enable communication which is selective both in terms of area and the number and types of vehicle served;
(3) it should have the capacity to communicate with a large proportion of vehicles;
(4) it should be standardized so that a vehicle equipped for use in one country can receive and decode messages in other countries;

(5) it should transmit messages in digital form so that they can be selectively decoded either as internationally recognizable symbols or in the native language of the driver;
(6) it should have reserve capacity to enable later expansion and enhancements.

A communication link which meets these requirements could be achieved in a variety of ways. Following the work of COST 30 [1985] and OECD [1988], they can be considered in terms of four classes of systems as follows:

Class 1—area broadcast systems;

Class 2—local roadside transmitter systems;

Class 3—mobile radio systems; and

Class 4—local roadside transceiver systems.

Each of these four classes of potentially dynamic route guidance systems can be achieved by communication between a traffic control center and an autonomous, or so called class 0, system shown in Figure 6.4 and discussed below.

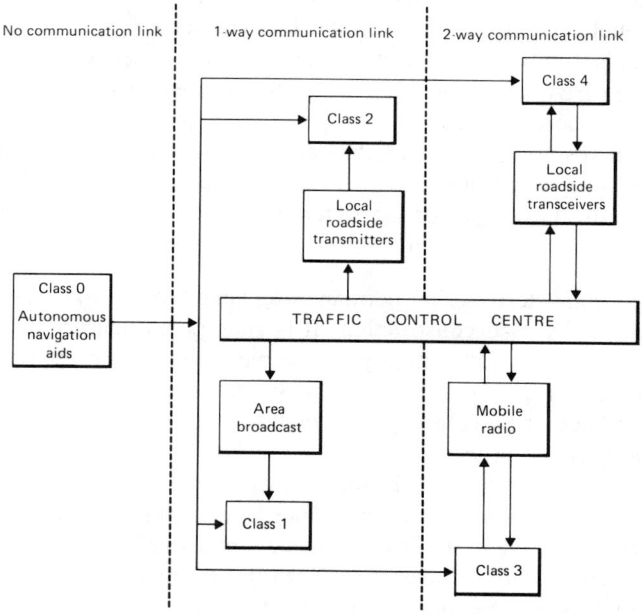

Figure 6.4 Classes of route guidance systems.

6.5 DYNAMIC ROUTE GUIDANCE SYSTEMS

6.5.1 Class 1—Area Broadcasting

Area broadcasting by radio enables traffic messages to be broadcast over a wide area, typically ten to a hundred or so kilometers in diameter. (See Chapter 5 for more information on traffic information broadcasting.)

Such a system provides one-way communication, from the transmitter to the vehicle, and could be realized in a number of ways. However, problems with frequency allocation mean that it is unlikely that a suitable frequency could be found for international use, and following the recommendations of COST 30 [1985], efforts have been directed at sharing facilities with an existing digital broadcasting system. Possibilities include the proposed pan-European cellular radio system or proposed satellite broadcasting systems.

The most promising technique for Europe at the present time, however, is to use the facilities of the RDS defined by the EBU [1984]. This system, which is described in detail in Chapter 5, has now been adopted by several major broadcasting companies. It enables digital messages to be superimposed on normal VHF radio program broadcasts. The next generation of car radios will be equipped to decode them. The channel capacity is around 1200 bits per second, and some 10% of this could be available for traffic messages. The system is seen to have considerable potential, and the EBU is cooperating with the European Conference of Ministers of Transport to define a standard.

The system resulting from the standard will be designed to transmit update information for vehicles equipped with Class 0 Map Display and Route Guidance Aid type of system. Two types of messages will be transmitted for the area served by a transmitter:

(1) *Network Data*—to describe permanent changes to the road network caused, for example, by road construction. It is anticipated that drivers would renew their "electronic maps" annually so that messages need only describe changes which occurred during the current year. Network Data messages would be broadcast once or twice a day, and could be received and stored, even when a vehicle was not in use.
(2) *Hazard Warnings*—to describe major traffic incidents such as accidents, roadwork, and bad weather conditions. These data can be integrated into the electronic maps carried in a Class 0 device, or displayed directly on a cheaper hazard-warning-only receiver unit. Hazard warning messages would be broadcast in real time. The means for gathering information and the criterion for selecting messages would be similar to those used in existing speech broadcasting systems (see, for example, [OECD, 1987]).

An initial system could employ a single transmitter to cover a whole country or region. More transmitters could be introduced later, if required, to give more detailed information in particular areas.

Individual messages could be directed at particular types of vehicles, or even individual vehicles, by suitably coding them. Area selectivity would be automatic for vehicles equipped with a Class 0 unit; if the incident fell in the area covered by the electronic map, it would be incorporated, if not, it would be ignored. Area selectivity could be achieved in a hazard-warning-only unit if the driver keyed in an area code so that his receiver would only accept messages carrying the code for that area. So that different types of vehicle unit can understand messages, and know how to deal with them, they must be packaged in a standard way. The format currently being considered comprises the following elements:

(HEADER) (VEHICLE TYPE) (LOCATION) (HAZARD TYPE) (EFFECT) (DURATION) (ADVICE)

Examples of Class 0 systems developed to incorporate data messages transmitted over the RDS are CARIN, EVA, and the TRRL NAVIGATOR. The Eureka project CARMINAT (see Section 6.8.1), also proposes to use RDS.

A system called AMTICS [Koshi, 1988] is being tried by the National Police Agency in Japan. AMTICS uses a system developed by the Japan Ministry of Posts and Telecommunications known as *Teleterminals*. These are digital radio transmitters operating at 800 MHz. They provide data rates of 4800 or 9600 baud, and are arranged in a network to provide a reception area of about 3 km in radius. In a pilot experiment conducted in Tokyo in 1988, dynamic traffic information was transmitted to suitably equipped vehicles and superimposed on a map display unit.

6.5.2 Class 2—Local Roadside Transmitter Systems

These systems represent a logical extension of the previous Class 1 area broadcasting systems to provide a higher density of transmitters, ultimately perhaps to equipping individual junctions. In this configuration they are more versatile however, and may be used in one of three ways:

(1) *Hazard-warning-only transmitters*—to broadcast information about an incident in the immediate vicinity. HAR, as used in the US and Japan, is an example. In the US, two frequencies, 530 kHz and 1610 kHz, are allocated for use by HAR. They respectively fall just below and just above the standard AM broadcast band, and can therefore be received on most car radios. Local transmitters, using one or other of these frequencies and with a power such that reception is confined to a radius of up to about 2 km, are sited on the

approaches to hazards where they broadcast prerecorded messages from a continuous loop tape. A fixed sign is needed to inform drivers when transmitters are in use, and to tell them which frequency to tune to. HAR is not a digital broadcast system, but development of a digital AHAR system using a frequency of 45.80 MHz has been reported by Mammano [1984]. HAR systems are discussed further in Chapter 5.

(2) The second use for local roadside transmitters is as *simple location beacons,* as used in public transport vehicle location systems (see, for example, [OECD, 1987]). A wide range of techniques has been tried or proposed, including: a coded pattern of buried magnets; short range radio broadcasts; buried loops; leaky coaxial cable; microwave; and infrared transmitters. Current Japanese work uses radio transmitters to broadcast position codes which can be detected and used for location fixing, and automatic reinitialization, by a range of Class 0 Autonomous Navigation Aids (see Shibata [1986]).

(3) The third use for local roadside transmitters is as *complex beacons* (or, more correctly, as *Electronic Signposts*) to download hazard warnings, network updates, and route guidance advice for use in a Class 0 Map Display or Route Guidance Aid unit. AUTO-SCOUT (an early version of ALI-SCOUT developed by Siemens in Germany) is an example of such a system. In this, an infrared communication link is used to transmit network and guidance data into a special vehicle unit which is a hybrid of the route guidance and simple directional aids. When the driver sets off on a journey, his or her vehicle unit acts in autonomous mode (i.e., as a simple directional aid). When encountering a "beacon," the driver receives data which can be interpreted by the vehicle unit to guide him or her over a "best" route to the destination (the vehicle unit acts as a route guidance aid). The driver then follows this advice until either encountering another beacon and receiving more advice, or leaving the guidance network—when the vehicle unit reverts to autonomous mode and acts as a simple directional aid again.

Vehicle selectivity can readily be achieved in local roadside transmitter systems by suitably coding the messages transmitted. Area selectivity is determined by the transmitter power and technology, but can be confined in most systems to cover a single junction, or even the individual approaches to a junction.

The criteria for transmitting messages, and updating them, will depend on requirements. Simple beacons transmit position information only, continuously—the information never changes. Hazard-warning-only transmitters broadcast traffic messages as, when, and where required. They are generally deployed only at the sites of major long term incidents, such as roadwork, but they are, in principle, no less versatile than the Class 1 Area Broadcasting systems, and could be controlled from a control center and made to transmit traffic messages in real time. The electronic signpost systems can either be preprogrammed to provide historic guidance data, or automatically updated from a control center to account for hazards and changing traffic conditions, again in real time. The criterion for real time updating would be similar to that currently used for determining when a traffic broadcast should be made over radio (see *hazard warnings* in Section 6.5.1).

6.5.3 Class 3—Mobile Radio Systems

These involve the use of a two-way communication link between a control center and the vehicle. Several mobile radio systems are commercially available, but because they are usually limited in terms of the number of vehicles they can serve, public radio telephone, and in particular cellular radio, systems are generally preferred. (See Chapter 3 for more details on cellular radio.)

Cellular radio systems exist, or are proposed, in several countries, and a pan-European system is planned for Europe. They involve dividing up the coverage areas (i.e., a country, or region) into abutting "cells" whose radii vary from 16 km in rural areas, down to 2 km in busy urban areas. Each cell is then served by a fixed transceiver unit which has available several duplex (i.e., two-way) radio channels over which communication with mobile units can be set up. The cells are generally grouped in clusters of seven, which share the several hundred channels allocated to the system by the relevant telecommunications authorities. The clusters are then repeated throughout the coverage area. Channels are therefore reused, but not by immediately neighboring cells. Interference effects are thus minimized, while the use of the available channels is maximized.

The transceivers for each cell are interlinked and computer controlled, so as a vehicle crosses the boundary from one cell to the next, a "hand-off" procedure occurs automatically. The vehicle unit is then allocated a new channel from those available to the transceiver serving the next cell without any apparent break in communication. Moreover, the system interconnects with the PSTN. Cellular systems, therefore, effectively extend the public telephone network into vehicles, and anything which can be achieved on the public network can, in principle, be achieved on the cellular system.

Proposals for using cellular radio to provide drivers with digitally encoded hazard warning and guidance advice have been formulated by several countries (see, for example, [Walker, 1985]). These range from providing service information and route plans on demand to downloading hazard warnings and up-to-the-minute electronic maps for Class 0 route guidance aid type devices.

The most recent proposals are from CALTRANS in the US (see French, [1987]) which proposes to use cellular radio in order to transmit details of traffic incidents. These will then be superimposed on the network shown by an ETAK map display unit. The driver must decide for himself whether to try and avoid the problem area. The two-way aspect of the communication link would also enable the driver to make emergency calls and to report on his or her position and status from time to time. This is seen as a valuable addition, particularly for the operators of large vehicle fleets. These data will also enable some estimates to be made of journey times on the individual roads making up the network, and should therefore prove useful for traffic management purposes as well as for indicating where and when new messages need to be transmitted.

Vehicle and area selectivity are inherent in the system because communication is set up between a transceiver which serves a known area and individual vehicles. Data gathering and message updates would be handled in a way similar to that described for the Class 1 and 2 systems above. No international standard exists for Cellular radio systems, but the CEPT countries are currently working to define a standard for a digital pan-European system (working in the 900 MHz region) in the 1990s.

6.5.4 Class 4—Local Roadside Transceiver Systems

Early examples of these systems include ERGS from the US, CACS from Japan, and ALI from Germany. They use a two-way communication link between roadside units placed on the approaches to major intersections, and special in-vehicle units. The roadside units are controlled from a central computer, and respond with a guidance instruction, upon receipt of a destination code transmitted from a passing vehicle. The guidance instruction is then displayed for the driver, usually on a visual display which mimics the layout of the junction ahead.

All of these early systems used inductive loop technology to provide the communication at a precise location along the road. Later developments, such as the initial proposals for AUTOGUIDE in the UK [Department of Transport, 1986]; see also Figure 6.5 and Section 6.8.2), have also favored buried loops: mainly because the technology is proven, and because the loops themselves can act as vehicle detectors—both to interrogate equipped vehicles to learn their trip times since crossing the last loop, and to count unequipped vehicles. Loops can therefore be used to collect most of the information needed in order to detect problem areas and give up-to-the-minute guidance advice (i.e., in real time).

However, recent work on the ALI-SCOUT system from Siemens and Blaupunkt in West Germany favors infrared, while similar work in Japan favors a radio or microwave communication link. The ALI-SCOUT system is designed to be evolutionary in three phases. The first phase uses a Class 0 simple directional aid to provide autonomous navigation. The second phase involves the addition of Class 2 complex beacons, as described in Section 6.5.2. These beacons transmit location and route guidance information to the vehicle over an infrared communication link. The third phase involves the addition of an infrared link for communicating from the vehicle back to the roadside beacons, so participating vehicles can communicate their trip times to the infrastructure. These data can then be taken into account by

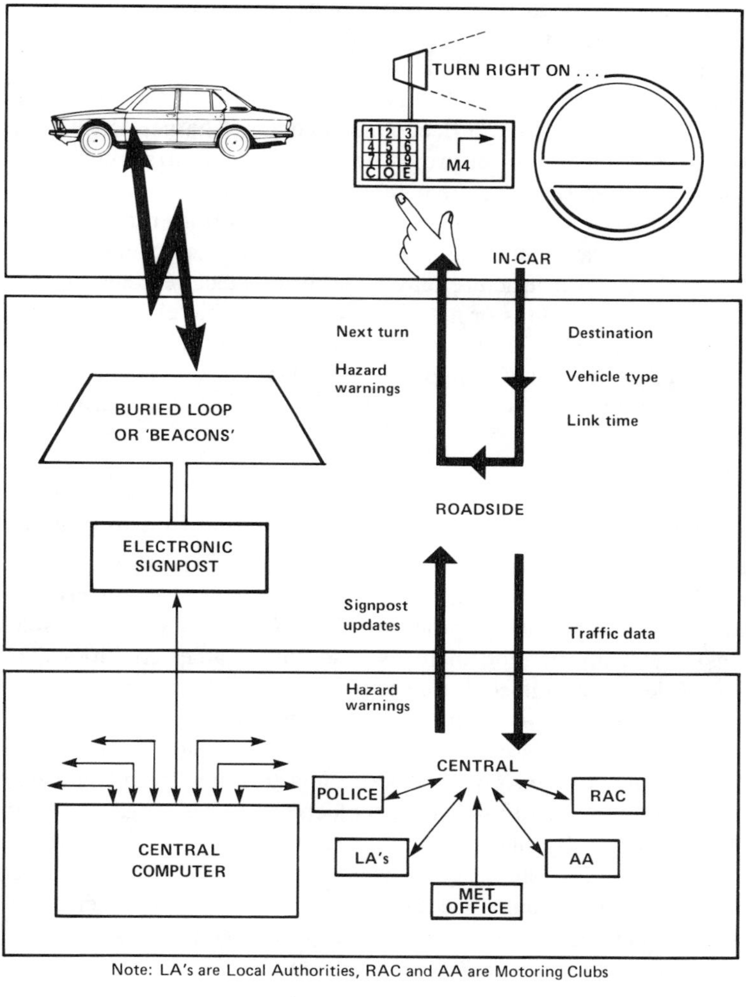

Note: LA's are Local Authorities, RAC and AA are Motoring Clubs

Figure 6.5 Elements of the AUTOGUIDE system.

a control center to update the guidance advice given by the roadside units in real time.

Japanese work has favored Class 0 map display devices as the basis for vehicle units, with location updating provided by Class 2 simple (radio) beacons, as described in Section 6.5.2. Current developments include work on microwave links as an alternative to radio beacons, and an enhancement to provide two-way communication between the roadside and vehicle units. This two-way microwave link would be used similarly to the two-way infrared link employed in ALI-SCOUT. However, the Japanese propose that drivers could also use the link to report their location and status to, and receive messages from, a fleet control center. Charges for this service could be imposed to cover the costs of providing the roadside infrastructure.

Vehicle and area selectivity are inherent in these systems because communication is set up between a transceiver at a known position in the road network and individual vehicles. Data gathering is also inherent in many of the proposed systems because the roadside infrastructure may be used to detect vehicles (whether they are equipped with in-vehicle units or not), and equipped vehicles may be asked to report details of their travel times on the roads leading to the transceiver. In the most recently developed systems that employ this last technique, great care has been taken to ensure that individual vehicles cannot be identified without permission.

6.6 SYSTEM FUNCTIONS

The main functions that can be achieved by the different Classes of system include:

(1) *Strategic information* (i.e., information of use for trip planning purposes, which can either be *historic* ("learned" from past experience), or *predictive* (taking into account forecasts of problems due to roadwork or bad weather conditions). Such information includes road and weather conditions, route plans, time of departure advice during holiday periods, and car parking availability.
(2) *Navigation information*—for location and route guidance purposes. Route guidance advice may either be historic, or dynamic (traffic responsive).
(3) *Hazard warning*—information describing transient problems caused, for example, by unexpected roadwork, accidents, and bad weather conditions. Such warnings generally include: (a) a description of the hazard; (b) its location; (c) its effect (e.g., the delays caused to traffic); and (d) any restrictions (such as speed limits) in force. Hazard warnings are generally received by the driver while driving; they are essentially dynamic, though they may also be predictive.
(4) *Emergency* or *service calls* (i.e., a means by which the driver can summon assistance in an emergency), or information of a service nature, such as hotel accommodations.

Plus, for the traffic management authorities, there is:

(5) *Traffic management information,* such as data on traffic speeds, flows, vehicle classification, and distribution on the network; all of which is useful for assessing both the present situation and the effects of various traffic control measures.

The extent to which each of the five Classes of systems is capable of fulfilling these functions is indicated in Table 6.1. In summary:

Table 6.1
Systems and Their Functions

System Class:	0	1	2	3	4
Function	Autonomous Navigation aid	Area Broadcast	Local Roadside Transmitter	Mobile Radio	Local Roadside Transceiver
Strategic Information					
Historical	x	x	x	x	x
Predictive		x		x	
Navigation Information					
Historical	x	x	x	x	x
Dynamic		x	x	x	x
Warnings		x	x	x	x
Calls for Emergency or Service				x	x
Traffic Management Information				x	x

(1) The Class 0 autonomous navigation aids are capable of providing strategic and navigation information, but only of the historic kind.
(2) The Class 1 and 2 systems, which use a one-way communication link, enable digital information to be transmitted to the vehicle. The information can be used directly to provide hazard warnings, or update electronic maps so that the driver obtains dynamic navigation information.
(3) The Class 3 and 4 systems enable two-way communication between the road and the vehicle. The road-to-vehicle direction of the communication link provides all the features of the Class 1 and 2 systems, while the addition of the vehicle to road direction enables traffic management information to be collected and emergency and service calls to be made.

The Class 3 and 4 systems are the only ones capable of collecting traffic information for themselves. This is a very significant advantage: it means they can monitor the effects of any control action they take on traffic and continuously update and modify that action to utilize the road network in the most efficient way.

Each of the system Classes 1 through 4 can be achieved directly from a Class 0 Autonomous Navigation Aid by adding communication with a traffic control center (see Figure 6.4). However, from the previous discussion, it is clear that developments in many countries are directed at providing systems which can evolve: from Class 0 through 1 or 2, and into Class 3 or 4 systems. These trends, and the reasons for them, are discussed further in Sections 6.8 and 6.10.

6.7 POTENTIAL BENEFITS FROM IMPROVED GUIDANCE SYSTEMS

Some of the benefits that can be expected from improved guidance systems are intangible and cannot be quantified. For example, it is difficult to estimate the value of increased reassurance to drivers that they would be guided over a good route and would not get lost, or that they would be informed in good time of incidents likely to affect their journey. Nevertheless, major sources of quantifiable benefit can be associated with the various system functions.

6.7.1 Historic Guidance Advice

Analysis of times and distances involved in a sample of journeys made in the UK [Jeffery, 1981] suggests that about two percent of all driver journey costs could, in principle, be recovered by an efficient route guidance system which based its advice on historical information.

For the typical private car driver in the UK, the benefit would be made up from savings or waste incurred on about 60 journeys each year that are longer than 5 km in length and take one over unfamiliar roads. On these journeys, the average inefficiency is about 20%, and gives rise to additional costs averaging $1.20 per trip, (a total of $70 per year). For the average commercial vehicle, the avoidable waste arises on about 45 journeys per year, where excess costs averaging 25% and $5 per trip make a total of about $220 per year. In national terms, the potential benefits available to an efficient "historic" route guidance system would seem to be about $1330 million per annum, of which about 15% is contributed by direct fuel costs, about 40% from other vehicle running costs, and the remainder from a valuation of vehicle occupant's time. If we assume that accidents and road maintenance are in proportion to vehicle mileage, then another $75 million a year may be saved from reduced accidents and $73 million from road maintenance, thus bringing the total benefit figure to over $1460 million a year in the UK alone. The potential in Western Europe as a whole could be ten times this figure, while in the US the total costs of navigational waste in noncommercial travel are estimated by King and Mast [1987] to be worth a staggering $45 billion.

6.7.2 Hazard Warnings

The UK TRRL [TRRL, 1979] has estimated the benefits of a system that could assist drivers by conveying information about congestion caused by significant traffic incidents. (In 1976, such incidents, causing at least 12 minutes of delay, were estimated to occur 17,000 times a year on UK roads—nearly 7000 on urban roads, about 10,000 on rural roads and about 300 on highways.) An information system enabling a significant proportion of drivers to divert and avoid such traffic incidents would yield benefits of around $50 million per year at present-day values.

6.7.3 Dynamic Guidance Advice

However, such a system would still not serve drivers on familiar trips (e.g., commuters, following routes largely chosen by themselves). Analysis has shown that when all trips are included, the potential available to a guidance system could be worth about 7% of all distance driven.

Moreover, traffic incidents usually cause additional delays, and while extra distance may be incurred in taking a diversion route, a net saving in time can often be achieved. An experiment in Tokyo with the CACS route guidance system is reported by Tsuji *et al.* [1985] to have achieved savings worth between 9% and 14% of the journey times between pairs of origins and destinations.

The two results can be combined to suggest that a route guidance system operating in real time should achieve savings (for those who chose to buy and use the system) averaging about 10% of journey costs. On the basis of these figures, Jeffery, Russam, and Robertson [1987] estimate achievable benefits worth around $1700 million per annum for a dynamic system of route guidance in London.

6.7.4 Traffic Management Information and Emergency Calls

Benefits here relate mainly to the efficient deployment of the police and other traffic management resources, and are difficult to quantify. However, substantial direct costs are incurred throughout Europe each year in collecting traffic data for purposes of updating traffic signal control plans, and for advance planning of new work and maintenance schedules. These costs amount to around $0.2 million per year in London alone where the costs of driving are estimated at around $5000 million per year [Jeffery, Russam, and Robertson, 1987]. A system that could provide the traffic data automatically, and at the same time enable a small percentage of improvement in the effectiveness of advance forward planning, should therefore achieve significant benefits for road users. Further benefits would be available to systems that provide automatic incident detection.

6.7.5 Other Sources of Benefit

Additional sources of potential benefit mainly from dynamic guidance systems include: (1) savings in congestion and delays, and hence time and vehicle operating costs for nonequipped vehicles because equipped vehicles are diverted out of their way (estimated by Jeffery, Russam, and Robertson [1987] to be worth around $17 million per year in London); (2) increased reassurance for drivers that they will be guided over a good route and warned in advance of any traffic problems they might encounter; (3) reduced anxiety (e.g., that they might get lost and arrive late), and hence reduced accident risk; and (4) improvements to the environment, including reductions of heavy traffic in environmentally sensitive areas, noise, and pollution.

6.7.6 Achievable Benefits

The level of potential benefit available to a particular system will depend on the extent to which it can provide each of the individual benefits discussed above. In general terms, and as we see in Table 6.1, the potential increases with Class number, and hence with the range of facilities provided. The largest single source of quantifiable benefit is expected to come from providing individual drivers with improved navigation information. As indicated above, this is estimated to be worth between about 2% of journey costs for a good static (Class 0), system, and about 10% for a traffic responsive Class 4 system operating in a large and busy city.

However, the actual level of benefit achieved in practice will be determined by the extent to which a system can penetrate the market (i.e., on the numbers of drivers who are prepared to buy and use it). This will depend not only on the range and attractiveness of the facilities provided, but also on cost and the credibility of the information given.

In consideration of these factors, European work on route guidance systems is advancing on two main fronts, as discussed below.

6.8 CURRENT DEVELOPMENTS IN EUROPE

Investigations reported by Jeffery [1988] confirm that drivers generally have a low perception of the value of navigation information unless it incorporates dynamic (up-to-the-minute) warning information. All systems must rely on communication with a traffic control center to provide this dynamic information. Much effort is therefore being directed to developing Class 0 autonomous systems which can take advantage of a communication link when it becomes available.

As a first step, this enhancement would produce a Class 1 area broadcast or a Class 2 roadside transmitter system. The Class 2 systems involve a much greater cost for providing the infrastructure. For Europe, therefore, a Class 1 system based on the existing RDS should have a better chance of success.

6.8.1 The Radio Data System

See Chapter 5 for a detailed description of the European RDS. A good example of developments on this front is provided by the EUREKA sponsored CARMINAT project, which effectively integrates the development of Renault's ATLAS and Philip's CARIN systems (see Dobias [1987]).

Both systems rely on a CRT to display information that a driver can recall from a mass memory unit carried in the vehicle. However, whereas the original emphasis for CARIN was on providing navigation information, for ATLAS it was on providing gazetteer, vehicle status, and diagnostic information.

Under CARMINAT, the two systems will be combined and, with the collaboration of the French television company TDF, adapted to receive traffic messages broadcast using the RDS. The resulting Class 1 system will provide a driver with a comprehensive in-vehicle information center that could, for example, warn him or her about malfunctions in the vehicle, show when service is due, provide maps showing roads, garages and hotels, show his or her location on the map, warn of weather and traffic problems in real time, and guide the vehicle over an efficient route to the driver's destination.

A considerable measure of agreement has been reached on the data structures and protocols for RDS traffic broadcasts in Europe (see Chapter 5), and current work is designed to agree on a set of messages that will cover the requirements for traffic information broadcasts in all European countries, and to specify how the messages should be coded. These codes can then be interpreted for the driver by reference to a look-up table of messages stored in his vehicle. Vehicles equipped with a CARMINAT type system will carry this look-up table, and will be able to incorporate the broadcast messages into their database and display the information for the driver on a "need-to-know" basis (to show congestion on a map display, or to account for delays in determining a best route).

However, also proposed is that RDS traffic broadcasts be usable by drivers who are not equipped with a navigation aid. In this case, a less expensive modification will incorporate the look-up table directly into an RDS car radio receiver so that it can interpret the messages and "speak" them to the driver as they are broadcast. Drivers must then select from the broadcast messages by keying an area code into their receiver, thus ensuring that drivers only hear broadcasts about the area in which they are driving. Outside of CARMINAT, work is underway in several countries to develop prototype car radios that will incorporate both the RDS receiver and the look-up table, and trials of the system are proposed in France and on a cross-border route between the Netherlands and West Germany.

The main advantages of RDS as a means for broadcasting traffic information are: it exists as a standard and is currently being implemented by the major national broadcasting companies in Europe; and the broadcasts are digitally encoded, and can easily be converted to synthetic speech in the vehicle, in the language of the driver's choice. A system should therefore be relatively cheap to implement, and will be international (i.e., a driver can continue to receive messages in his own language as he travels throughout Europe).

However, the system employs only a one-way communication link. Drivers can learn about changed road conditions, but no traffic management information is collected. Consequently, all drivers receive the same information, so the opportunities for traffic control are limited, and the information itself must be collected by conventional means, which are often not reliable or timely [OECD, 1987]. Nevertheless, RDS should be of considerable value in warning drivers of impending bad road and weather conditions, and of serious tie-ups such as occur on main roads and highways from time to time.

6.8.2 LISB and AUTOGUIDE

In congested urban areas however, relatively minor incidents can cause a rapid build-up of congestion over a wide area. In order to achieve credibility in these circumstances, a route guidance system itself must collect the information on which to base the guidance advice on a minute-by-minute basis.

A two-way communication link is therefore required. Class 3 systems were considered (see, for example, [Walker, 1985], and [Jeffery, Russam, and Robertson, 1987]), but were rejected in favor of Class 4, which involves a higher cost for the infrastructure, but the least expensive vehicle units, and hence the greatest potential for market penetration.

Two trials of Class 4 automatic route guidance systems are planned in Europe: LISB in West Berlin [Hoffman *et al.*, 1987], and AUTOGUIDE in London [Belcher and Catling, 1987]. (See also Chapter 9.) As a result of work by a bilateral working party established by the British and West German Governments to agree a standard for the road-vehicle communication link, the systems are similar. In concept, they both involve equipping the main road junctions in a city with beacons that can communicate with in-vehicle units via an infrared communication link (see Figure 6.6).

The beacons transmit data which describes the road network in the area surrounding the beacon, together with route guidance advice on the best routes to take to reach distant destinations. The driver must key a code for his destination into the in-vehicle unit. On passing a beacon, the in-vehicle unit (which is shown in Figures 6.7 and 9.6, and has a layout similar to that shown in Figure 6.3), will then receive the data, deduce the route which must be taken, and guide the driver over it until he or she reaches either the destination or another beacon.

The in-vehicle units also transmit data back to the beacons. These data describe the route taken by a vehicle, and the times spent transversing the individual lengths of road since passing the last beacon. The data can then be transmitted from the beacons to a traffic control center where the units can be monitored to see which roads are congested and which are free. Delays are therefore monitored in real time, and the route guidance advice given by the beacons can be changed within minutes to divert traffic around congested areas.

Figure 6.6 Infrared beacons installed on traffic signals, as used in AUTOGUIDE and LisB.

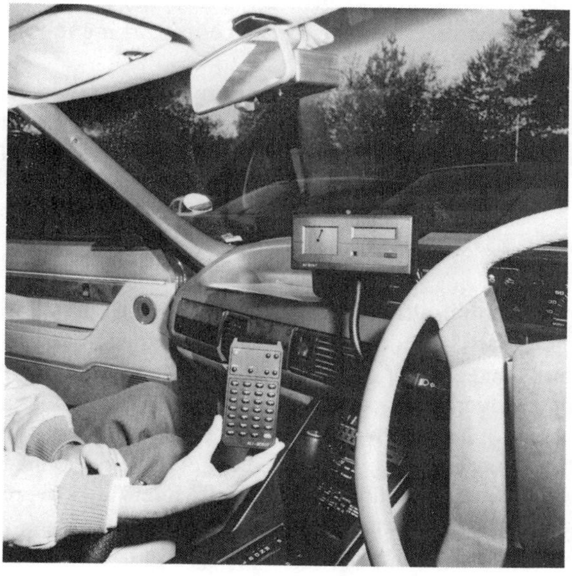

Figure 6.7 In-vehicle units as used in AUTOGUIDE and LISB.

In Berlin, some 230 junctions are being equipped with beacons, and 700 vehicles with in-vehicle units. Field tests began in September 1988, and involved the selection of a structured sample of vehicles to be equipped including: private cars, commercial vehicles, taxis, and hired cars. The use made of LISB by these vehicles will be monitored to show: (1) drivers' acceptance of the system and willingness to follow guidance advice; (2) the usability of the system and its effects on safety; (3) the feasibility and reliability of the system concept and equipment; and (4) to confirm the costs and benefits. (Results were expected by the middle of 1989.) The funds for LISB amount to some $9 million, and are shared equally between government and a collaboration of German industries.

In London, proposals for a trial on a scale similar to LISB are under consideration [Belcher and Catling, 1987]. This would provide a first step on the way to a commercially based system covering all of the London area out to, and including, the M25 Orbital Motorway. A full system could involve some 800 beacons and cover an area of around 2500 km^2. In the meantime, a demonstration system, involving five equipped junctions and some 12 equipped vehicles, has been installed to demonstrate the principles of automatic route guidance on a corridor of routes between central London and Heathrow Airport to the west.

The three main aims of this London demonstration are: to demonstrate automatic route guidance to decision-makers and the media in a real environment, and to learn more about the market for a system; to provide practical hands-on-experience working with the technology; and to give the numerous government and industrial partners, who must be involved, some experience in working together to finance and organize a system.

We should anticipate that the demonstration system would remain operational for as long as necessary, which could be for two years or so until the proposed large scale trial became operational (see also Department of Transport [1988]).

6.9 THE MARKET FOR ROUTE GUIDANCE

To learn about the potential attractiveness of AUTOGUIDE, a firm of consultants was employed by the TRRL to design and conduct a questionnaire survey to investigate driving patterns and current navigation practices, the appeal of AUTOGUIDE, based on an outline product description, and aspects of purchasing sensitivity to price and design features. Copies of the questionnaire were mailed to a sample of over 5000 private and business addresses. Thirty percent of the questionnaires were returned, and subsequently analyzed. In addition, in-depth interviews were conducted with some 70 respondents who had completed questionnaires.

The results indicate a generally positive response to AUTOGUIDE with likely sales to about 25 or 30% of vehicles in and around London for an in-vehicle unit costing some $340 plus a $40 annual license fee. Drivers were particularly attracted to the potential of the system for reducing the difficulties of city driving, and for providing greater reliability for meeting delivery times and appointments. The potential safety aspects resulting from reduced stress, tiredness, and frustration were also acknowledged and welcomed. Drivers covering high mileages for business purposes in the area were particularly attracted to the system.

There was a widespread view that historic route guidance, without the ability to avoid congestion, would not be sufficiently attractive to warrant purchase, and high value was placed on terminal guidance (i.e., guidance at journey's end to the specific destination address). There were some doubts expressed about the ability of the system to do everything advertised, and respondents wanted to see it demonstrated.

Various other attempts have been made to estimate the market for route guidance and some results are presented in OECD [1988], but other results are commercially confidential, and can only be surmised with regard to the types of systems under development, the time scales in which they are expected to become available, and the forecast cost of the vehicle units to drivers.

6.10 TRENDS IN ROUTE GUIDANCE

Many of the world's leading electronics companies and motor manufacturers are actively developing, and can demonstrate, route guidance systems. They range from autonomous navigation aids and systems that rely on satellite or radio transmitters to provide updated information to systems that rely on an extensive infrastructure of roadside equipment for providing both route guidance and improved opportunities for traffic control.

Whatever systems are eventually preferred, it is clear that route guidance will be with us in the 1990s; and a communication link will be needed so that drivers can be warned of changing road conditions.

For Europe, the major push is toward information systems based on RDS and, in congested urban areas, for traffic-responsive route guidance systems such as LISB and AUTOGUIDE. The consensus view is that these systems will coexist. The EUREKA sponsored PROMETHEUS project (see Chapter 9), which is concerned, among other things, with developing electronic systems for the next generation of vehicles, will no doubt consider how these two different systems can be integrated. As a complementary activity, the EEC's proposed DRIVE program should further integrate information and traffic management functions in ways that account for government concerns.

In the meantime, the justification for developing the various schemes is based largely on forecasts and estimates of the likely benefits. These cannot be confirmed until experimental results become available from the trials and demonstrations.

REFERENCES

Belcher, P., and I. Catling, "Electronic route guidance by AUTOGUIDE: The London demonstration," Traffic Engineering and Control, London, November 1987.

COST 30 *bis*, Electronic traffic aids on major roads—Final Report. EUR 9835. Commission of the European Communities, Luxembourg, 1985.

Department of Transport, UK, AUTOGUIDE—a better way to go? London, 1986.

Department of Transport, UK, AUTOGUIDE—pilot stage proposals, London, 1988.

Dobias, G., *"La voiture intelligente, un plus pour la securite routiere," Revue de l'INRETS,* No 15, pp. 13–18, Arcueil, France, September 1987.

European Broadcasting Union, Specification of the radio data system RDS for VHF/FM sound broadcasting. Document Tech. 3244. EBU, Brussels, 1984.

French, R.L., "The evolving roles of vehicular navigation," *Navigation: Journal of the Inst. of Navigation,* Vol. 34, No. 3, Fall. 1987, pp. 212–228.

Heed, V. and K.D. Kricke, "GPS Satellite Navigation in the urban environment," *Proc. NAV 85 Land Navigation and Location for Mobile Applications,* Conference of the Royal Institute of Navigation, York, 1985.

Hoffman, G., *et al.,* Large scale experiment: Berlin guidance and information system. Translated reprint from Strassenverkehrstechnik, No. 2, Siemens, Munich, 1987.

Jeffery, D.J., The potential benefits of route guidance. TRRL Report LR 997. Transport and Road Research Laboratory, Crowthorne, 1981.

Jeffery, D.J., K. Russam, and D.I. Robertson, Electronic route guidance by AUTOGUIDE: the research background, *Traffic Engineering and Control,* London, October 1987.

Jeffery, D.J., "Driver route guidance systems: State of the art," *Proc. Int. Symposium on Telematics—Transportation and Spatial Developments,* The Hague, April 1988.

Jenezon, J.H., J.J. Klijnhout, and H.C.G. Langelaar, Motorway control and signalling, *Traffic Engineering and Control,* Vol. 28, No. 6, pp. 349–355, London, 1987.

King, G.F., and T.M. Mast, "Excess travel: causes, extent and consequences," *Proc. 66th Annual Meeting of the Transportation Research Board,* Washington, DC, 1987.

Koshi, M., "An overview of area control system and motor vehicle navigation/route guidance developments in Japan," *Proc. Roads and Traffic 2000,* Berlin, 1988.

Mammano, F.J., "Speech synthesis for a motorist information system," *Proc. OECD Seminar on Microelectronics for Road and Traffic Management,* Tokyo, November 1984, pp. 182–191.

Mooney, F.W., "Terrestrial evaluation of the GPS standard positioning service," *Navigation: Journal of the Institute of Navigation,* Vol. 32, No. 4, Winter 1985–86, pp. 351–369.

OECD, "Dynamic traffic management in urban and suburban road systems," OECD, Paris, 1987.

OECD, "Route guidance and in-car communication systems," OECD, Paris, 1987.

Robertson, D.I., "International perspective on traffic signal systems—Great Britain," *Proc. 67th Annual Meeting, Transportation Research Board,* Washington, DC, 1988.

Shibata, M., "Development of in-car information system," *Annual Report of Roads,* Japan Road Association, Tokyo, 1986, pp. 51–53.

TRRL, Report of the working group on the broadcasting of traffic information. Department of Transport, UK, TRRL Report SR 506, Transport and Road Research Laboratory, Crowthorne, UK, 1979.

Tsuji, H. *et al.,* "A stochastic approach for estimating the effectiveness of a route guidance system and its related parameters," *Transp. Sci.,* Vol. 19, No. 4, November 1985, pp. 333–351.

van Vliet. D. "D'esopo—a forgotten tree building algorithm," *Traffic Engineering and Control,* London, July–August 1977.

Walker, J. "The navigation and command and control aspects of the Mobile Information Systems (MIS) Alvey Demonstrator Project," *Proc. NAV 85 Land navigation and location for Mobile Applications,* Conference of the Royal Institute of Navigation, York, 1985.

Chapter 7
Urban Traffic Control

K.W. Huddart

Traffic Engineering Consultant

7.1 INTRODUCTION

Urban traffic control comprises the use of computers at central offices to control traffic from moment to moment. The control is exercised by traffic signals at road junctions. The signals are connected to the computers by various forms of data transmission, usually involving telephone-type circuits. Currently, UTC is the only comprehensive means of applying traffic control policies to motorists in real time, although other techniques, especially those described in Chapters 5 and 6, are related.

The UTC example indicates how the provision of additional information at the right time and place can confer very large benefits to the community. It also indicates the difficulty of arranging that the beneficiaries pay for the systems required to provide the information.

UTC systems also demonstrate the possibilities, difficulties, and the type of techniques required to provide widely dispersed equipment in the street for interaction with members of the public. The systems show that some of the greatest advantages are obtained by simplifying the problem, and providing only the information required. In many cases, this is achieved by processing the information so that only precalculated plans are available to the equipment which displays the signals to the motorist. In this way, the complexity of providing fully functioning online equipment to deal with all eventualities can be avoided. Some modern systems are complex, however, and provide greater benefits in suitable circumstances.

UTC systems communicate with motorists through traffic signals, with which they share opportunities for, and limitations of, control. The signals are provided only at selected intersections in the network, except in sections of cities, mainly in North America, where streets lie in regular grid patterns, and all intersections can be signalized. The control exercised by UTC is therefore selective, and in general, the achievement of significantly different objectives requires the installation of special traffic signals for that purpose.

Because road traffic is increasingly a part of all commercial and social activity, the importance of its control increases. Similarly, as traffic flows increase, a greater proportion of the traffic capacity of roads is used, and it becomes more important to provide control systems which preserve this capacity at all times. In all cases, to provide additional roads to carry traffic is expensive and, sometimes, particularly in urban areas, it may not be practical. To optimize use of existing roads, therefore, traffic control systems are increasing in number, size, and sophistication.

Urban traffic control (when not abbreviated as UTC) is sometimes interpreted to cover any means of controlling urban traffic as a whole. A wide range of techniques is applied, including numerous traffic regulations, tolls, and taxes on roads and vehicle usage, and planning controls on the extent and nature of land use. In many cases, the objectives of these control techniques include wider issues such as raising revenue and controlling the environment. These techniques are dealt with in this chapter only if they interact with control by traffic signals or real-time motorist information.

Special purpose roads, such as freeways and motorways (UK),* physically separate motor traffic from other highway activities, and spatially separate conflicting traffic streams. Such highways achieve important benefits in accident prevention, journey time, and efficient use of space for traffic movement. While they are generally considered in interurban terms, the more sophisticated control systems on highways are in urban areas and thus interact with other roads. Current systems comprehensively control corridors comprising both freeways and general-purpose streets.

7.2 HISTORY

Traffic signal lights made their first appearance near the Houses of Parliament in London in 1868 [Anon, 1868]. They were used to control the conflict of traffic (literally horse-powered) and pedestrians. They were gas-lit and blew up, so their use was discontinued. Electrically lit traffic lights came into use (first in the US) in the twentieth century. Initially, they operated on a fixed-time cyclic basis, but traffic actuated versions became available in the 1930s, which was also the time when the subject began to attract research interest.

From that time, coordination was provided between traffic signals at neighboring intersections. Coordination by central digital computers started with the Toronto

*US and UK terminology differ; a comparison of the two is given in the glossary at the end of this Chapter.

experiment [Cass and Casciato, 1960]. Microcomputers have been used for the control of individual intersections since 1972. The range and importance of functions which are carried out by traffic signal systems has progressively increased over the years, so that local and central computers are now being used to monitor the performance of traffic signal controllers.

Freeways were built in the 1930s, but interest in traffic control on them did not emerge until after the war. Lighthill and Whitham [1955] established the fundamental theory of unidirectional traffic flow. From the outset, US work addressed the control of on-ramps to ensure that highway traffic continued flowing smoothly, with the street system being used to store or handle overloads. The work progressed to providing transit priority. In contrast, UK motorway control (from the 1960s) provided systems for motorist aid and police reaction to incidents; this significant difference in approach still applies.

7.3 TRAFFIC SIGNALS ON STREETS

7.3.1 Traffic Signal Heads

The traffic signal aspects within the signal heads are usually mounted on posts by the roadside (Northern Europe) or suspended over the intersection (US and Japan). They are the primary means of providing motorists with real-time information, specifically instructions as to when to proceed or stop to avoid conflicts with other traffic or pedestrians.

The signal aspects may be arranged vertically above one another (US and Europe), in which case red is at the top and green at the bottom, or horizontally (Japan). The third (intermediate) yellow (UK amber) aspect gives warning that the red is about to appear and (primarily UK) in conjunction with a red that the green is about to appear. In some European countries, the amber is accompanied by green. Additional information is given by restricting the green lens (possibly by providing an extra green signal) to show an arrow, and in some countries this also applies to the amber and red. A symbol of a cyclist may also be used. Red crosses and white arrows may be used over traffic lanes and at toll booths to indicate which lanes are currently in use for each direction of traffic flow.

For the signals to be visible to motorists is vital, both on the approach to the intersection and when stopped at the marked stop line. To provide duplicate signals to meet these objectives is usual, except in some Western European countries, where only one signal head is provided. The aspect design varies, UK light output requirements being more demanding and being met by using low-voltage quartz halogen lamps. On the other hand, US practice uses long-life (5000 or 8000 hours) higher voltage lamps.

7.3.2 Fixed-Time Operation

Originally the traffic signals were programmed to provide fixed periods of right-of-way to the various approaches. Once each approach had been served, the cycle repeated, cycle times generally being under two minutes. A small electric motor (see Figure 7.1) drove a dial provided with levers to indicate when the light changes should occur; a stepping motor with cams then operated the switches for the lamps. Such a dial and cam controller might have several dials and have different motor speeds which would be activated by a time-switch at different times of day or days of the week. Such controllers were cheap, small, easy to install and maintain, and could still give good service.

Coordination with neighboring intersections could be provided by pulses from a master controller or central computer. One type of pulse released the dial movement after it had stopped at a defined point in the cycle; another actually stepped on the camshaft. Subsequently, such fixed-time operation has been provided using relays, contactors, other electronic circuitry, and microcomputers.

Satisfactory fixed-time operation depends on the repeatability of traffic patterns. These are observed using surveys of all directions of traffic flow at the intersection at different times of day. Saturation flows (the maximum instantaneous traffic flows on each approach) also must be observed, and usually occur when a line of traffic is being discharged. Then one can deduce the correct period to be provided for each traffic movement.

The basis of many such calculations is from Webster [1958]. This accounts for the time lost as the right-of-way changes from one traffic stream to another. To

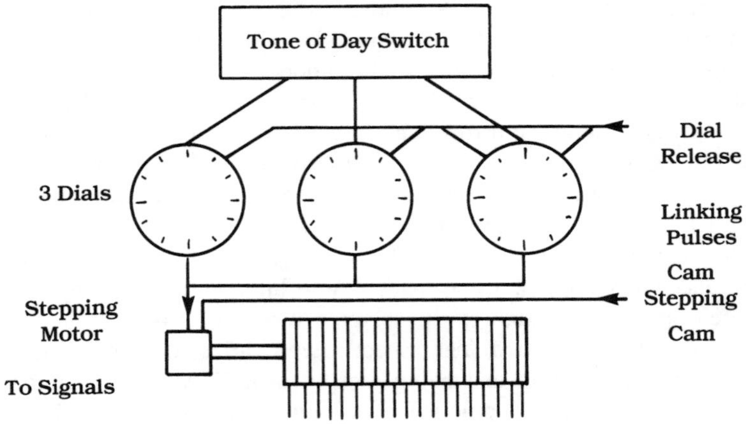

Figure 7.1 Dial and cam controller.

maximize traffic capacity, the cycle time must be raised so that the time available to traffic is dominant. At times of light traffic, the reverse applies; cycle time must be reduced so as to minimize the average delay waiting for the signals to change. Should the actual conditions differ from those on which the calculations were based, control would be suboptimal. To guard against the possibility that traffic capacity may not be adequate, to set the cycle time longer than optimum is prudent, even at the cost of extra delay. With very light traffic, the delays can be frustrating; to avoid this, the signals may be turned off or switched to flashing amber or yellow at night.

7.3.3 Traffic Actuation

Detection of actual traffic overcomes some problems of fixed-time operation. The signals provide right-of-way to the approach on which traffic is detected. Opposing traffic streams have to wait (subject to maximum time settings) until no more vehicles are detected in the stream being served. When there is continuous traffic on all approaches, each traffic stream receives right-of-way for the maximum prescribed time, which is calculated in the same way as for fixed-time operation.

At simple intersections, *semiactuated* (US practice and terminology) operation is possible. In this approach, the main road receives right-of-way until traffic is detected on the side road. There is no need to detect vehicles on the main road. The

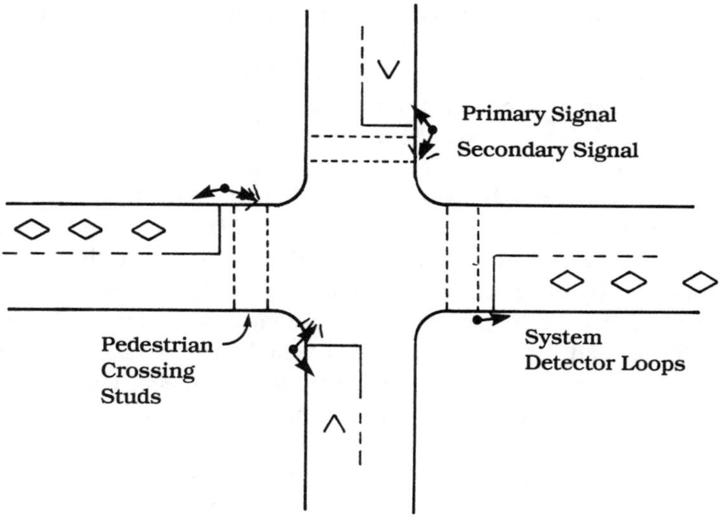

Figure 7.2 Typical UK signalized intersection.

layout of the more comprehensively controlled intersections standard in the UK is shown in Figure 7.2.

European (except UK) and Japanese signal controllers commonly use a basic fixed-time cycle. The detectors are used to extend the periods in the cycle while relevant traffic is detected. If no traffic is detected on a particular approach, the period can be skipped altogether. A limitation of such a system is that the traffic streams have to be serviced in the order prescribed in the signal controller. UK controllers [Department of Transport, 1984] are intended to serve traffic streams in any order, depending on the traffic actually detected. These international differences in standards pose real difficulties of communication when traffic engineers meet.

Traffic actuation requires drivers to be alert and obey the signals. The sequence and duration of particular traffic movements may vary considerably from cycle to cycle. Drivers who assume a fixed pattern may be taken by surprise, even to the extent of wrongly starting to enter the intersection after another movement is finished. The traffic engineer must ensure that sufficient clearance time is allowed between each movement. Engineers must also be cautious if traffic actuation is suddenly introduced after a long period of fixed-time operation. They should at least increase the clearance times for any new transitions which may occur.

7.3.4 Vehicle Detection

Traffic actuated control, and UTC in general, require that vehicles be detected and some traffic flow parameters such as speed be measured. There are numerous types of vehicle detector; some are listed in Table 7.1. They all work imperfectly, they are liable to faults, they require considerable attention if they are to give satisfactory service, and the careers of engineers and the prosperity of companies have been jeopardized in attempts to improve them.

Induction loop detectors are used with increasing frequency. They comprise loops of insulated wire through a slot cut into the surface of the road and connected to electronic circuitry. The loop is energized at about 100 kHz. The resultant electromagnetic field is disturbed by any metallic body; the resultant change of loop parameters is detected by the electronics, usually as a change in resonant frequency of a circuit incorporating the loop. The circuitry compensates for gradual changes in climatic conditions and for events such as a vehicle semipermanently parked over the loop. Loops can be used in pairs to measure vehicle speed, including the special case of vehicles traveling across the loops in the wrong direction.

Transit vehicles may be detected by examining the variation in loop inductance change as the vehicle passes across. The pattern is known as a *signature;* the signatures can be stored and classified so that different vehicle types may be distinguished. In Europe, traffic lanes are less well defined, so the signature may be erratic. More often transit vehicles are fitted with transponders, which communicate

Table 7.1
Detector Types and Their Characteristics

Detector Type	*Placement*	*Detected Vehicles*	*Comment*
Induction loop	"Surface"	All	Very adaptable
Transponder	"Surface"	Equipped	Selective
Miniloop	"Surface"	All	Very localized
Vertical probe	Buried	All	Very localized
Magnetometer	Buried	All	Localized
Ultrasonic	Overhead	All	Localized
Microwave	On Signal	Moving >5 mi/hr	Directional
Infrared	Side	All	Experimental
Rubber tube	Surface	Moving	Temporary
Tape switch	Surface	Moving	Temporary
Triboelectric	Buried	Moving	Specialist
Piezoelectric	Buried	Moving	Specialist
Optical	Vehicle	Equipped	Selective

with special electronics connected to the loop. When transit vehicles are detected, they may be accorded greater priority at the traffic signals.

Induction loop detectors are susceptible to damage to the road surface, but this excuse is overused. Their faults can be reduced by working to a sound specification for supply and maintenance (UK), or by placing the loop wire and feeder cables in special small ducts (US). Modern electronic circuits using digital techniques overcome many of these faults, which may be attributed to inadequate performance of the circuitry.

Microwave or radar vehicle detectors are post-mounted at the intersection facing the oncoming traffic. The wave reflected from a moving vehicle differs in frequency from the emitted radiation. They are useful if loops cannot be installed (e.g., for traffic signals at sites of road construction). They detect vehicles which may not be as precisely positioned over a loop, but the zone of detection is poorly defined, which adversely affects traffic control logic at an intersection. Stationary vehicles are not detected; hence, the traffic signal controller has to deduce which vehicles in a previous signal cycle actually passed through the intersection.

Ultrasonic detectors are favored in Japan. They are mounted over the approach to an intersection. The time of return of ultrasonic waves from the road or vehicle surface is measured. Clearly, they operate only at a single point, and require physical support, normally on a post. Wind deflects the ultrasonic waves and the detector itself; traffic noise also contains components which interfere with the return waves.

7.3.5 Linking Traffic Signals

There are advantages in interconnecting closely adjacent traffic signals. Traffic passing through one intersection may then be given undelayed right-of-way through the adjacent intersection. The process may be continued along an expressway. Linking has been practiced in the UK since the 1930s, usually with cables interconnecting the signal controllers.

If only two signal controllers are to be linked, a wide variety of techniques may be employed. Essentially, an event at one of the signal controllers is used to exert an influence at the adjacent controller. The beginning or end of right-of-way for a traffic stream at one controller may be used to force a similar condition for the same traffic at the adjacent intersection, after a delay reflecting the traffic travel time between the intersections. A recent more standardized example is midblock pedestrian crossings of divided roads in the UK. At these, the push button for pedestrians on one half of the road is disabled while the crossing invitation (a green walking man) is shown to pedestrians on the other half of the crossing. The effect is to provide linking for pedestrians, who can then cross the whole road without significant delay in the middle.

When larger numbers of signal controllers must be linked, one of the controllers (or a special device) is nominated as a master controller. The master controller creates a pattern of electrical signals, which is sent to all the interconnected controllers. This pattern, perhaps in conjunction with time settings in the individual controllers, ensures that the controllers cooperate in the optimum way. The timing patterns can be varied at different times of day. The algorithms used for the timing patterns have been subject to much research and controversy, which is more fully discussed in Section 7.4 in the context of UTC with central computers.

Most recently, linking has been provided without interconnecting cables. The controller contains an accurate clock. This may normally be in synchronism with the power station (mains) supply on the assumption that all the controllers are connected to the same mains supply system, a quartz clock being provided to cover limited periods of mains disconnection. Synchronism is specified to within two seconds per year. Alternatively, a national radio station with a coded time signal may be used, in which case synchronism can be re-established automatically after a break. In either case, every signal controller in the system must have stored sets of timings (known as *signal plans*) to provide the required coordination, and must have clock settings to know at which times of day or week to introduce different plans.

Local linking plans have always suffered from difficulties of maintenance. Lack of synchronism between intersections and other faults may have serious effects on traffic, but the reason for this is not immediately obvious to motorists, the police or other casual observers. Hence the faults may persist unreported. Local linking is perhaps most effective nowadays as a back-up option for a centrally controlled system, or in conjunction with some form of central monitoring.

7.3.6 Standards for Modern Signal Controllers

The increase in complexity of signal controllers has accelerated since the introduction of microprocessors in 1972. To incorporate numerous facilities is now feasible. There are considerable difficulties in ensuring that these facilities can be understood well enough by traffic engineers who program them, and that they do not interact in unexpected ways. Various standards have been promulgated, a discussion of which is confused by inconsistent terminology between countries.

The UK standard [Department of Transport, 1984] is based on *traffic phases* (groups of traffic streams that always receive identical indications), which occur in different time intervals called *stages*. The specified facilities include:

Determination of stage sequence (16 or more stages);

Parallel stage streaming (independent control of parts of an intersection or of a neighboring intersection);

Phase characteristics for vehicular traffic and pedestrians (16 or more phases);

Interphase clearances and extended clearances—shuttle working;

Operating modes and mode changing (possible every cycle);

Phase timing—calls, extensions;

Detector logic including continuous fault monitoring;

Queue (line) detection logic (for turning traffic);

Speed timing logic (increased clearances on high-speed roads);

Fixed-time mode (optional);

Manual control mode (optional);

"Hurry" call facilities for emergency vehicles;

Transit priority logic, with compensation for other traffic;

Master time clock system—8 timing plans, time-of-day clock;

Control of associated traffic signs;

Urban traffic control mode and interface;

Three-tiered monitoring of signal conflicts;

Special conditioning logic (i.e., 8 kb of logic available for user specification); and

Plug-in facility for operator handset or portable computer.

The main US standard [NEMA TS1-1983] is based on interconnecting units, which can be obtained from different sources and fitted in the same controller. The units are based on phases that combine the features of UK stages and phases. In the larger controllers, eight phases may be configured for "dual-ring" operation. This is particularly useful for the standard "HI" (for *high capacity*) type intersection in which separate off-side (left turn) pockets are provided in the intersection layout. The logic for this, known as *dual-ring control*, is shown in Figure 7.3; it allows decisions to be made in parallel, provided synchronism is established twice per cycle at the barrier line. On a given axis, say north-south, there are two left turns for which conflicts with their respective ahead movements are independently adjudicated. While various specialized units can be added, the standard controller features include:

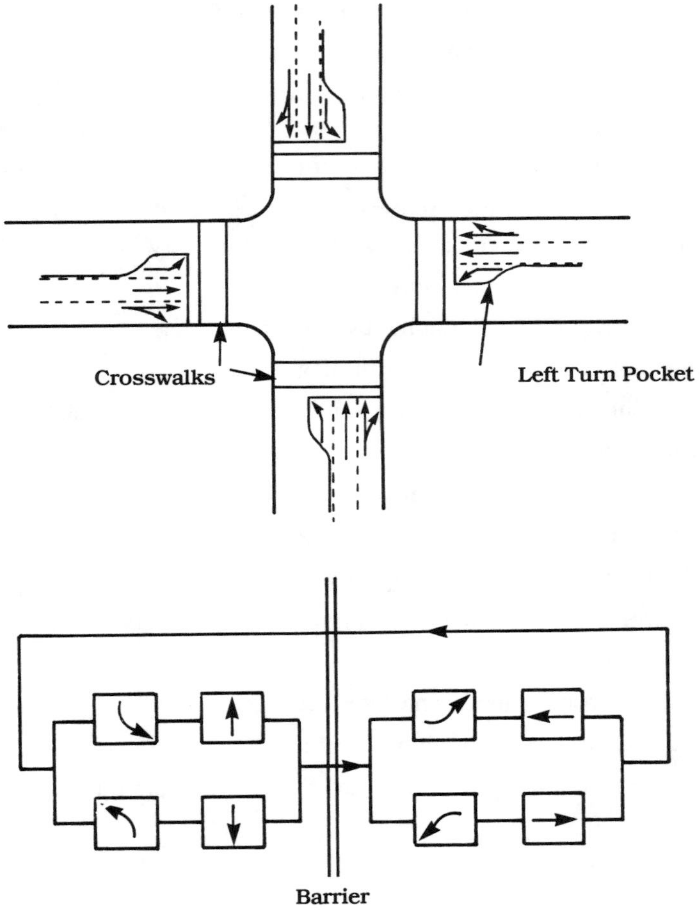

Figure 7.3 (a) US "HI" type (high traffic capacity) junction; (b) dual-ring control logic for the junction.

Up to eight traffic phases, each of which may control a selection of separately signaled traffic streams;

Organization of phases as single-ring or dual-ring operation;

Overlap timings between phases;

Pedestrian facilities to be associated with each phase;

Recall of pedestrian facilities in a long phase;

Vehicle actuation;

Reducing gap logic for terminating traffic phases;

Variable initial green based on red time detections;

Red time extension;

Independent conflict monitor unit;

Solid-state load switches;

Solid-state flasher units; and

Standardized pin connections between units.

Japanese controllers [Takahashi *et al.*, 1977] have a relatively rigid structure with a maximum of 16 steps (a step must be used for *every* period, including flashing periods and yellow periods). Very high reliability (around 50,000 hours MTBF [*mean time between failure*]) is claimed for these controllers (as compared with 10,000 hours expected elsewhere). Subsequent developments [Yamamoto *et al.*, 1988] have first replaced the discrete components with integrated circuits and then with microprocessors. Numerous operating modes are available, including analysis of traffic congestion, different traffic-dependent ways of determining step length, and remote control operation.

A great variety of signal controller types is available in Europe. Traditionally controllers have been based on steps, large numbers of which were available so that the times of different signals at large intersections (often including trams) could differ by a few seconds. Traffic actuation worked by omitting or varying the duration of selected steps. Now, controllers have appeared in which traffic streams (usually defined as *"signal groups"*) are allocated to phases; the phases are expected to appear in any order, depending on traffic demands. The operation of these controllers can be complex, so simulation programs are used to understand them (e.g., FLEXSYT [Middleham and Van Zuylen, 1986]).

Generally, microprocessor-based signal controllers from any country can now mimic the control logic common in other countries, although they do not necessarily pass the relevant approval procedures. Thus, the UK's parallel stage streaming feature can readily imitate the US type of dual-ring control.

7.4 CENTRAL COMPUTER CONTROL

This section describes the history and implementation of UTC. The sections on control algorithms and estimation of benefits follow.

7.4.1 Early Experiments and Objectives

The use of central computers lies at the heart of modern UTC. The first such computers were introduced in Toronto in 1960 [Cass and Casciato, 1960]. At that time, computers were expensive and treasured. One computer was responsible for basic communication with the local signal controllers. The other, which had to be returned to nontraffic duties outside the traffic peak hours, was responsible for applying the main traffic control algorithms.

By 1967, *area traffic control* (ATC) systems (as they were then known) had been introduced experimentally in San José, Munich, Barcelona, west London, and Glasgow. The system in San José was advanced in that detectors were used not only for the traffic control algorithm, but also to evaluate traffic performance with and without the computer in operation. In west London and Glasgow, evaluation was particularly thorough, using moving car observers to produce statistically dependable estimates of benefits [Williams, 1969; Holroyd and Hillier, 1969].

The main perceived purpose of ATC was to provide more satisfactory coordination between signals on an area basis. Previous coordination had primarily been linear, along particular expressways or other main roads and, as mentioned above, was not entirely satisfactory. By having the timing plans stored centrally, they could be adjusted and applied or not as required. The computer would apply them reliably at the correct times of day and would monitor the traffic signal controllers for correct responses. In the experimental systems, it was particularly useful to be able to interchange the optimum signal plans for those ruling before the introduction of computers, so that measurements could be made of the different traffic performance.

The measured and published benefits of these experiments in terms of journey time savings to road users were used as the justification for introducing numerous similar schemes.

7.4.2 The UTC Computer

The arrangement of a UTC system is shown in Figure 7.4. The core of the system is the computer. The computer should be suited to process control and programmable in an appropriate language. The features required include the ability to process various peripherals, including data communication, to attend to specified tasks at specified times, and to adjust individual data bits. With the increase of capability of computers of all types, the distinctions between computers designed for different

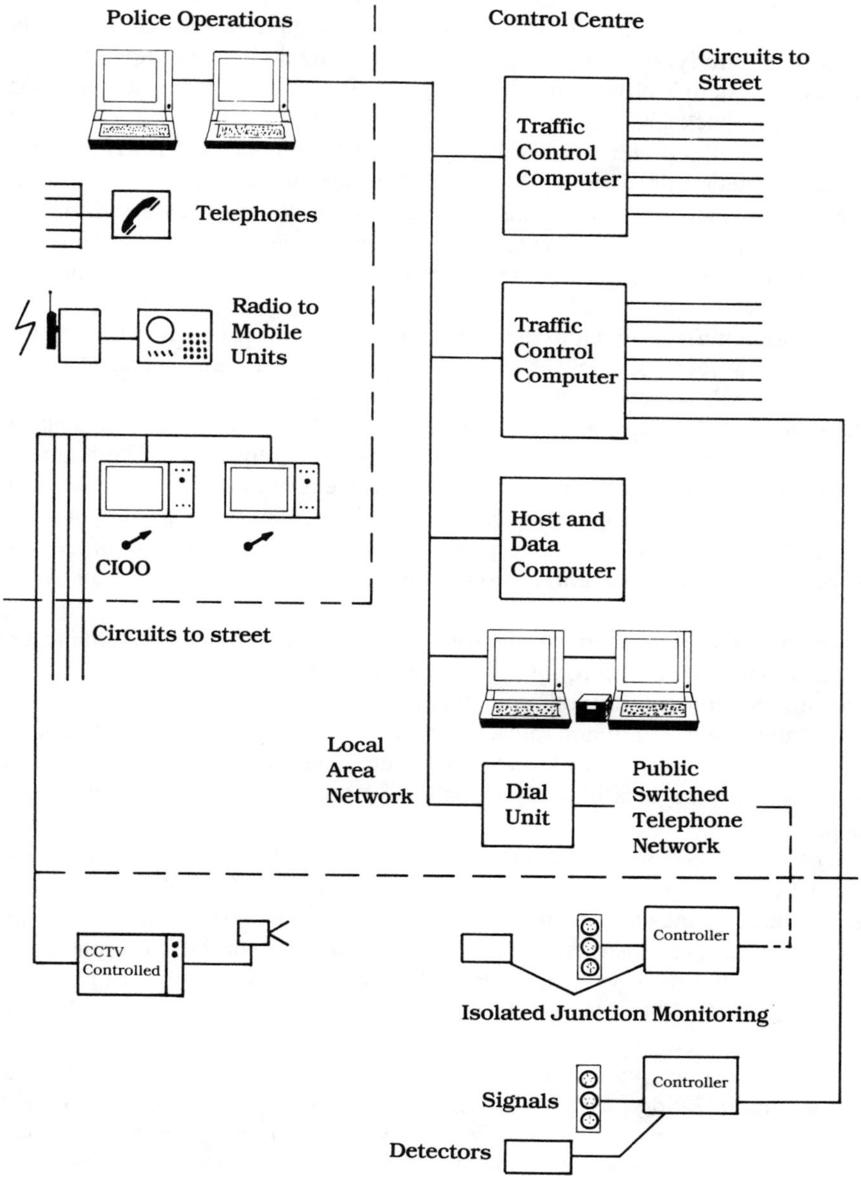

Figure 7.4 Urban traffic control systems.

purposes are becoming less pronounced, and it may be possible to press any computer into service. For example, UTC systems have been implemented using personal computers. Similarly, the once demanding task of time division multiplexing the data transmission system is now more effectively handled by readily available application-specific circuit packages.

There is still a preference for duplicate central computers to provide backup in case one computer fails. Strictly, this is not necessary, because good backup facilities have to be provided in the local signal controllers to cover for data communication failure. However, backup computers are not expensive, and there is considerable merit in having an identical computer in the office so that the local team can develop modified software.

Attention must be given to the nature of the backup. An on-line UTC computer updates various records, such as the current versions of signal plans. On a large system, there will be several changes to such plans each day, and it is desirable for change-over to a backup computer to continue implementing the same plan. More importantly, the computer will have records of the current state of communication with all intersections, including the status of current attempts to establish or reestablish communication. The current process must be either tidily aborted or correctly continued. Finally, the computer keeps various clocks, such as a count of the number of seconds since some moment to reset to zero (possibly midnight) for all the signal plans, and these ought to be transferred.

Overall, these requirements demand either that continuous communication is maintained with the backup computer throughout normal operation, or that some area of data storage common to both computers is regularly updated.

An alternative is a hierarchical arrangement in which all relevant data is distributed in a local area network and can be retrieved by a backup computer. However, the signal plan data is likely to be too voluminous and must be more locally available.

The computer contains or has prompt access to all the signal plans which are normally required, or which are required as alternatives for police use or for the system to react to unusual circumstances. On receiving an instruction for plan changing, the first changed instructions should be dispatched within five seconds. Completion of the change is a traffic matter, depending on safe sequences of signals at the local controller. The change should be complete within a signal cycle, even if simultaneous changes were required throughout the network.

The signal plans will normally be produced, with computer assistance, using one of the optimization programs described in Section 7.5. It should be possible to run these programs in the control computer, in the backup computer, or in another computer linked to the control computer. Once the plans have been properly authorized, transfer into the control computer should be automatic. When they have been experimentally applied on streets under skilled observation, they can be authorized for regular use.

7.4.3 Data Transmission and Circuits

Data communication must be provided to the local controllers. In the past, one-way systems such as radio were used. These are not now normally acceptable, because a significant function of UTC is monitoring of the local controllers.

Design and procurement of the data communication network can be complex. In the absence of television, the requirement is for continuous (once per second) two-way communication between the computer center and each local signal controller. The rate of data transfer is modest (i.e., 200 bs/s), but it must be free of breaks in transmission. Normally, the task of supplying these circuits would be left to the telephone company. Unfortunately, this has led to some unsatisfactory experiences in the US because of increased rental charge imposed shortly after the systems have been introduced. In one case, a UTC system was prematurely abandoned because the highway authority would not pay the grossly increased line rental fee. The telephone company may also fail to provide the level of maintenance required.

An alternative is for the traffic authority to provide its own lines. The major cost of doing so is in supplying the ducts for the cables. During the construction of major highways, a prudent option is for such ducts to be provided at the same time (e.g., two four-inch ducts on each side of a freeway, with frequent cross-linking ducts). Otherwise, the cost of ducts may be reduced by renting ducts, by using disused facilities (hydraulic mains have been used), or laying cables alongside railroad tracks with other communication cables. If extensive cabling between intersections is required or has been provided, spare capacity may be obtained in those cables or ducts. Care must be taken not to expose the circuits to damage, or to place them where access for repair may be difficult.

If the highway authority provides its own circuits, it has to maintain them, and may not have the correct organization for this. If rented telephone-type circuits become faulty, it is probable that the telephone company has spares which can be used for the time being. Similarly, if there are excavation works on the cable route, the telephone company may be better placed than the highway authority to ensure that the circuits are properly protected or rerouted.

The structure of the communication network needs analysis. For short distances, a separate circuit is used to each intersection. In some US systems, separate circuits are used for each data item at each intersection. Such an arrangement is no longer recommended, because data transmission circuitry is so readily available and inexpensive, and because any extra circuit requirements create a risk of having to pay high rentals if the telephone company services are used.

Several intersections may be connected to one circuit, as shown in Figure 7.5. One system, known as multidrop, may be used at all distances. In this system, one circuit is connected to all its intersections. Such an arrangement is not usually satisfactory because of the difficulty of finding faults. It is desirable for each line to pass only between two points, for access to be available to it at those points for

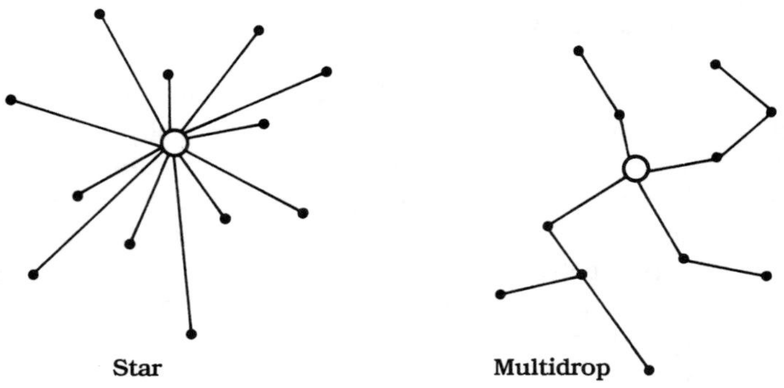

Figure 7.5 Alternative data transmission circuits.

testing, and for remote procedures to be able to identify the faulty section of line. If multidrop seems to be the economical solution, separate lines to each intersection will probably not be expensive, and thus are preferable.

A star arrangement with trunk and local circuits is more likely to be needed. In this system, a relatively high capacity circuit is taken from the computer center to the midpoint of a group of intersections; from this point, the circuit feeds individual lines to the intersections. The multiplexing should be arranged so that a fault on one of the lines does not affect the others. In this case, identification of the intersections that have faulty communication unambiguously identifies the faulty section of line; this identification can be done automatically by the central computer.

One problem with line sharing is that a fault in the shared line affects several intersections, and thus impairs traffic control over a substantial area. If the intersections are in a tightly grouped traffic network, this does not matter, because loss of control to one intersection makes it necessary to return all the intersections to a coordinated fallback plan. Otherwise, there is a case for duplicating the shared line, possibly by an alternative route, if this can be arranged, and providing for automatic or remotely switched changeover.

7.4.4 Police and Operator Interfaces

Historically, the police have been responsible for traffic control. As sophistication has increased, for them to take action to preserve the efficiency of control has become more difficult. Many actions, even responses to abnormal situations, can be more reliably handled automatically. Consequently, there is now less of a case for large control rooms with animated wall maps. Therefore, the provision of these depends as much on their value for publicity and demonstrating the status of the authority and its members (which may be vital to secure funding) as on their functional use.

The case is made often and correctly that a wall map indicating main road green signals or levels of congestion can be interpreted to show the style of traffic control in use. However, this requires alert observers. The automated alternative is for the computer to set threshold levels on the queue detectors, to display alarms when really necessary, and otherwise to introduce algorithms to react automatically to the situation. Some algorithms are described in Section 7.5.

There will be some occasions when conditions are far beyond the limits of design, in which case police intervention may be essential. For these circumstances, display consoles should be provided. Modern graphics are largely self-supporting and readily driven by the computer (perhaps through a dedicated slave computer), so that these displays can be made meaningful and colorful. There should also be associated means for the police to instruct the computer, probably by selecting alternative signal plans from a library.

Similar displays are required for highway authority engineers. They can be useful for commissioning the system, and for commissioning the many changes that are required in any traffic system. Displays are also required for fault reporting. On a small system, one terminal may provide displays for many purposes. On a large system, special displays should be provided to staff with responsibility for analyzing and processing fault reports.

7.4.5 Closed Circuit Television

The police role is to take action under abnormal circumstances, and for this they need further information. Some of this will be gleaned by radio communication with police traffic units. Two-way radio also provides the means for the traffic units to take action on the street (this could include encouraging motorists to obey the signals or not block intersections, or may involve switching off the signals so that the police themselves can control traffic).

Television has the advantage that the cause of any troubles can be identified before a traffic unit has time to visit the site. For example, different reactions are required to deal with a traffic accident, as compared with traffic jams caused by excessive demand. Different diversion routes and signal plans may be needed. Although television can be very useful for these purposes, once the operator's attention has been drawn to it, to expect an operator to scan numerous pictures is unrealistic, whether they are simultaneously displayed or not. Therefore, any television system must be supplemented with a detector and alarm system designed to identify those pictures which are likely to be worthy of further inspection. The alarm system need not be as foolproof as one on which automatic action is to be taken.

Economy in the use of television is achieved by careful siting of cameras so that each can view several intersections or likely trouble areas. This involves mounting on high masts or buildings (the former are usually preferred for ease of access for installation and maintenance).

Communication circuits for television are extremely expensive, and less readily available than telephone circuits. There is, therefore, a greater case for them to be provided especially for the UTC system. Data transmission techniques have improved substantially in the past decade, so it is now possible for relatively conventional circuits (such as twisted pairs) to be used over many miles. If circuits have to be specially laid, there is now the possibility of using fiber optic transmission. Many circuits may be provided within a cable either as separate fibers or by multiplexing, and these circuits can be extended on a lower performance basis to the individual sites once the main distance has been covered. In suitable terrain, microwave radio transmission may be used.

7.4.6 Priority for Special Vehicle Classes

Greatest priority must be given to emergency vehicles. A UTC system achieves this by providing "green waves" (UK; in Germany *"grunnewelle"* apply to all traffic). In a small city with one fire station, the crew presses a button when leaving the fire station, perhaps indicating the intended route. The UTC computer then forces the signals on the green wave to favor the fire apparatus after a time based on observations of its normal travel time to the intersection. This is usually, but not necessarily, to give a green signal to the fire apparatus; however, it may be preferable to stop other traffic and allow the emergency vehicle to overtake on the wrong side of the road.

Traffic congestion may make the fire engine's travel time uncertain, so to track it and adjust the timing is useful. This is done by EVADE (*Emergency Vehicle Automatic Detection Equipment,* UK), for which considerable benefits have been reported [Griffin and Johnson, 1980].

If the fire engine's route is unknown, some more general assistance must be provided. This can be on a localized basis, by a system such as the OPTICOM system (a proprietary product of the 3M Company). In OPTICOM, a flashing strobe light is fixed on the roof of the apparatus and is detected at the intersection. The detector causes the local traffic signal controller to change immediately to favor the emergency vehicle. The beam is strong enough to detect an apparatus up to 300 yards away, thus giving time for the traffic signal controller to react.

Transit priority is required more often. At a local level, this is achieved by detecting transit vehicles and causing the signal controller to react favorably. Detection may be by a loop detector covering a prescribed area of road that is only reached by transit vehicles, or in which only transit vehicles are large enough to fill the loop and give the required response. In some cases, the pattern of response in the loop detector as the vehicle travels over it can be used to identify the vehicle class.

In the UK, the problem is complicated by the general variation in road width and consequent poor lane discipline; this is particularly the case when a bus leaves a bus stop, which is a commonly required position for the detector. Accordingly, the more successful system is to provide a transponder on the vehicle, which interacts with special electronics connected to a loop. Such a transponder may also contain information about the vehicle, indicating its future route at the intersection (which may be relevant to select the correct response). Such a system is in use for buses in London [Chandler, 1984], for trams (trolleys) in Melbourne [Cornwell *et al.*, 1986] and for light rail (commuter train) in Hong Kong [Bodell and Huddart, 1987]. The major difficulty with buses is to ensure that all of them are correctly fitted, but there is potential for substantial benefits (see Figure 7.6) in reduced delays and, particularly, reduced *variation* of delay.

Once priority has been given to the transit vehicle, care must be taken not to unduly handicap other road users. The priority is usually given by instigating the traffic phase to favor the transit vehicle (or continuing the phase if it is already

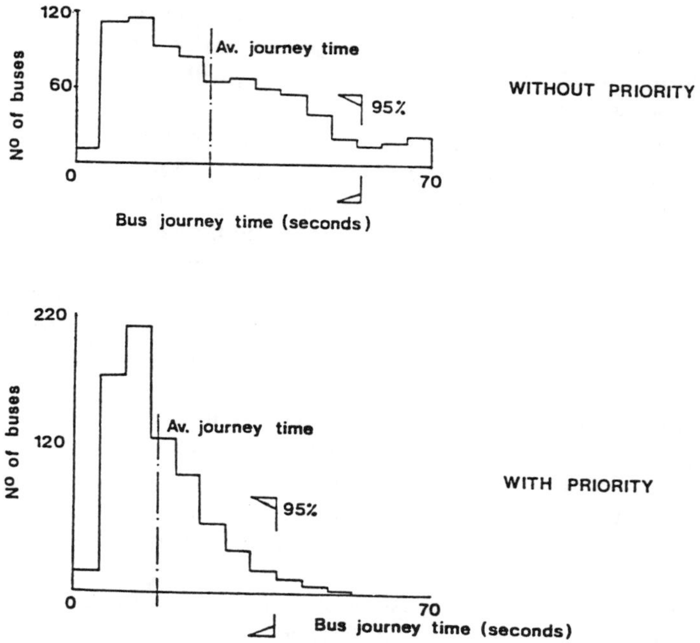

Figure 7.6 Effect of bus priority using vehicle identification [Chandler, 1984].

active). To compensate for this at heavy traffic intersections, additional time can be given to other specified traffic phases in the next signal cycle. To inhibit further priority action may also be useful for a period, such as one signal cycle.

Within a UTC system, transit priority is more difficult to provide. The system derives its main advantage by coordinating traffic signals in a network. Disruption of the signal timings for an individual vehicle destroys this coordination to an extent which may be difficult to repair. The Melbourne system [Cornwell *et al.*, 1986] avoids this by having the UTC system provide a time window, during which priority could be given to the tram or specified other traffic can flow. Critical cross traffic timing is not substantially affected, but the tram does not always receive priority. An alternative approach is to set the coordination so that it more nearly favors transit operations. In essence, the travel time used between intersections applies to the transit vehicle (hopefully, without needing to allow stopping time to pick up passengers) rather than other traffic. This is done in the TRANSYT algorithm [Robertson and Vincent, 1975].

7.4.7 Variable Message Signs and Car Park Status

The UTC system can be used as an interface with any other traffic control device. Variable message signs may be used to advise motorists of changes in routing. One example is lifting river bridges in the traffic network, with the possibility of rerouting over a neighboring bridge.

When the bridge is to be lifted, the local signal controller is switched to provide a different traffic control mode, usually including red signals for the route across the bridge. There may be variable message signs close to the bridge, and these can be activated either by the signal controller or by the bridge opening controller. The variable message signs to indicate an alternative route are further away, probably at another intersection. To activate these signs, feedback from the controller at the bridge to the UTC computer is used. The computer may send the relevant messages directly through a special outstation at the variable message sign, or to a conveniently placed traffic signal controller.

A common requirement for variable message signs is to indicate car park (parking lot) usage. Careful advance traffic engineering design is required. The car parks selected for consideration must have adequate counting equipment based on vehicles detected entering and leaving, or better still, drivers taking tickets at an entry barrier and paying on leaving. Threshold levels are set, which may vary by time of day, to indicate when the car park is considered to be full or nearly full. At suitable points in the network, signs are provided listing the car parks, and against each is set the legend "spaces," "nearly full," or "full," as appropriate. These signs can again be driven directly by the UTC system or through a nearby signal controller.

7.4.8 Remote Monitoring of Traffic Signals

We have established that the problems arising from traffic signal faults are substantial, and can outweigh the benefits to be obtained by more sophisticated control systems. Straightforward faults like failed lamps may be reported and rectified promptly. They do not, however, contribute much to the causes of traffic delay. These are usually attributed to faulty timings. Examples include vehicle detector failure, as a result of which a signal remains green for the maximum prescribed time, thus holding up other traffic; the signal controller may balance the delays with other traffic movements, but may not be able to eliminate them.

Traffic signals connected to a UTC system are increasingly monitored. Independent investigations of faults at them [Oastler, 1985] showed them and their related intersection equipment to have ten times fewer uncorrected faults than those not connected to the system.

With the availability of microcomputers, it is possible to provide one within a traffic signal controller to monitor the performance of the controller and associated equipment. The monitoring unit is connected to the controller as if it were a UTC connection. It also has connections to the detector input terminals and to special lamp monitoring circuits. The latter connections group lamp leads together and passes them through simple transformers. Up to 16 leads may be served by one transformer and monitoring unit. The monitoring unit relates the current measured at the transformer to the pattern of current expected from the lamps intended to be switched by the traffic signal controller. For each pattern of controller switching, there should be a standard pattern of lamp current changes. When all is well, this pattern is memorized by the monitoring unit, so that any lamp failure can be detected as a change in the pattern.

When this or any other fault is identified by the monitoring unit, the autodialing unit associated with it uses the PSTN to report the fault to the central office. At specified intervals, the central office calls up all its outstations and causes them to start a self-checking routine. Data on the traffic signal controllers' logs can be transferred at this time.

The system has several advantages in terms of economy and reliability. The public telephone circuit rented only extends from the signal controller to the nearest telephone exchange. It is, therefore, much less expensive than a dedicated circuit all the way to the central office. For the rest of the route, use of the switched network means that charges for the circuit are only incurred during the limited period when it is in use. At other times, the circuit is used by other telephone users. If, at any time, the circuit becomes faulty, this is likely to be detected by other telephone users or by the telephone exchange equipment, which will normally provide alternative circuits. Hence, the fault reporting is most unlikely to be delayed for repair to this part of the circuit.

An additional advantage of remote monitoring equipment is that it has the capability of transferring data to the traffic signal controller. In principle, the central office can recalculate traffic signal settings or react to a temporary traffic situation by sending altered signal plans to the signal controller. Because these plans could include coordination with other intersections by using clock synchronization (itself confirmed on the dialed link), the effect would be as if a UTC computer were in direct communication with the traffic signal controller. This is known as *dial-up UTC*. Regrettably, it has problems. One is that remotely retiming signal controllers may ignore specific local traffic interactions, unless thorough relevant site observations have been made, or the site is under television surveillance. Another problem is in synchronizing the start of a new plan between intersections. Although traffic signal controllers already have the facility to start the plan at a predetermined instant, this is to no avail if the telephone call cannot be established to one or more controllers at a group of intersections.

7.5 TRAFFIC CONTROL PROGRAMS FOR UTC

7.5.1 The Range of Available Programs

The main quantifiable advantages of UTC may be attributed to coordination between adjacent intersections. Such coordination is based on the travel speeds of vehicles between those intersections. Coordination is also influenced by traffic flows, because these have a primary effect on the intersection cycle times necessary to pass the traffic. However, once a signal plan has been set up, changes to individual intersections or links in the network are likely to be difficult to make without destroying the entire optimization. Therefore, the plan is usually used for some time (i.e., hours), as long as the traffic conditions are predicted or measured to be suitable. During this time, all signal switching instants can be rigidly fixed on a cyclic basis. This is known as a *fixed-time plan*.

Fixed-time plans have obvious deficiencies, which become apparent as traffic patterns change. The research justification clearly shows that to maintain the plan is better than trying to match the traffic variations in an unstructured way. However, the deficiencies have attracted research attention, and several solutions have been proposed. The more successful ones are classified and described in Sections 7.5.2 to 7.5.7.

Considerable further research and development is being done, and numerous other programs are available. Selection of the examples is based on the extent to which they apply to large numbers of intersections or have enjoyed some success beyond national boundaries.

There is—or rather there was until the 1973 oil crisis—a division between the US and UK approaches to coordinating traffic signals. The US approach (and many manual methods) started from a desire to produce a visible green band on the arterial road. The manual methods used "time-distance" diagrams like that in Figure 7.7. The UK approach was to calculate a strictly optimal set of timings for all traffic on the network, including turning traffic. This was partly attributable to the less structured nature of UK traffic networks. Clearly, both approaches have merit, and their modern development seeks to include the more important advantages of both.

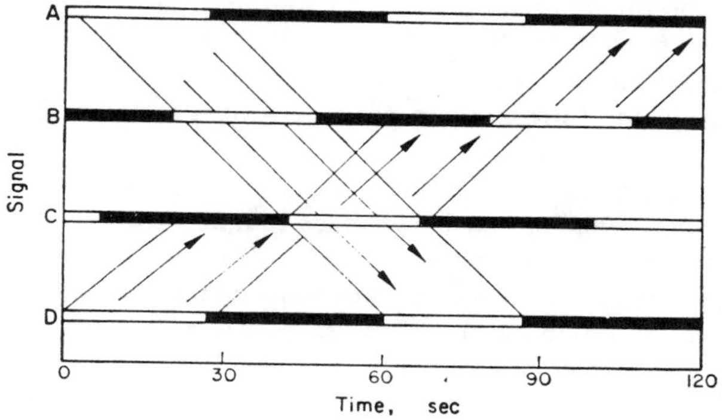

Figure 7.7 Time-distance diagram showing green-band through four signalized intersections.

7.5.2 Fixed Time Plans Derived from TRANSYT

TRANSYT (Traffic Network Study Tool) is a UK-developed program for off-line analysis and calculation of signal timings for a network. There are no restrictions on network shape, although the sequence of optimization implies assumptions of the relevant importance of links in some closed loops in the network. The program models "give way" junctions in the network, including those related to traffic sources and "sinks" such as parking lots. The model assumes that all signaled intersections are operating on a rigid fixed-time plan. The required input to the model includes the lengths of links between intersections, travel times between intersections, traffic flows (including turning movements), and information on the signal timings at individual intersections (to account for pedestrian movements and safety clearances).

The model for each link (two-way links are treated as two one-way links) computes a traffic flow for each second of the signal cycle for each link entry, most of the entries being exits of signaled intersections, based on the average traffic flows and the signal settings. Allowing for variation of travel speed between individual vehicles, the model deduces the second-by-second arrival pattern at each intersection approach, and hence the disbenefit created by traffic there in terms of stopping and being delayed. The degree of disbenefit depends on the relative timing of the two intersections at the ends of the link, known as the *offset*. The optimization model treats intersections in the network in turn, making large or small changes to their offsets, in an attempt to produce a set of offsets that minimizes the disbenefit. The optimization does not claim to be mathematically rigid; indeed, the attraction of TRANSYT is that it uses judgment and traffic engineering knowledge to minimize computation requirements. There is evidence that the real optimum is seldom or never missed if it shows an improvement greater than one percent.

As well as optimizing offsets, TRANSYT adjusts "splits," the balance between timings provided for different traffic flows at individual intersections. The model provides information and procedures to permit the selection of suitable cycle times, and to avoid traffic line formation on particular links that may block neighboring intersections.

TRANSYT's "performance index" (or measure of effectiveness) accounts for the likelihood of accidents caused by vehicles required to stop, the value of time, fuel consumption, and environmental pollution. The user may cause the model to attach extra importance to particular links or movements. The program provides, on request, a full report, link by link, of the factors contributing to the performance index, together with pictograms of the traffic flow during a signal cycle on each link. The traffic engineer is therefore able to judge where the problems are predicted, and perhaps to make changes (such as rerouting some traffic streams) beyond the scope of the TRANSYT model.

Recent versions of TRANSYT give settings that provide transit priority [Robertson and Vincent, 1975]. The model of relevant links is altered to treat separately buses and other traffic. Hence, the effect of their different travel times and passenger occupancy can be judged. Experiments in Glasgow demonstrated that BUS TRANSYT provided 8% improvements in bus journey times, with negligible detriment to other traffic.

The benefits attributed to TRANSYT have been thoroughly researched in Glasgow and many other UTC implementations. The model is used as a reference, and is applied worldwide, including its adaptation as version 7F to retiming of all linked signals in the US following the oil crisis of 1973. The program is normally used off-line. When the traffic engineer is satisfied with the signal settings calculated by TRANSYT, they are entered as input to the control computer for application on the street.

7.5.3 US Bandwidth Maximizing Models

US thinking on optimization has been directed to linear systems and the maximization of throughput on a two-way arterial road. The objective is usually to maximize bandwidth (i.e., the time within a signal cycle during which traffic proceeding along the arterial road at its normal speed can do so without being stopped at any red signals).

The procedure for such models starts by setting up the signaling structure of each intersection (i.e., the times that must be allowed for the various movements, the sequence in which they occur, the clearance times between them, and the way in which different movements are combined). From this, the red and green periods for the two directions of travel along the arterial road, and the relationship between them, are specified. Speeds between intersections are assumed. The model then adjusts the timing of the signals relative to one another to maximize the width (in time) of the band of traffic that can flow along the arterial road without stopping. The model can be set to maximize the bandwidth in one direction, to be equal in both directions, or to be at some intermediate proportion. An interesting phenomenon is that maximum equal bandwidths in both directions occur when every intersection is either exactly in synchronism (center instant of green time is the same), or exactly antisynchronous with every other intersection in the arterial road.

Apparent these calculations give no regard to traffic volumes. Although the reduction of data input may be thought to speed computations, in practice the requirement to consider all combinations of signal timings regardless of their importance means that bandwidth maximizing models need relatively large computer powers. Some more recent ones investigate the effect on bandwidth of changing the signal structure of intersections, particularly with respect to whether the arterial road's turning traffic flows before or after the cross street traffic.

These ideas have been embodied in US Federal Highway Administration products such as PASSER. Evaluation of their performance is frequently by use of the detailed and comprehensive NETSIM model. Unlike TRANSYT, this model uses random traffic generators, and also does not include penalties for stopping vehicles in its performance index.

Recent developments [Gartner, 1987] are making these models sensitive to traffic volumes and to the network implications of turning traffic. Similarly, the US versions of TRANSYT now include more specific modeling of the HI type of intersection common on arterial roads.

7.5.4 Classification of Control by Traffic Responsiveness

The fundamental ground rules and terminology for US on-line network control were set in the *Urban Traffic Control System* (UTCS) program in Washington, DC between 1967 and 1976 [FHWA, 1976]. In this program, several competing strategies were compared with one another and the normal means of creating fixed-time signal plans by the District of Columbia. At that time, it was envisaged that the sophistication of traffic control would increase progressively, so the expected generations of system were defined as follows:

- *first-generation control* (1GC) uses the precalculated timing pattern that most closely approximates the current level of demand. Demand assessment includes both traffic volume and occupancy (the proportion of time for which a vehicle detector is occupied by a vehicle).

- *Second-generation control* (2GC) strategy is an on-line, real-time traffic signal optimization model that computes signal timing plans as a function of current and predicted traffic conditions. A five-minute prediction period was used.
- *Third-generation control* (3GC) strategy is similar, except that it does not require that intersections within each coordinated subnetwork have the same signal cycle time. Different algorithms were planned for uncongested and congested conditions, but only the former could be developed to a satisfactory state for evaluation.

We note that none of these control strategies included "vehicle actuation" in the UK sense and as applied at isolated intersections in the US, whereby traffic signal phases are called and extended in response to detection of individual vehicles. These strategies might be better defined as *traffic selected* (or adjusted) *fixed-time systems*. Even so, the clearest finding of the evaluation was that 3GC was the most expensive and least effective of the control strategies.

First-generation control was found to be slightly (up to 4% less delay) better than the original three-dial control, but only marginally better than computer-produced plans selected by time of day. Second-generation control was not found to have significant advantages. The finding that disturbing a fixed-time plan to react to traffic would be counterproductive was consistent with other results, such as those in Glasgow [Holroyd and Hillier, 1969]. Consequently, current US systems are normally described as 1GC or 1.5GC.

Other subordinate control strategies were tested. *Critical Intersection Control* adjusts the balance of lines of traffic on the arms of selected intersections within the coordinated network. This strategy was found to have limited value in conjunction with time-of-day plan selection and 1GC. Current systems use the strategy in this way. A *Bus Priority System,* in which bus flows affected the signal settings, provided benefits only if bus flows were not too high (not above 400 buses per hour), and there was no traffic congestion; in the latter case, later versions of the program temporarily disabled the bus priority feature.

7.5.5 Current Control Systems—US

In the US, there are some 400,000 traffic signals, of which up to 30% are coordinated. Of course, these are coordinated in a variety of systems, ranging from local master controllers linked to a central personal computer, to systems in which every intersection is directly controlled from a substantial computer center.

An alternative which attempts to achieve a flexible compromise [Gardner, 1987] is *Series 2000*. In this system, normal coordination is operated by local masters, but they are in second-by-second communication with the central computers. Many fault diagnostic programs are run, and the results are immediately reported. This and other graphical presentations to the operator are a significant feature of the system. The central computer is responsible for recording new plans and downloading them to the local masters. Developments include running TRANSYT on-line to produce updated plans and make traffic-dependent selections. However, in the event of circuit failure, the local masters can continue to operate a coordinated signal plan and change it by time of day. This development, therefore, parallels the dial-up UTC developments in the UK, except that dedicated lines are used.

7.5.6 Current Control Systems—SCOOT

Only a limited number of the many available control systems can be described. SCOOT (*split, cycle time and offset optimizing technique*; see [Hunt *et al.*, 1981]) is chosen because of the thorough documentation of its performance during development, its use as a national standard in the UK, and its increasingly international use.

Like 1GC and 2GC, SCOOT uses traffic-adjusted fixed-time plans. Its advantages include:

- Adjustments are made very frequently, at the next signal change for split and offset optimization, and at the next signal cycle for cycle time optimization.
- The adjustments are small, the new plan always being a marginal change on the existing one, so that there is no disruption of traffic flow for plan changing.
- The adjustments are implemented as the optimization proceeds, so that there is no delay in plan implementation; hence, the traffic data is up to date.

The SCOOT system must have a central computer in permanent communication with its signal controllers and vehicle detectors, as shown in Figure 7.8. The detectors are placed at the beginning of every link approaching a traffic signal; within a network, the detectors therefore usually are on the exits from intersections. The detectors report vehicle occupancy every quarter second; this information is used by the computer model to create a second-by-second pattern of vehicle arrivals within the signal cycle for every link. The model uses previously supplied data for travel time on each link and link storage capacity, in addition to the actual timings of the traffic signals at the exit to assess how the traffic will line up at the signals. This may be displayed to the operator in the form of Figure 7.9. The model stores the number of stops and the delays suffered by the traffic in each link for the current signal timings, and for conditions if the timings are altered by a few seconds.

When a signal change at an intersection is imminent, SCOOT sums these patterns for all approaches and determines whether advantage will be gained by advancing or delaying the signal change by a few seconds. If so, the change is made immediately. This has the effect of optimizing for both offset (the relative timing of

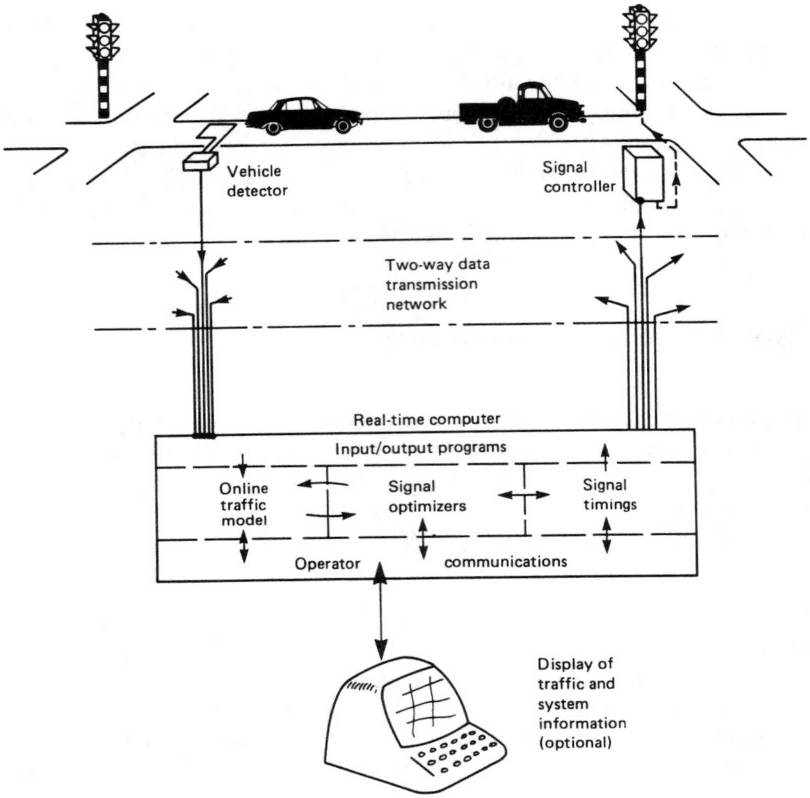

Figure 7.8 The flow of information in a SCOOT urban traffic control system [Hunt, *et al.*, 1981].

adjacent intersections) and split (the balance of time allocated to traffic on competing approaches to one intersection).

Additionally, every signal cycle, the model estimates the lowest practical cycle time for each intersection in the network. The highest of these cycle times is then taken as the required network cycle time. The common cycle time must apply to all intersections, although the model tests whether advantage will be gained by having some intersections or pedestrian crossings operate at a submultiple (a half or a third) of the common cycle time. If the required cycle time differs from the current cycle time, a small change is made for the next cycle. The changes for increasing cycle time are greater than those for decreasing it, so that the system can respond rapidly to sudden increases in traffic without risking the formation of traffic jams which may be slow to clear.

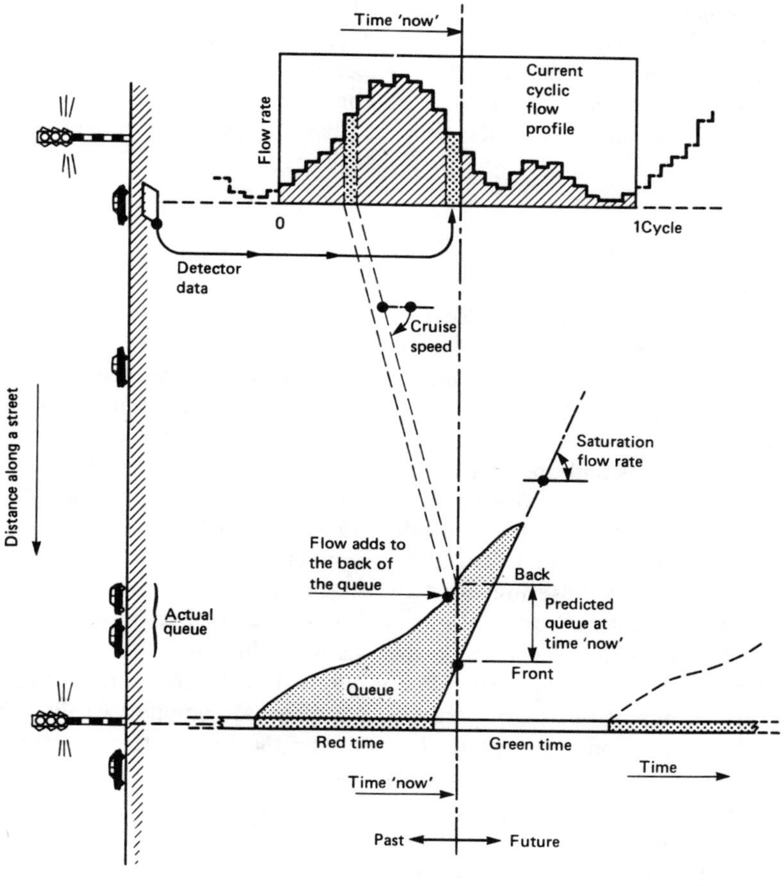

Figure 7.9 Principles of the SCOOT traffic model [Hunt, *et al.*, 1981].

In performance terms, simplistically, a 10% rule can be applied. Coordinated systems with precalculated plans using TRANSYT show a 10% journey time improvement over the previous vehicle-actuated isolated systems, and SCOOT shows a 10% journey time improvement over TRANSYT. The delays at intersections are improved by twice as much (20%), but, of course, these relationships depend on network features such as the length of individual blocks. Coordination benefits links within a system, but does not benefit traffic on the entry links, because the common system cycle time will probably be longer than that possible at the entry intersections. SCOOT retrieves some of these losses by using lower cycle times whenever traffic conditions permit.

The advantage of an adaptive system, such as SCOOT, over precalculated fixed-time systems is enhanced by the difficulties of deploying adequate resources to keep the fixed-time systems up to date. A plan selection system such as 1.5GC naturally falls halfway between the two, because its constituent plans and selection criteria need to be updated, but at least the time of plan change is dependent on current traffic patterns.

7.5.7 Current Control Systems—SCATS

The *Sydney Coordinated Adaptive Traffic System* (SCATS) is used in most major cities in Australia and New Zealand at nearly 5000 signal installations. SCATS's development started in 1975. Intersection controllers, all equipped with microprocessors, are subject to second-by-second control from regional master computers located in unattended cabins. Central office computers exercise overall coordination by providing databases and operator interfaces, including police interfaces. Data for a given computer in the system may be input from any other point, including each of the intersection controllers and a terminal in a helicopter.

The traffic algorithm [Lowrie, 1982] is essentially pragmatic. For each important intersection and those immediately associated with it, several control plans are devised for different traffic conditions. This is similar to specifying the acceptable sequences of traffic phases at a single intersection, plus any coordination requirements with the immediately associated intersections. Local traffic information is used to adjust the timings of this plan, partly by the regional computer varying the plan that it sends every second, and partly by allowing the local intersection controller freedom to interpret certain periods within the plan. Limits to many timings are specified and often reached in peak traffic conditions, so the timings applied are entirely predictable.

Coordination between adjacent important intersections is enabled or disabled in accordance with traffic conditions. Coordination requires the adjacent intersection controllers to use a common cycle time; the regional computer assesses whether this would require one of the controllers to change from its current cycle time, and whether the disadvantage of doing so would be adequately compensated by the advantages of coordination. Several coordination plans are available for balanced two-way traffic flow, or in favor of one direction of flow, depending on the network geometry and current traffic flows. Such bilateral links in suitable circumstances may spread throughout the network.

For the system to work effectively, part of the computing is undertaken locally. In particular, vehicle detector actuations are converted into a more usable proprietary format, indicating average flow and occupancy, before dispatch to the regional computer.

The difficulty of evaluating the system is that it has such a large number of adjustable variables. In the main comparative test [Luk *et al.*, 1983], the system was significantly better than any other for times and sections of the network where the skilled designer, network geometry, traffic flows, and objectives set in the evaluation were in harmony. However, there were circumstances in which the results were more erratic and subject to dispute. Neither on that occasion nor any other has there been an effective comparison between SCATS and systems using traffic-adjusted fixed-time plans, such as SCOOT or the Federal Highway Administration's UTCS software (1GC).

7.5.8 Current Control Systems—Japan

Of the 122,000 traffic signaled intersections in Japan, some 35,000 were (*circa* March 1987) in 74 central area control systems, including some 8000 in one system in Tokyo. The design of these fixed-time systems [Nakahara *et al.*, 1970; Yamamoto *et al.*, 1984] uses a multicriterion approach. Numerous detectors are used to assess the state of traffic and congestion, and an appropriate plan or control algorithm is applied. In principle, this approximates the US 1.5GC systems. Japanese traffic control is entirely the responsibility of the police; success is therefore judged by the operator facilities made available to the police and the effect that they have on the traffic, rather than by the maximization of quantified measures of effectiveness.

7.6 FREEWAY CONTROL IN URBAN AREAS

7.6.1 Capacity Optimization

Freeways in the US have relatively sparse instrumentation. Attention has mainly been given to optimizing the capacity and distributing this capacity equitably between different classes of road user, particularly by allocating priority to transit vehicles and other high occupancy vehicles.

As has been well established [Lighthill and Whitham, 1955], traffic flow is unstable. Figure 7.10 indicates that, as flow increases, the speed progressively falls until the capacity is momentarily reached when speed falls suddenly to a very low value, possibly zero. The instantaneously stationary traffic condition then propagates as a shock backward through the traffic stream, thereby reducing the effective traffic capacity of other sections of road. The section of road with locally lower capacity which initiates such an unstable reaction is one where motorists hold back slightly. This may be at an on-ramp, where they need greater space to deal safely with the more complex movements or at the bottom of an upgrade (such as in a cross-river tunnel), where they are slow to compensate for the decelerating effect of the gradient.

Effective control of the traffic input to such sites so that the critical condition is not reached can achieve traffic capacity increases of up to 10%. The most widespread application is to traffic merging at on-ramps, as in Figure 7.11. Here, the

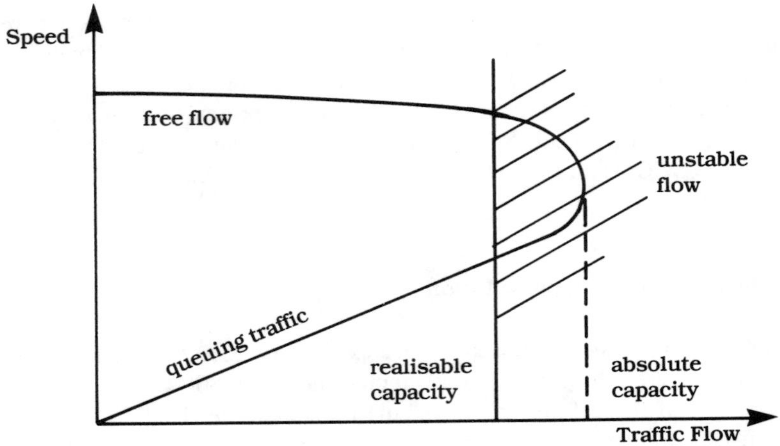

Figure 7.10 Speed-flow relationship for a uniform road.

traffic capacity downstream of the merge point is either known to the system, or calculated on-line by using detectors to measure speed and flow, and the computer algorithm to derive the resultant speed-flow relationship. The detectors also measure the main line traffic input upstream of the merge point. The difference is available for on-ramp traffic, which must be metered.

On the on-ramp, when traffic metering is in operation, traffic signals periodically turn green for short durations to permit one or more vehicles to pass. The equipment measures the actual number of vehicles passing and adjusts the traffic signal settings to achieve the metering rate required by the system. A special lane

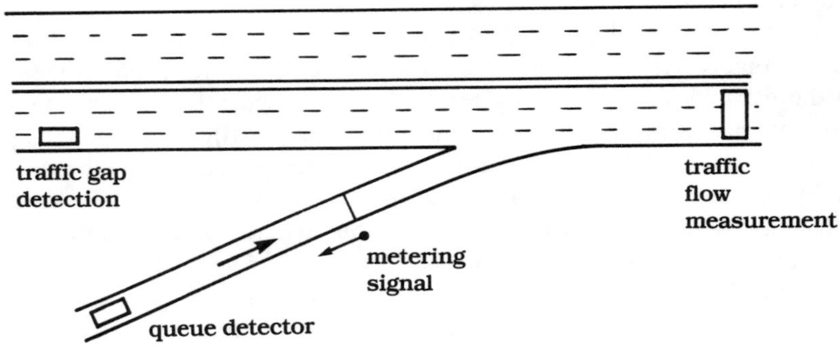

Figure 7.11 Arrangement to meter traffic onto a freeway.

may be provided on the on-ramp for transit or other high-occupancy vehicles, which gain priority by bypassing the signals.

In certain cases, such as the Oakland Bay bridge [Scott Maccalden, 1984], the operation is dominated by the needs of high-occupancy vehicles. Some of the lanes are reversible, so the control system must set signals indicating which lanes are to be used and whether vehicles need to pass through the toll booths. High-occupancy vehicles are allocated special lanes on the approach. Their flow is measured and the other traffic is metered to conform to the bridge capacity derived from on-line speed and flow measurements.

7.6.2 Incident Avoidance

European motorways are more heavily equipped with instrumentation. Motorist aid telephones are always provided at intervals of less than two miles. On many motorways, such as in the UK, standard variable message signals are provided to indicate recommended speeds and lane closures. At many sites, closed circuit television is provided for information to the police, who also have communication to emergency services and broadcasting stations (see also Chapters 5 and 6 for details of traffic information systems). The communication is controlled by computers; one arrangement of the current UK specification is shown in Figure 7.12.

Large-scale trials in the Netherlands [Klijnhout, 1986] illustrate the most sophisticated and successful systems. Detector stations and signals are provided every 500 m. Low speed is interpreted as congestion. The signals at the previous set of signals are then set to a marginally lower speed. This has the effect of avoiding accidents by reducing the speed differentials between vehicles. The fast reaction of the system (within 28 s) ensures that the instructions are relevant. Consequently, a much higher level of driver obedience to the signals is achieved than in systems where the signals are set by police. The system has been particularly successful in fog conditions, because the absence of varying vehicle speeds reduces the likelihood of collisions, while still maintaining relatively high recommended and actual speeds. For the cases of snow and black ice, special additional signals must be provided to drivers, because they will not otherwise know of the hidden danger for which a lower speed is required. The system was found to reduce accidents overall by 25% and secondary accidents by 46%. Capacity past obstructions was also slightly increased.

In Japan, very elaborate variable message signs are used. On occasion, these can be used to close sections of the motorways. More attention, however, is given to advising motorists about the extent of and reasons for delays ahead than to requiring any driver reaction (probably none is feasible anyway). Similarly, experiments on in-vehicle communication within these systems seem primarily addressed to enabling drivers to communicate with their offices and hence to be better able to organize their time.

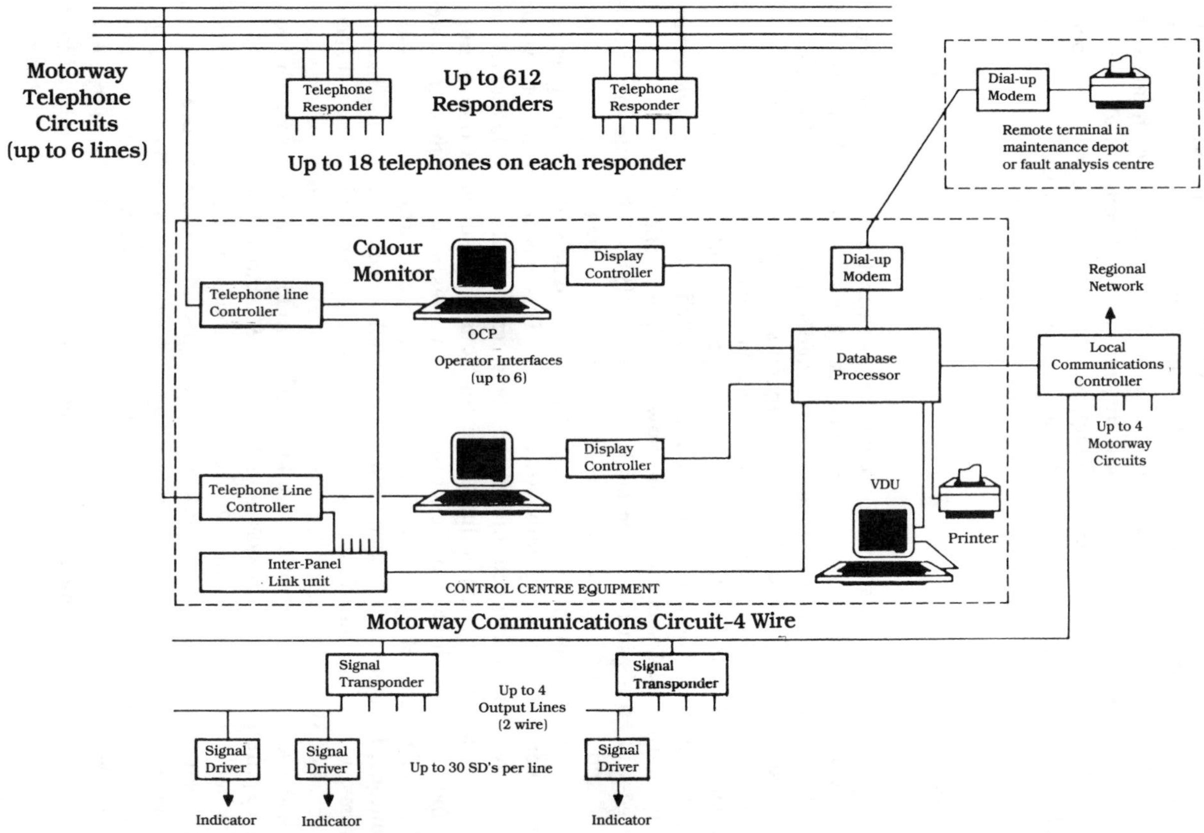

Figure 7.12 UK motorway control system. (Source: Ferranti Computer Systems Ltd.)

7.6.3 Corridor Control

In urban conditions, freeway control can be extended to control of the whole corridor, including neighboring streets. This is particularly applicable in US cities with strong street networks running parallel to the freeways. Motorists naturally tend to take advantage of the higher journey speed on the freeways, so that normally the streets are less heavily loaded. In the event of traffic being metered, or of freeway incidents, surplus traffic must be diverted onto the street system. With increasing traffic levels, the need for such diversions is increasing, and the ability of the street systems to handle them is decreasing.

Limited corridor control systems have been implemented during the past two decades. A modern system is the "SMART" street demonstration project adjacent to the Santa Monica freeway (see Figure 7.13) in Los Angeles [LACTC, 1987]. The techniques to be deployed include:

- Automated traffic surveillance with detectors on the freeway, in the streets and in turning lanes to identify the opportunities for traffic diversions;
- Human interaction with the *Automated Traffic Surveillance and Control System* (ATSAC), in reaction to the freeway information;
- Direct access, real-time information to commuters by telephone at home or on the road;
- Improved emergency response by the police.

7.6.4 On-line Collection of Data

The increasing use of detection devices in modern traffic control systems offers opportunities for automatically collecting data on traffic performance. Although the detectors are positioned for control purposes, they are increasingly being set far back from intersections so the signal controllers can model traffic behavior to produce an optimum response. The controllers or the coordinating computers can use this modeling to estimate delays, stops, fuel consumption, and other measures of effectiveness.

Some systems are capable of tracking individual vehicles. For transit priority systems, detection of transit vehicles can be done by transponders which identify the vehicles. If this information is passed to central computers, the journey times of the vehicles can be deduced. In some conditions, this may indicate the journey time of other vehicles in the traffic stream.

Navigational systems, such as AUTOGUIDE in the UK, exchange information between roadside and vehicle. Such systems derive some of their benefits from providing motorists with up-to-date information on best routes. This information is obtained from the journey times currently being achieved by vehicles on those routes (see also Chapters 6 and 9).

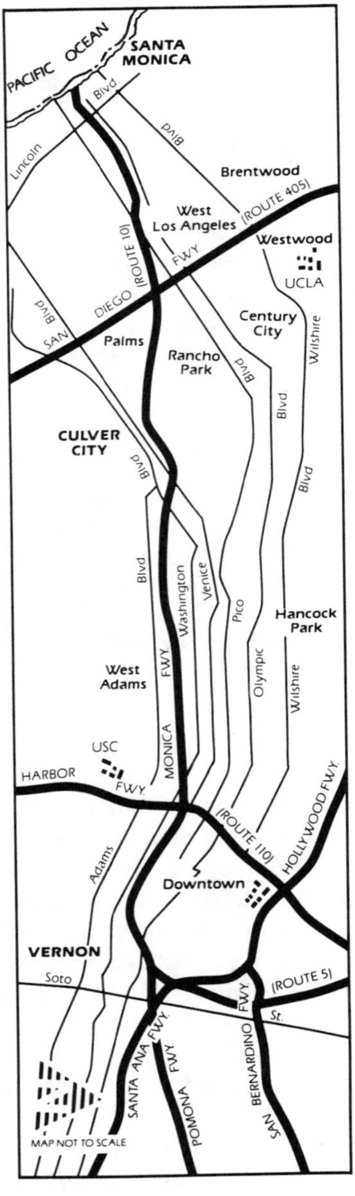

Figure 7.13 Santa Monica freeway, Los Angeles: site of the SMART street corridor control project [LACTC, 1987].

7.7 BENEFITS

7.7.1 The Main Classes of Benefit and the Beneficiaries

The benefits achieved by better traffic control systems are experienced by motorists, vehicle owners and other members of the public. However, the costs of providing the systems (except on toll roads) fall on public authorities. The situation is identical to that of building the roads themselves. The framework for justifying the building of roads may therefore be used to justify implementing traffic control facilities. The classes of benefit are discussed in this light:

- Journey time savings form the largest part of the benefits. For particular traffic situations, the time saved benefits drivers and passengers. The rates allocated must take into account the difference in the value of time when people are on business or on leisure activities. In some cases, different rates apply in different areas. Journey time savings are also a component of the operating costs of vehicles, particularly commercial vehicles which are used for business a large part of the time.
- Vehicle stops are seen as a problem by motorists. Decision makers are favorably influenced by traffic control systems that reduce the number of stops to perceived important movements. This has caused the US preoccupation with bandwidth optimization. Stops are also thought to correlate with accidents; roads where vehicles need not stop, such as freeways, have much lower accident rates than roads on which they must stop.
- Fuel consumption roughly correlates with the previous factors, although several models have been developed to refine the relationship. Current versions of TRANSYT include fuel consumption in their performance index, so that the signal timings chosen will reduce the amount of braking and reacceleration required. Normally, fuel savings are valued at lower levels than journey time, but it was nevertheless sufficient on its own to justify the retiming of all US signals. Benefits attributed to fuel savings have the advantage that they represent a reduction of use of real resources, and are not dependent on estimates of the value of people's time.
- Pollution is reduced if fuel is saved. Noise pollution is reduced by reducing the number of vehicle stops and reacceleration. However, there is a conflict of interest which has not been properly addressed; good UTC may attract traffic through towns instead of encouraging the use of bypass routes. In this case, good traffic control may have an adverse environmental effect.

7.7.2 The Magnitude of Benefits

Generally, the potential benefits from a traffic control system are very large compared with its cost. Not uncommonly, the annual benefits exceed the cost of the system (i.e., the net rate of return exceeds 100%, or the benefits pay for the system in less than one year). The main difficulty is in being sure that the benefits materialize.

By way of example, a recent evaluation of connection of 1370 signals in New York to a controlled system [Kell, 1987] found:

System cost	\$16,000,000 capital
Total annual cost	\$3,750,000 annually
Benefits from stops and delays	\$25,000,000 annually
Fuel consumption reduction	3,500,000 gals annually
Vehicle emissions reduction	2,500,000 lbs annually

A coordination program, such as TRANSYT, may produce journey time savings on the order of 16% [Holroyd and Hillier, 1969]. However, the magnitude of the benefits depends on the nature of the network [Robertson and Hunt, 1982]. In particular, benefits are greater if traffic flows are greater, if traffic flows smoothly between intersections, and if intersections are closely spaced. Thus, one expression for the benefit of coordination of a link to an intersection is

$$B = 11fv/(1 + t)$$

where

B is the annual benefit in monetary units,
f is the flow in vehicles per hour,
v is the value of time in monetary units per vehicle hour, and
t is the travel time in minutes along the link.

An allowance for increasing the cycle time of most junctions in the network to achieve the common cycle time necessary for coordination must be deducted from this figure.

Additional benefits of traffic adjusted coordination plans such as SCOOT derive from:

- The aging of fixed plans, considered to reduce journey time advantages by 4% per year;
- The difficulty of recruiting and retaining sufficient skilled staff to update the plans;
- Better coordination by adapting to traffic variations. The demonstration projects suggest that this amounts to a 12% reduction in delays and traffic jams;
- Providing an advantage in terms of the selected cycle time because optimization keeps cycle time lower than it would otherwise have to be set, so that even uncoordinated approaches receive some benefits.

SCOOT coordination therefore may be taken to have overall benefits that are 25% greater than fixed coordination plans. Much of this advantage remains if the plans are used in some traffic-based selection program such as 1GC or 1.5GC. While

SCOOT costs more than some locally linked versions of fixed coordination plans, it can readily be supported at no greater infrastructure cost than many current UTC systems (such as Series 2000).

7.8 RELATIONSHIP TO OTHER MOBILE INFORMATION SYSTEMS

7.8.1 Available Information Channels

Traffic signal systems, and to a lesser extent traffic signs, provide a means of conveying information to drivers and influencing their behavior in real time. Traffic signals in particular rely on a very simple message—"stop"—, which can be learned until it becomes instinctive. This provides a response time and certainty that is difficult to match with other means of communication that also depend on the driver as a link in the system. More complex messages require more elaborate signals or "variable message signs," which are becoming more common.

More complex signing systems may require the driver to assimilate and accumulate information from several sources, including previous knowledge. Thus, to benefit from a signing system advising on the occupancy of various car parks in a town, drivers have to know whether those car parks are suitable for their destination, duration of stay, *et cetera*. Moreover, they must process this information in a short time, on the move, in safety-critical circumstances. Emerging information systems seek to provide the necessary processing within the vehicle, thus in due course simplifying the driver's task to the original responses of stopping, going and turning as directly instructed.

The UTC example indicates how the provision of additional information at the right time and place can confer relatively large benefits on the community. The example also indicates the difficulty of arranging that the systems required to provide the information are paid for by the beneficiaries.

UTC systems demonstrate the possibilities, difficulties, and in general the type of techniques required to provide widely dispersed equipment in the street for interaction with members of the public. UTC systems also show that some of the greatest advantages are obtained by simplifying the problem and providing only the information required. In many cases, this simplification is achieved by processing the information so that only precalculated plans are available to the equipment which displays the signals to the motorist, thereby avoiding the complexity of providing fully functioning on-line equipment to deal with all eventualities. The plans are also regularly tested by use, and any imperfection is eliminated.

However, the complexity of UTC systems is increasing, and systems such as SCOOT have demonstrated the additional advantages obtained by retaining the information and processing it on-line. This offers greater scope for integration with other types of mobile information systems. The next research phase of SCOOT is intended to identify means of passing information from the system to other users. This should benefit those using automatic navigation systems (see Chapter 6), and should provide messages for broadcasting via traffic information systems (e.g., through RDS; see Chapter 5).

7.8.2 Driver Reaction to Information

We already know that radio announcements of incidents on clearly defined roads with a readily available diversion route are effective in transferring traffic promptly between those routes. Improved techniques such as cellular radio and RDS (see Chapter 5) will increase the proportion of drivers responding. In some cases, this will mean that too many drivers respond too fast, creating worse problems on the alternative routes; those problems may persist after the original difficulty has been cleared. If there is no good alternative route, backed up traffic should be kept on its original road even though many drivers would prefer to try their luck elsewhere. The people living and working in the vicinity then suffer.

The police already restrict the release of information to radio outlets to account for these issues. Regrettably, however, other information sources, such as helicopters reporting to radio programs, are not so disciplined; there is pressure for their accounts to be newsworthy, so they may be embellished and may include the pilot's view of what he or she ought to be seeing at places not directly in view.

Even supplying good information may be counterproductive. A good in-vehicle navigation system will enable drivers to navigate easily and successfully on minor roads, which they would otherwise have avoided. In North America, in places where the street system has spare capacity for through traffic, this could increase network capacity, but at the expense of people living and working near those streets. In other situations, the effect may well be seriously disruptive, and the authorities are likely to react by applying traffic management measures to restrict such through traffic.

The current enthusiasm of those promoting improved information systems will predictably elicit an adverse reaction from those who see their environment threatened; this reaction will include the imposition of restrictive regulations. Such a situation has existed for many years in the field of road pricing; transportation planners readily demonstrate that some form of traffic restraint has overall benefits, and are assisted by the increasing availability of technology such as the road pricing system demonstrated in Hong Kong [Dawson and Brown, 1985]. Each new demonstration serves to confirm that charging for the use of existing roads is "politically" unacceptable (i.e., that the public will not stand for it).

7.9 FUTURE APPLICATIONS

Advanced street signaling systems are expected to spread to most signaled intersections. For those that are not closely spaced, this may be in the form only of remote monitoring of faults. However, this will provide a widespread data network which may be of use for other traffic-related purposes. Those providing coordination, however, will be expected to cover only a proportion of total signaled intersections (i.e., up to 40%). These will be distributed evenly between the addition of more intersections to existing systems, and the implementation of new systems in towns and cities only now reaching the level of urbanization and traffic flow necessary to take advantage of them.

At present, there are probably 800,000 traffic signaled intersections worldwide, half in North America and 15% in Japan. The increased facilities available, and the way in which these bring advantages without suffering associated disadvantages, enable the number of signaled intersections to increase without any sign of saturation. Demands for more facilities from each of them means the real cost of each installation tends to remain constant or rises, despite the dramatically better performance available from newer products.

The benefits from better control systems are so large as compared to the costs of alternatives (such as the construction of new or improved roads), that their use will continue to expand. In many cities, sophisticated systems such as SCOOT, originally associated with substantial signalized networks, are proving useful at small groups of closely spaced critical intersections; often, as few as three critical intersections are connected.

As mobile information systems increasingly provide data processing facilities within the vehicle, signaling systems may become more sophisticated. There are suggestions that the signals may be located within the vehicle, as they are increasingly on high-speed trains. However, the private ownership of vehicles and the long delays required for innovations to become universal within the vehicle population will severely inhibit such developments.

However, signaling systems are rapidly making greater use of information fed to them, including information on vehicle position and movement. Such data collection from the traffic stream can be expected to increase. In this case, to achieve useful benefits frequently is possible when only a proportion of vehicles is fitted. For example, an in-car navigation system such as AUTOGUIDE (see Chapters 6 and 9) receives a report of the local traffic network's performance every time an equipped vehicle passes a beacon. This information can be used to assess speeds in that part of the network, and to reassess traffic factors such as average flow and the proportion of turning traffic. We expect that SCOOT and similar systems will be developed to make use of this extra information and achieve better optimized signal settings.

REFERENCES

Anon., "Signal for regulating street traffic," *Engineering,* 18 December, 1868, p. 537.

Bodell, G., and K.W. Huddart, "Tram priority in Hong Kong's first Light Rail Transit system—Part 1," *Traffic Engineering and Control,* 28, 9, September 1987, pp. 446–451, 470.

Cass, S., and L. Casciato, "Centralized traffic signal control by general purpose computer," *Proceedings of 30th Annual Meeting of the Institution of Traffic Engineers,* 1960, p. 203.

Chandler, M.J.H., "Remote monitoring, SCOOT experience and bus priority in London," *Institution of Electrical Engineers Colloquium on road traffic signaling,* London, May 1984.

Cornwell, P.R., J.Y.K. Luk, and B.J. Negus, "Tram priority in SCATS," *Traffic Engineering and Control,* 27, 11, November 1986, pp. 561–565.

Dawson, J.A.L., and F.N. Brown, "Electronic road pricing in Hong Kong 1. A fair way to go?" *Traffic Engineering and Control,* 26, 11, November 1985 (continued in three subsequent issues), pp. 522–529.

Department of Transport, "Microprocessor-based traffic signal controller for isolated, linked and Urban Traffic Control installations," specification MCE 0141, London, 1984.

FHWA, "Evaluation of control strategies, final report," Federal Highway Administration, Washington, D.C., August 1976.

Gardner, C., "Current status of operating traffic control systems," *Engineering Foundation conference on management and control of urban traffic,* Henniker, New Hampshire, June 1987.

Gartner, N., "Modifying MAXBAND by disaggregate link volume weights," *Engineering Foundation conference on management and control of urban traffic,* Henniker, New Hampshire, June 1987.

Griffin, R.M., and D. Johnson, "Northampton fire priority demonstration scheme—the "before" study and EVADE," *Traffic Engineering and Control,* 21, 4, April 1980, pp. 182–185.

Holroyd, J., and J.A. Hillier, "Area traffic control in Glasgow. A summary of results from four control schemes," *Traffic Engineering and Control,* 11, 5, p. 220, September 1969.

Hunt, P.B., D.I. Robertson, R.D. Bretherton and R.I. Wilton, "SCOOT—a traffic responsive method of co-ordinating traffic signals," Laboratory Report LR 1014, Transport and Road Research Laboratory, Crowthorne, Berks, England, 1981.

Kell, J.H., "Topics III, VTCS expansion, summary of evaluation," James H. Kell and Associates, New York, January 1987.

Klijnhout, J.J., "Motorway control and signaling—operational experience in the Netherlands," *International conference on road traffic control,* The Institution of Electrical Engineers, London, April 1986, pp. 31–34.

LACTC, "On the way to the year 2000, Highway plan for Los Angeles County," Los Angeles County Transportation Commission, August 1987.

Lighthill, M.J., and F.B. Whitham, "On kinematic waves, II. A theory of traffic flow on long crowded roads," *Proc. Roy. Soc.* (Series A), 229, 1955, pp. 317–345.

Lowrie, P.R., "The Sydney Co-ordinated Adaptive Traffic System—principles, methodology, algorithms," *Proc. I.E.E. International conference on road traffic signaling,* London, 1982.

Luk, J.Y.K., A.G. Sims and P.R. Lowrie, "The Parramatta experiment—evaluation of four methods of area traffic control," A.R.R.B. Res. Report No. 132, 1983.

Middleham, F., and Van Zuylen, "Inca and Flexsyt: computer-aided design, evaluation and testing of control schemes," *2nd international conference on Road Traffic Control, The Institution of Electrical Engineers,* pp. 127–132. London, April 1986.

Nakahara, T., N. Yumoto, and A. Tanaka, "Multi-criterion area traffic control system with feedback features," *IFAC/IFIP 1st International symposium on traffic control,* Versailles, Paper 3.9, June 1970.

NEMA TS1-1983 (and updates), "Traffic control systems," National Electrical Manufacturers Association, Washington, DC.

Oastler, K.H.S., "Maintenance of traffic signals in London," *Traffic Engineering and Control,* 26, 3, March 1985, pp. 104–108.

Robertson, D.I., and P.B. Hunt, "A method of estimating the benefit of coordinating signals by TRANSYT and SCOOT," *Traffic Engineering and Control,* 23, 11, November 1982, pp. 527–531.

Robertson, D.I., and R.A. Vincent, "Bus priority in a network of fixed-time signals," Laboratory Report LR666, Transport and Road Research Laboratory, Crowthorne, England, 1975.

Scott Maccalden, Jr, M., "A traffic management system for the San Francisco-Oakland Bay Bridge," *ITE Journal,* May 1984, pp. 46–50.

Takahashi, H., K. Nakamuro, and M. Okabe, "A highly reliable and compact traffic signal controller series," *Sumitomo Electric Technical Review,* 17, October 1977, pp. 59–71.

Webster, F.V., "Traffic signal settings," Road Research Technical Paper No. 39, H.M. Stationery Office, London, 1958.

Williams, D.A.B., "Area Traffic Control in West London—Assessment of first experiment," *Traffic Engineering and Control,* 11, 1969, p. 125.

Yamamoto, T., T. Okamoto, K. Iwane, K. Sawai, S. Ueda, and T. Kitagawa, "Advanced arterial street control system," *Sumitomo Electric Technical Review,* 23, January 1984, pp. 125–135.

Yamamoto, T., S. Yamaoka, Y. Eikawa, M. Doi and T. Higashi, "Advanced local traffic signal controller for urban traffic control systems," *Sumitomo Electric Technical Review,* 27, January 1988, pp. 97–103.

GENERAL READING

Institute of Transportation Engineers, *Transportation and Traffic Engineering Handbook,* 2nd Ed., W.S. Homberger, ed., Prentice-Hall, Englewood Cliffs, NJ, 1982.

Institution of Highways and Transportation, "Roads and traffic in urban areas" (particularly Chapters 20, 24, 26, 41 and 43), Her Majesty's Stationery Office, London, 1987.

APPENDIX 7.A
GLOSSARY

If a term is not defined as being US or UK, the US definition applies in this glossary.

Crosswalk (US) or *pedestrian crossing* (UK): Part of pavement (carriageway) marked to provide for pedestrian movement across the highway. The terms are not synonymous, because the US usage applies at intersections, and the UK usage applies away from intersections (*midblock* (US)).

Expressway (US) (no real UK equivalent): A road, mainly for motor vehicles, with one axis on which all crossing and off-side turning traffic conflicts are spatially separated or controlled by traffic signals.

Freeway, parkway, highway, turnpike (US) or *motorway* (UK): A road provided especially for motor vehicles from which pedestrians and nonmotorized traffic are banned, and on which conflicting traffic movements are separated spatially.

Far side traffic signals: Traffic signals installed on the far side (in the direction of travel) of the intersection from that being approached by the motorist.

Interchange: The system of ramps (slip roads) used to connect two or more freeways (motorways) while maintaining spatial separation of conflicting traffic streams. May also apply to facilities to interchange between different transport modes, such as car, bus and rail.

Intersection (US) or *junction* (UK): An area of road in which traffic streams diverge, merge or cross one another.

Near side: Toward the curb (right (US) or left (UK)) of the vehicle or traffic stream. May also be used to distinguish from far side.

Off side: Away from the curb (left (US) or right (UK)) of the vehicle or traffic stream.

On-ramp, off-ramp (US) or *entry, exit slip road* (UK): Roads connected to a freeway (motorway) which enable traffic to join or leave without significantly interfering with through traffic.

Pavement (US) or *carriageway* (UK): Part of highway constructed for vehicular traffic.

Sidewalk (US) or *pavement* (UK) = *footway* (both): Part of highway provided for pedestrian movement along the highway.

Traffic actuated (US) or *vehicle actuated* (UK): Where traffic signals respond to individual vehicles in the traffic stream.

Traffic responsive (US) or *traffic actuated* (UK): Where a traffic control system responds to current traffic levels. Because of the potential confusion with these current technology systems, the US usage is preferred.

Transit (US) or *bus or rail passenger transport* (UK).

Chapter 8
CARIN—A Car Information and Navigation System

M.L.G. THOONE and J. WALKER
PHILIPS CONSUMER ELECTRONICS — RACAL RESEARCH LTD.

8.1 INTRODUCTION

In a car, the passenger sitting next to the driver usually must navigate: plan the route, read the map, relate actual position to the map, and give directions to the driver. The passenger performs these tasks with varying degrees of success, and any inadequacies of performance can lead to disharmony. Perhaps more serious is that mistakes in planning and following a route can mean that greater distances than necessary are traveled. For professional road users, such as police, firefighters, delivery and freight services, ambulances, and taxis, economic route planning and good vehicle guidance can yield substantial savings (see Chapter 6). This is certainly so if the guidance helps drivers avoid traffic delays by an early change of route. All this can become reality if the tasks of the navigator are assumed by an electronic computer-controlled system. Electronic navigation and information systems are well established in marine and aviation applications. Car and electronics manufacturers are now trying to develop systems that are suitable for automotive applications.

Almost all of the car navigation and information systems now in use have the disadvantage that they lack a suitable medium for storing topographical information. Some of the systems also require substantial investments in "infrastructure," such as beacons for radio rangefinding. Neither are most of the systems very user-friendly, because the driver must enter the starting position and desired destination as a map grid reference [Zimdahl, 1984; Friedl, 1975; Jeffery, 1985].

A car navigation and information system without these disadvantages is being developed by Philips [Thoone and Breukers, 1984]. The system, called CARIN

Figure 8.1 The interior of a car fitted with the CARIN vehicular guidance system.

(*Car Information and Navigation System*), plans the best route, guides the driver with the aid of a speech synthesizer, periodically determines the position of the vehicle, selects an alternative route if there are coded digital radio signals reporting traffic obstructions, and even gives tourist information (see Fig. 8.1). The special feature of CARIN is that it uses a highly efficient medium for storing digital data, the *compact disc*.

The main benefits of the CARIN system are efficient route planning and a contribution to traffic safety. Research has shown [Jeffery, 1981] that drivers in big cities can plan their routes approximately 20% more efficiently with an effective route guidance system. In the UK, automatic route guidance systems could save $900 million annually; an effective traffic information broadcasting system could save another $500 million per annum. Furthermore, traffic safety could be improved considerably if drivers were not distracted by trying to read maps; also, the fewer miles a driver travels, the less likely he or she is to have an accident.

This chapter gives an outline description of the CARIN system. Some aspects of the storage of digital map databases on compact disc are subsequently discussed, followed by an explanation of the vehicle-locating system. Of course, much attention must be paid to the design of the human interface, which involves ergonomics, traffic safety, perception, and styling. Because consumer-electronics products must adhere to standards, the CD-I concept will be described separately, as will the control software aspects of the CARIN system and the use of the radio data system.

8.2 THE COMPONENTS OF A CARIN SYSTEM

Figure 8.2 shows a block diagram of a CARIN system; it allows various configurations, so that a broad range of products can be defined with different displays,

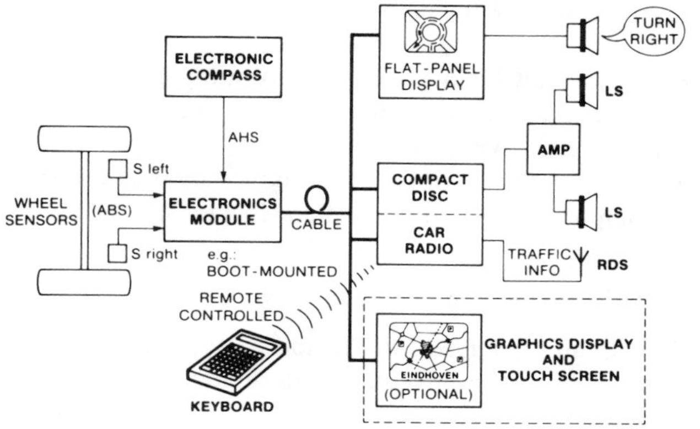

Figure 8.2 Block diagram of a CARIN-II prototype system.

input devices, and extensions. The basic configuration comprises the following parts:

1. An *automotive compact disc player,* which can play back digital audio discs and read compact discs with computer data. A *compact disc interactive* (CD-I) can store up to 600 Mbytes of computer data, which allows the storage of 150,000 A4-sized (equivalently, US legal-sized) pages of typewritten text. In CARIN, CD-I discs are mainly used for the storage of digitized road maps, traffic and tourist information, telephone listings, *et cetera.* A single disc can contain the detailed road map of a typical medium-sized country such as France, the Federal Republic of Germany, or the UK. (See [Thoone and Breukers, 1984] and [Benning, 1986].)
2. An *electronics module* includes the on-board computer, which can read the relevant parts of the map database from the CD-I disc and calculate an efficient route to the required destination. Furthermore, the module can calculate the current position of the vehicle by using the navigation sensors, and determine the necessary instructions for route guidance.
3. A *simple hand-held keyboard* for entering the required destination via common identifications (e.g., street name) and for issuing commands to the system; the keyboard comprises the full alphabet, and integers 0 to 9, and several function buttons.
4. A *speech synthesizer unit* for giving the driver directions.
5. A *small, flat-panel display* (e.g., dot-matrix LCD) can be used for visual support if the speech-synthesized instruction is not well understood. (For traffic safety reasons, however, preference has been given to spoken guidance instructions.) The display shows a simplified map of the intersection ahead as well as the recommended route. The display is preferably mounted in the center of the instrument panel, so that the picture shown can be interpreted at a glance.

The display is also used for communication with the driver while the vehicle is stationary.

6. An *electronic compass* for measuring the heading of the vehicle relative to the terrestrial magnetic field, and odometers (e.g., the wheel-sensors used for *anti-skid braking systems* (ABS) for determining the distance covered by the automobile). The vehicle's position is determined by the onboard computer's processing the outputs of the sensors. (See Chapter 10 for more information on vehicle sensors.)

The basic configuration can be extended with several units. For instance:

1. A car radio with an *RDS decoder*. The radio data system (see Chapter 5) allows the transmission of traffic information in digital form via existing FM broadcasting transmitters without interference to or interruption of the normal radio program.
2. A *graphics display* for the visualization of outline maps, the planned route, tourist information, *et cetera*. Because of road safety, this display can be used only while the car is stationary.
3. An interface with a *dedicated engine management* or body computer for implementing vehicle status monitoring functions (i.e., oil pressure, diagnostics, servicing *et cetera*. (See Chapter 10 for more details of such status monitoring and diagnostics in the "intelligent car.") The voice synthesizer and the displays can be used for a multitude of functions.
4. A standardized computer interface for future extensions. Mobile radio equipment (see Chapters 2 and 3) can be connected to the system so that a communications link can be established with the dispatcher of a fleet of vehicles. This allows *Automatic Vehicle Location* (AVL) and fleet management.

In addition to these physical components, several software modules (applications programs, firmware) are required for operating the system. These include navigation algorithms, programs for automatic route planning, route guidance, communication with the user, information retrieval programs, and control programs.

8.3 HOW THE CARIN SYSTEM IS USED

In a typical scenario, the CD-I disc with the map database of the geographic area concerned is put in the car's CD player. The required destination is entered via the keyboard, using common identifications such as name of town, street, and possibly address. Then, the best route is planned and stored in the working memory of the computer. The driver now begins the journey and replaces the CD-I disc with a music disc. During the drive, the music signal is interrupted by the speech synthesizer each time a direction is given.

The driver can only consult the map data on the touch screen when the car is stationary. The keyboard is used for entering the starting position and destination, and any special requirements. The driver does not have to enter a map reference, but can simply specify the names of countries, towns and streets, because lists of such names are stored with their map references on the compact disc. The computer loads the required application programs from the disc, calculates an efficient route from the starting position to the destination, and ensures that the information required for following this route is read from the disc and stored in the main memory. Then the player can be used for playing music again. During the journey, the driver is guided by spoken directions from the speech synthesizer, with directions such as "take first left turn," "bear right," "take first exit at rotary," and "we have arrived." The information on the small flat display gives extra support. In CARIN, the emphasis is on speech guidance for the driver, for reasons of traffic safety.

The principle of the vehicle-locating procedure is as follows: sensors that measure the number of revolutions of the nondriven wheels provide information about the distance traveled and changes in the direction of the vehicle. An electronic compass, which measures the direction of the vehicle in relation to the Earth's magnetic field, gives information about the heading. The computer uses both kinds of information to calculate the location of the car. This is compared with the topographical information in the main memory. Because the car should be on a road whose data is stored in the memory, this comparison determines the part of the road where the car is located. The positioning error with this method is about 20 m.

The map database of the CD-I disc is a static set of data, representing the road network as it exists when the digital map was edited. Traffic situations, however, are not static. They change continuously: for the worse if an accident blocks a highway, for the better if a new section of motorway opens. Updating CARIN with this dynamic traffic information would greatly improve its usefulness.

The EBU has recently reached agreement on RDS, a system that can be used to transmit digital data at a rate of 1200 b/s superimposed on the FM broadcast signal. RDS is currently being introduced in many Western European countries. In future versions of CARIN, if the system receives an RDS message about traffic delays or road obstructions, it will calculate the time to reach the affected position. If the received message says that the obstruction will have disappeared within that time, the route will stay unchanged. If the obstruction remains, CARIN will then calculate a route that bypasses the obstacle and brings the car to its destination in the shortest possible time.

8.4 COMPACT DISC INTERACTIVE (CD-I) FOR CONSUMER ELECTRONICS PRODUCTS

CARIN is a product range aimed at the consumer electronics market on a worldwide scale. This imposes at least two requirements:

- the product must be reasonably priced,
- the system must be based on international standards.

Two standards relating to compact discs have currently been set by Philips and Sony: *compact disc digital audio* (CD-DA), announced publicly in 1982; and *compact disc read-only memory* (CD-ROM), mainly applied as a professional computer peripheral. CARIN systems are based on a third standard, the single-medium concept CD-I for the storage of information on the compact disc.

In May 1986, the draft CD-I specification was issued, and the final full functional specification was released to the licensees in 1989. The specification is for a multi-media, interactive information system, which combines audio, still video pictures, and text with powerful data processing facilities. CD-I thus allows the realization of a broad variety of interactive applications.

A CD-I disc can combine the following types of information on a single medium:

1. *Audio data*. The CD-I player can reproduce existing CD-DA discs. Furthermore, within the CD-I standard, there are four quality levels of audio: CD-DA, LP record, FM radio, and AM radio. The possibility of computer-generated phonetic speech is also foreseen.
2. *Video data*. Three quality levels of resolution are defined: first, the best resolution achievable for pictures on present-day television receivers (normal resolution—384 lines of 280 pixels); second, the best achievable for characters displayed on the same receivers or color monitors (double resolution); third, the best achievable with the upcoming enhanced or digital television receivers (HDTV) as well as high-resolution color monitors. Due to the rate at which data can be read from the disc, restoration of a full-screen, high-resolution picture takes 0.6 s. Recently, the CD-I standard has been extended with full-motion video, thanks to the progress in video compression techniques.
3. *Ordinary data*. Apart from the audio and video, there are three types of CARIN data which can be put on a CD-I disc:
   - The executable object code (68000 processor); this is the application software for CARIN (e.g., navigation, route planning).
   - System text; this is text required by the system to transmit messages to the user.
   - Data to be processed by the application—the map database and related information.

A CD-I disc containing digital map databases and related application programs can be employed for various functions, both in the car and at home (see Table 8.1).

The success of the CD-DA guarantees the availability of compact disc drives at low prices. Moreover, from the general acceptance of CD-DA as a world standard, a similarly favorable reception for CD-I can be expected.

Table 8.1
Availability of Some Map-Related Functions on Home and Car CD-I Systems

APPLICATION SW PROGRAMS	SYSTEM CONFIGURATION	
	IN CAR CD Drive Automotive CD-I Computer Simple Display Audio Output Nav. Sensors	AT HOME CD Drive CD-I Computer Video Display or TV Audio Output
Navigation	▨	
Route Planning	▨	▨
Driver Guidance	▨	
Show Map		▨
Interactive School Atlas		▨
Interactive Travel Guide		▨
Tourist Information	▨	▨

8.5 MAP DATABASES ON COMPACT DISC

An indispensible part of the CARIN system is a reliable and accurate electronic road map stored on a compact disc. Figure 8.3 depicts the steps in producing such CD-I discs.

Due to the time-consuming nature of data collection, specialized companies need to become involved in supplying data files. Data must be combined to produce a large, consistent, extendable, and widely applicable database. Data that have the necessary accuracy can be collected in phases: a basic set of data can be specified; later, new information can be added as it becomes available and commercially interesting. To ensure that the incoming data are complete, current, accurate, and consistent, various checks must be performed [Benning, 1986].

CARIN is one application that may extract this information, appropriately converting it. CARIN also requires special demands on the data structures. Conventional database designs are not suitable for data storage on compact disc [Cooke, 1986; Lambert and Ropiequet, 1986]. The CARIN database may be combined with other digital data, such as application programs, still pictures, and audio tracks, and subsequently formatted according to the CD-I standard [Philips, 1986]. The final step is the conventional compact disc mastering and replication process.

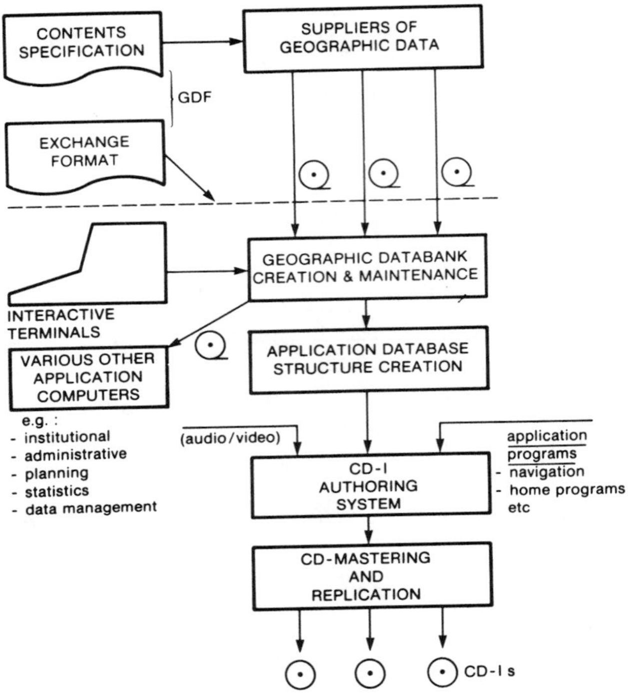

Figure 8.3 Flowchart of processes related to the production of CD-I discs with map databases.

For car navigation systems, the real road network is most suitably digitized by a *vector method,* rather than a *raster scanning method* [Thoone and Breukers, 1984; Benning, 1986; Burrough, 1986; Broome, 1984]. Each road or lane intersection is represented by a point within a particular coordinate system. Each road segment between two points is represented by a sequence of straight-line segments chosen to approximate the real curvature of the road, with endpoints referenced to the same coordinate system. Thus, road maps are modeled by points (nodes and intermediates), line segments (vectors), and polygons, as shown in Figure 8.4.

8.5.1 Digitizing the Map Data

In modern cartography, some of the topographical information is now processed digitally. Two methods are used, the raster scanning method and the vector method. In the former, a map is divided into picture elements (pixels) (i.e., 0.1 mm × 0.1 mm). The color of each pixel is represented by a digital code. To digitize the

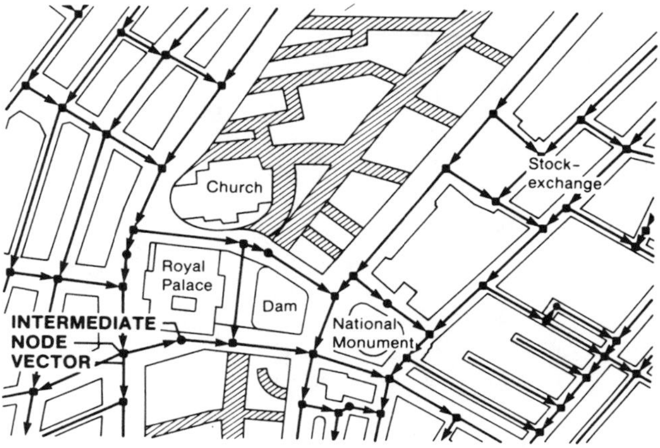

Figure 8.4 Vectorial representation of a road map.

map of Amsterdam (covering 14 km × 12 km), for example, on a scale of 1/15,000, a storage capacity of 375 Mb is required for the raster method [Thoone and Breukers, 1984]. In the vector method, as we use it, the axes of the roads are approximated by straight-line segments, which each represent a "vector." This method requires far less storage capacity. Digitized this way, the map of Amsterdam occupies 1 Mb. If other information is added to the purely topographical data, such as the names of streets and general information about filling stations and restaurants, another 1 Mb of memory is required. As we noted earlier, the compact disc has a capacity of 4800 Mb, so more than 1000 maps like that of Amsterdam can be stored on one disc.

Figure 8.5 illustrates the digitization procedure we use. The road network is translated into a graphic structure of points connected by straight-line segments. The position of each point is given as the coordinates of a map reference (i.e., as x,y coordinates in a rectangular system). (In a second-generation system, these could be geodetic coordinates.) There are two kinds of points, *nodes* and *intermediate points*. A node can be:

- a point where at least three segments meet (point 1),
- an intersection with the edge of the map or some other artificial boundary (point 2),
- an intersection with an administrative boundary (point 3),
- the end of a cul-de-sac (point 4),
- a point where one or more items of data change for a street (i.e., its name or the type of road surface; point 5).

All other points are called intermediate points. They are found at a bend in a road or are used to approximate the curvature of a road by a number of segments,

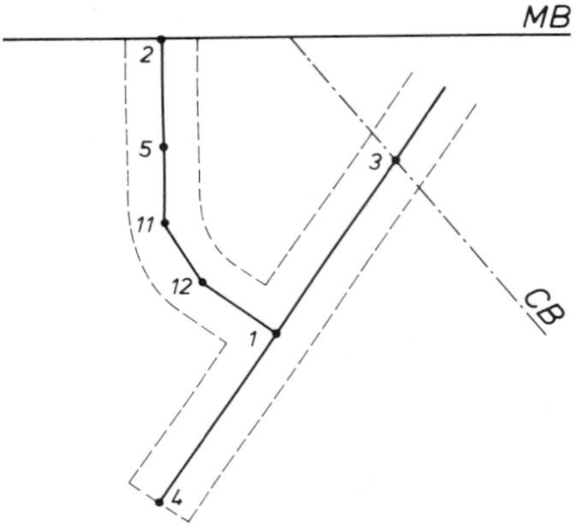

Figure 8.5 Vector method for digitizing information from the map. The axes of roads are approximated by line segments that connect points, the position of which is fixed by the coordinates of map references. 1 to 5 are nodes. At node 5, the condition of the road surface changes. 11 and 12 are intermediate points. *MB*-map boundary. *CB*-Local government boundary. The segment 1–4 is a cul-de-sac. Points 1, 12, 11 and 5 form a "chain."

like the points 11 and 12 in Figure 8.5. The succession of points in a sequence that starts and ends at a node is called a *chain*. In the figure, points 1, 12, 11, and 5 form a chain, as do 1 and 4, 1 and 3, and 2 and 5. The graphic structure formed by the various chains can be regarded as the *skeleton* of the digitized map. *Attributes* can then be added to the skeleton, such as:

- street names;
- classes of road—motorways, trunk roads, secondary roads, and limited-access roads;
- directions of one-way streets.

8.5.2 The Compact Disc as a Carrier of Digital Map Data

As mentioned earlier, we decided to use CD-I as the storage medium for the CARIN system. An international standard is being prepared for CD-I. The use of a standardized method of storage for the information has the advantage that the data carriers for different navigation systems are interchangeable, and savings can be made in research and development costs. The CD-I storage medium was developed for consumer-electronics applications by the licensors of the CD-DA system. A CD-I can store audio and video signals of different quality levels, combined with computer data. The proposed CD-I standard expresses detailed agreements relating to the storage of computer data, coding, and error correction systems; the operating system;

and the connection of peripherals. The standard thus allows for discs and systems from different manufacturers to be compatible.

In addition to digital map data, which will occupy most of the disc, a CD-I for geographical applications will soon contain applications software. Examples are parts of the navigation software, the route-planning program, and the driver-guidance program. Programs also can be included that will make the CD-I with its player in the home a kind of interactive atlas. The disc can also be used for teaching geography, if the topographical information is augmented by photographs, drawings, sound, and (with a few restrictions) moving video pictures. Possibly, depending on the options available with the hardware, we will soon be able to plan a route at home, study it on a video display, and make a printed copy.

The development of a geographical CD-I makes use of topographical databases. The rules to be observed in such a database also are established. The database will be regularly updated by publishers of electronic geographical data, and it can be used as well for other applications (e.g., for town and regional planning, keeping records of traffic accidents, and processing census data). The part of the database intended for use by CARIN is given the correct structure by a conversion program. The file thus obtained is combined with application software and possibly with other data. The result is then formatted to the CD-I standard. In simple terms, this means that the data are divided into blocks of 2 kbytes (16,384 bits), each with a *header,* synchronization bits, and bits for later error correction. After coding and modulation (*eight-to-fourteen modulation,* EFM), the data are in the correct form for writing to the disc [Heemskerk and Schouhamer Immink, 1982]. The blocks are converted into sectors, and the address is included in the header.

The result of these operations is that the successive sectors form a sequence of *channel bits*. Each 1 in the sequence of channel bits is a land-pit or pit-land transition in the spiral track of pits on the final disc [Carasso, Peek, and Sinjou, 1982]. The pattern of pits and lands is transferred to a photosensitive layer on a rotating disc with the aid of a laser. Development of this layer produces a *master,* and the molds for manufacturing the discs are made from the master in a number of intermediate steps.

As noted earlier, the storage capacity of a compact disc is 4800 Mbit. The main memory of the computer, however, has a capacity of only about 8 Mbit. While this is still a very respectable capacity for a semiconductor memory, the capacity is small compared with that of a compact disc. The time required for reading all the information from a disc is about an hour. The player used in the CARIN system is an ordinary compact disc player designed for music playback in cars. To move the optical pick-up in the player from the smallest to the largest radius of the useful area on the disc takes one or two seconds, although the procedure will take less time in the later generations of players. To determine beforehand exactly where the pick-up will arrive after a movement is not possible. Because one cannot read backward, when the pick-up is to be moved to a particular address on the disc, the reading of data must start well before the address is reached. This means that extra information must be read before the correct address is found, and this takes time. (Of course, this extra information is not transferred to the main memory.)

Because the access time is not negligible, the data required for navigation cannot be read from the compact disc at the exact moment that is required. At the start of a journey, therefore, all the topographical information required for navigation is (as far as possible) retrieved, read, and stored in the main memory. Because this must be done fairly quickly, to store the data on the compact disc in the most advantageous way obviously makes sense.

The topographical information is divided into *parcels*. The information in each parcel is stored as a *block* of data on the compact disc, and is distributed over an integer number of successive sectors. Also stored on the compact disc is a list of addresses, or a *directory*. This includes the address of the first sector where a block is stored, the number of sectors used for the block, and the coordinates of the lines that define the corresponding parcel on the map. A request for map data from the computer generally means that the data for a rectangular area are called. The directory is then consulted to find which blocks contain information about the requested area. The information can hence be retrieved from their addresses.

The time required for these operations can be short if two requirements are satisfied. The map must be divided into convenient parcels, and the corresponding blocks ought to be stored on the compact disc in a sequence that does not require too much movement of the pick-up. The first requirement means that all the blocks should contain about the same amount of data, which must not be so large that the blocks requested for a particular area take too much space in the main memory. Neither should there be so many blocks that too much space would be required for the directory or that too many movements of the pick-up would be necessary. The second requirement means that adjacent parcels on the map should correspond to blocks close to each other on the disc. For efficient management of the information read from the compact disc, it is important that the information be efficiently structured inside each block. At the start of a journey (and perhaps during the journey as well), some of the map data on the compact disc must be read and stored in the main memory. Therefore, the data must be partitioned and arranged on the compact disc so as to give short access times. The CARIN project group has made extensive studies of the problem of the optimum method of data storage [Thoone, 1987], because agreement on proper standards is of considerable importance to all parties involved: set-maker, CD-publisher, and end-user.

8.6 VEHICLE LOCATION

First-generation CARIN systems use dead reckoning, augmented by correlation with a digital road map. A compass based on a *magnetometer,* which measures the direction of the vehicle with respect to the earth's magnetic field, determines the heading of the vehicle. The distance it covers is determined by the odometers measuring wheel revolution. The odometers of the anti-lock braking system (ABS) can be utilized.

The difference of the outputs of the two complementary odometers (e.g., left and right wheel sensor on the nondriven wheels) give additional information on the *heading* (rate of change). The magnetometer reading is affected by disturbances of the terrestrial magnetic field (e.g., caused by iron objects), whereas the differential odometer signal contains "drift" components, due to varying tire pressure and wear. By combining both heading signals through an optimal estimation algorithm, which is implemented by software in the on-board computer, a more reliable measure of heading is obtained.

Figure 8.6 shows a functional block diagram of the navigational part of CARIN. On the basis of the optimal estimation of heading and the measured distance, the position of the vehicle is calculated by dead reckoning, but an increasing position error appears, due to cumulative measuring errors. To keep this error within acceptable bounds, the dead-reckoned position must be regularly updated by map-matching (i.e., the dead-reckoned track is related to a digitized road map). If statistically significant errors are detected, corrections are made, as illustrated in Figure

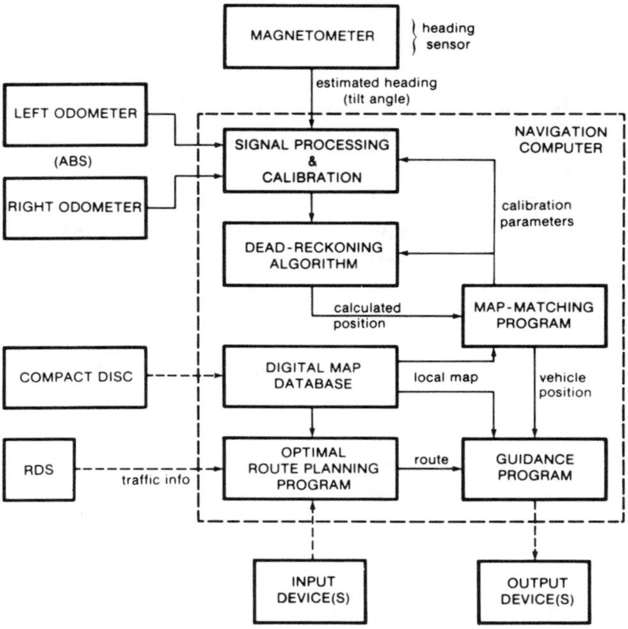

Figure 8.6 Functional block diagram of the navigation part of CARIN.

8.7. From these corrections, new estimates can be determined for several sensor parameters, such as the offset of the magnetometer and the scale factors of the odometers. The position and speed of the vehicle are used by the route guidance program, among others. In the 1990s, the CARIN system can be extended with a receiver for the NAVSTAR GPS satellite navigation system [Blanchard, 1986].

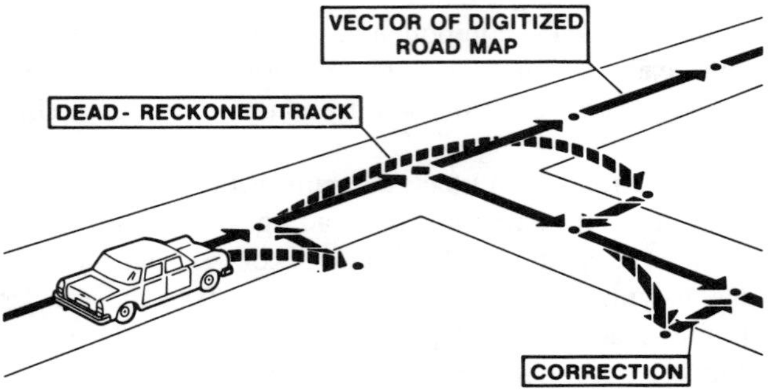

Figure 8.7 Explanation of the correction of the dead-reckoned positions using the digital road map.

8.6.1 General Equations for Location

The electronic navigation system of CARIN must obtain a "fix" for the vehicle at regular intervals (i.e., every five meters), by dead reckoning. This amounts to determining the center of the *vehicle location probability area* (VLPA), the area in which the vehicle is most likely to be traveling, with a probability of 95%, for example. The coordinates of this center are X_n and Y_n for the *n*th dead-reckoned fix. (The Y-axis points towards geographic north.) The navigation system takes the coordinates X_n and Y_n and calculates new coordinates X_{n+1} and Y_{n+1} from the equations:

$$x_{n+1} = x_n + d_n \sin \frac{1}{2} (\phi_{n+1} + \phi_n) \tag{8.1a}$$

and

$$Y_{n+1} = Y_n + d_n \cos \frac{1}{2} (\phi_{n+1} + \phi_n) \tag{8.1b}$$

where d_n represents the distance traveled between the *n*th and the $(n + 1)$th fix, and ϕ_n and ϕ_{n+1} represent the heading of the vehicle with respect to geographic north at the *n*th and the $(n + 1)$th fix.

The periodic dead reckoning of a fix requires periodic measurements of the heading and the distance traveled. As noted earlier, the car has sensors for this: an electronic compass and two-wheel sensors or odometers. These measure the number of revolutions of nondriven wheels about their axes. We shall now take a closer look at these sensors and the processing of the signals that they produce.

8.6.2 The Sensors

Both the electronic compass and the odometers contain transducers that convert a magnetic flux into a voltage. The transducers are made by Philips Valvo, type number KMZ10. They are based on the *magnetoresistance effect,* in which the resistance of certain materials depends on the angle between the direction of the current and that of the magnetization [Thoone, 1987]. The maximum sensitivity of such a transducer is about 0.1 mV per A/m at a supply voltage of 5 V.

The electronic compass contains three such transducers for the magnetic field. They measure the components of H, the strength of the terrestrial magnetic field, in three orthogonal directions. For the moment, only the components $H \sin\phi_n$ and $H \cos\phi_n$ are of interest. (Later we will use the vertical component to determine the gradient of the road.) H is about 15 A/m. Measurements of such a weak magnetic field are very difficult because of slow variations of the output signal, due to temperature changes. So, the output signal of the transducer is converted into a *square wave* signal. This is done by periodically changing the direction of magnetization in the transducer strips by energizing a coil around the transducer with alternate positive and negative pulses of current. Synchronous demodulation of the rectangular voltage then converts it into a direct voltage, with none of the low-frequency components that interfered with the original signal.

Transducers of the same type are used for measuring the number of revolutions of the nondriven wheels. Figure 8.8 shows how two transducers are mounted in relation to a strip bonded to the inside edge of a wheel rim. The strip is permanently magnetized so that the magnetic field is radial and alternately inward and outward, and so 26 periods of the magnetization, each of length p, correspond to a rotation of the wheel through 360°. The spacing of the transducers is $p/4$. The figure also shows how the transducer signal is electronically processed. When the wheel is turning, the transducers each supply an approximately sinusoidal signal, with a 90° phase difference between the two signals. The direction of rotation of the wheel determines which of the two signals leads in phase. The sinusoidal signals are amplified and converted into square wave voltages. These are transformed into a pulse train in such a way that one wheel revolution corresponds to $4 \times 26 = 104$ pulses. At the same time, the decision is made as to whether each pulse should be positive or negative. Finally, a binary number Z_l is produced, which is a measure of the angular rotation of the left-hand wheel with respect to the previous dead-reckoned fix. In the same

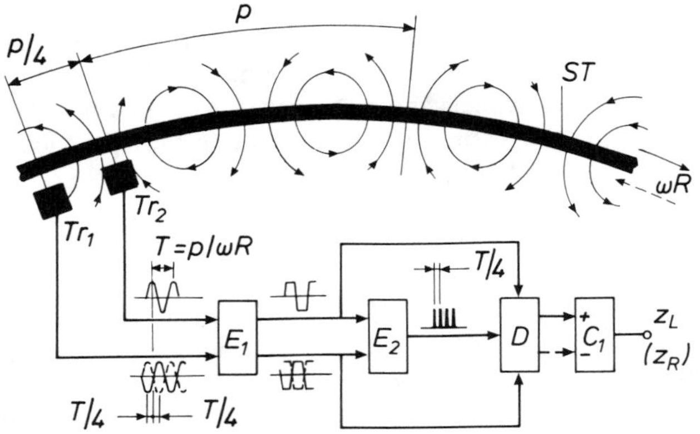

Figure 8.8 Principle of a wheel sensor: *ST* is a permanently magnetized strip; p is the length of the period of the radial magnetization; ωR is the peripheral velocity of the strip; ω is angular velocity; R is radius. Tr_1 and Tr_2 are magnetic transducers. The output signal is approximately sinusoidal with a period $T = p/\omega R$. The phase difference between the output signals from the two transducers is $T/4$, and is positive or negative, depending on the direction of rotation of the wheel. E_1 is a circuit that converts the signals into square waves, E_2 a circuit that derives pulse trains of period $T/4$ from the two square waves; D is a circuit that determines whether a positive or negative value should be assigned to the pulses; C_1 is a counter that supplies a binary number Z_l or Z_r, which is a measure of the angular displacement of the left-hand or right-hand wheel. The circuits are digital except for E_1.

way, a binary number Z_r is produced, which is a measure of the angular rotation of the right-hand wheel.

Values for the quantities d_n and ϕ_n (defined in equations (8.1a) and (8.1b)) are found from the measured results from the sensors. The distance traveled, d_n, and the change in heading $\phi_n^w - \phi_{n-1}^w$ from the wheel sensors are found from the equations:

$$d_n = \frac{(Z_l + Z_r)\pi R}{104} \tag{8.2}$$

and

$$\phi_n^w - \phi_{n-1}^w = \frac{(Z_l - Z_l)\pi R}{52\ S} \tag{8.3}$$

where R is the wheel radius and S is the track width. The heading measured by the electronic compass is denoted by ϕ_n^c. A weighted mean of the measured values for the heading is given by

$$\phi_n = \frac{A\phi_n^w + B\phi_n^c}{A + B} \tag{8.4}$$

where A and B are weighting factors. Now, we shall discuss a relatively simple way of obtaining a value for ϕ_n that is virtually free from interference.

8.6.3 Combining the Measured Values for the Change in Heading

We have seen how the signal from the electronic compass and the difference signal from the wheel sensors can provide a measure for the heading of the vehicle. The compass signal supplies an absolute value of the heading relative to magnetic north; the wheel signal supplies a relative value with respect to the initial value of the heading. The signals thus must be made comparable. Both signals are also subject to interference because of the way in which the signals have been obtained. The difference signal from the wheel sensors has a slow drift due to differences in tire pressure and wear, *et cetera*. The compass signal contains very fast fluctuations due to the magnetic fields of other vehicles, steel bridges, structural metal in buildings, and tramlines (streetcar, trolley). The compass signal also has a constant error due to the magnetic field of the steel body of the car; see Figure 8.9.

Figure 8.10 shows how the two signals for the heading are corrected and the interfering signals removed. In the computing unit C2, the successive numbers of

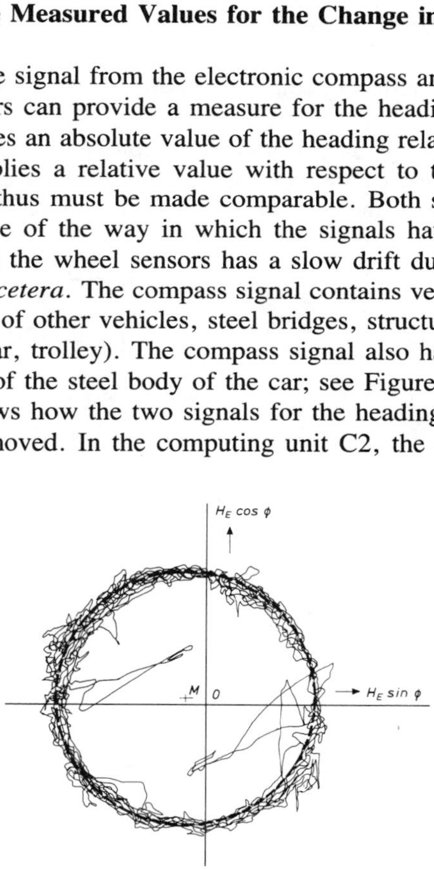

Figure 8.9 The component H cosϕ of the Earth's magnetic field H as a function of the component H sinϕ; ϕ is the heading of the vehicle. The graph is a recording, made during a journey, of measurements made with two magnetic transducers in the electronic compass. Theoretically, the measured points should lie on a circle with center zero. The best-fitting circle, however, has M as its center, because of the effect of the magnetic field of the vehicle. The other deviations are due to external magnetic fields from passing vehicles, steel bridges, electrical cables, *et cetera*.

pulses Z_l and Z_r from the left-hand and right-hand wheels are converted into a signal ϕ^w, which still contains an unwanted low-frequency component. In the computing unit C3, the effect of the car's own magnetic field is eliminated from the signals for $H \cos\phi$ and $H \sin\phi$. This is done by calculating the equation of a circle of best fit in the diagram for $H \cos\phi$ as a function of $H \sin\phi$ (see Figure 8.9) for a large number of measured results. Next, C3 calculates the signal ϕ^c for the heading. The values of ϕ^w and ϕ^c are each multiplied by a constant factor to make them comparable with each other in magnitude. The signal ϕ^w is then passed through a high-pass filter, and the signal ϕ^c is passed through a low-pass filter. Both filters have the same cut-off frequency f_c. The interfering components effectively are completely removed in the filters.

The special feature of our method is the continuous adjustment of the cut-off frequency f_c of the filters by the control unit R. The interfering components in ϕ^w and ϕ^c are measured in R during a certain period of time by subtracting the mean

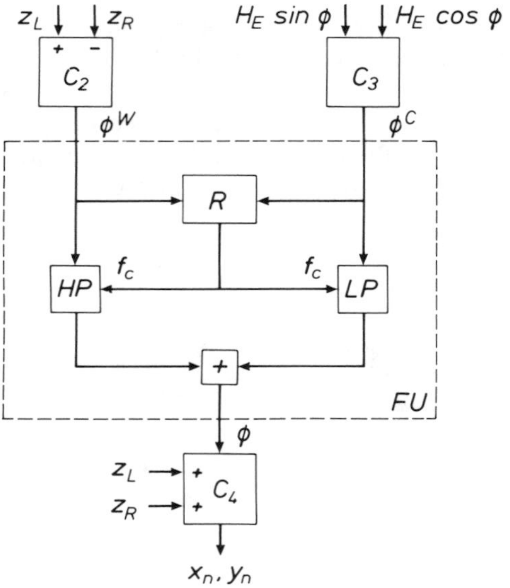

Figure 8.10 Processing the signals Z_l and Z_r from the wheel sensors, and $H_E \sin\phi$ and $H_E \cos\phi$ from the electronic compass. C2, C3, and C4 are computing units. *FU* is a filter unit. In C2, the signals Z_l and Z_r are converted into a signal ϕ^w for the heading. In C3, the compass signals are converted into a signal ϕ^c comparable with ϕ^w. *HP* is a high-pass filter, *LP* a low-pass filter, and f_c the cut-off frequency of the two filters. *R* is a control unit that continuously controls the magnitude of f_c by measuring the rms value of the interfering components in ϕ^w and ϕ^c. After addition, the resultant signal ϕ is periodically converted into values X_n and Y_n for the coordinates of position, with aid of the sum signal from the wheel sensors.

value from the signals. The root mean square (rms) value of each of the interfering signals is then determined. If the rms value of the interference in ϕ^w is large, f_c is increased; if this rms value is large for ϕ^c, then f_c is reduced. Finally, the two filtered signals are added. Setting the value of f_c also weights ϕ^w and ϕ^c as in equation (8.4). The computing unit C4 calculates coordinates of position X_n and Y_n from samples of the sum signal of Z_l and Z_r, which is a measure of the distance traveled, d_n. This calculation makes use of equations (8.1a) and (8.1b). The corrections described are not executed with samples of analog signals, but are derived from calculations with binary numbers, so that analog-to-digital converters are used where required. The calculations are made in a special microcomputer for navigation.

8.6.4 Improving the Dead-Reckoned Fix from Map Data

Errors in the fix determined by dead reckoning may be divided into *systematic* and *random* errors. The magnitude of systematic errors is essentially predictable, and they can be eliminated from the results of the measurements by calibration, whereas random errors cannot.

Once the system has been installed in a car, several calibrations are necessary. First, the car is driven around a closed loop to calibrate the compass. The effect of the car's magnetic field is then eliminated, and the transducers are matched to each other. The computer matches the transducers by fitting an imaginary circle (a regression circle) to a diagram resembling the one in Figure 8.9. Then, the car is driven along a straight line, so that the wheel sensors can be calibrated by setting the difference signal equal to zero and comparing the sum signal with the distance actually traveled. Such calibrations may also be necessary after new tires or snow chains have been fitted.

Calibration is also performed during a journey. The sum signal of the wheel sensors is then compared with the length of map segments, and the electronic compass is calibrated as described above. The remaining errors, those that cannot be eliminated by calibration, must be treated as random errors.

The magnitude of random errors can be estimated, because their standard deviation can be calculated with a fair probability from repeated measurements. The standard deviations of the random errors in the measurements of heading and distance can be obtained from a comparison of a large number of dead-reckoned fixes with the actual position of the vehicle. The magnitude of the random errors is expressed by the dimensions of the VLPA. In theory, this area is bounded by an ellipse, the dimensions of which are given by rules for the propagation of random errors.

The size of the VLPA increases continuously during the journey, until a position correction of the dead-reckoned fix is made, when the computer has recognized the pattern of a number of successive fixes by comparing them with topographical information in the main memory. The procedure requires a considerable amount of computation. To keep this within reasonable bounds, the VLPA ellipse is approximated by its smallest enclosing rectangle.

Figure 8.11 shows how the comparison of map data and dead-reckoned fix is made. Whenever a dead reckoning is made, the computer finds which segments lie inside the VLPA rectangle. A record of these segments is kept in the form of "trees" of possible routes, as illustrated in Figure 8.11(b). The lower-case letters in the trees refer to segments in directions that do not really correspond to the heading of the vehicle. If such a segment lies outside the VLPA in the next dead reckoning, it is removed from the tree. A position correction (an update) is made if the computer can decide with a good degree of certainty which of the possible routes has been followed.

In addition to position corrections, there is also a *map-matching procedure,* in which, at every dead-reckoned fix, the center of the VLPA is projected onto a segment inside it. If two or more segments are inside the VLPA, the choice of segment

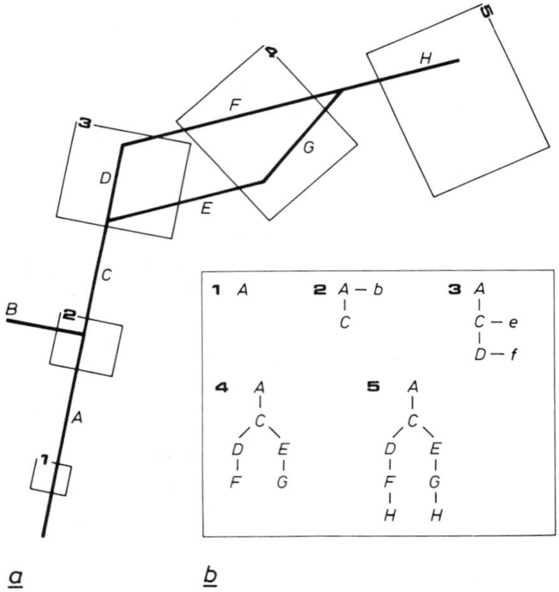

Figure 8.11 Comparison of map data with dead-reckoned fix. (a) The computer finds out which map segments lie inside a VLPA. (b) "Trees" of possible routes resulting from this comparison. Segments the direction of which does not correspond to the heading of the vehicle are shown in lower-case letters. The size of the VLPA steadily increases because of the increasing uncertainty of position. The tree at VLPA 5 shows that there are two possible routes. (In reality, the successive VLPAs overlap each other, because their centers are typically spaced at a distance of 5 m.)

is determined by factors such as whether the segment is on the planned route, or the correspondence between the direction of the segment and the heading. Map matching can be used, for example, to display the correct segments, nodes, and intermediate points on the LCD screen. The position resulting from map matching is always at the center of the display, which therefore always corresponds to a part of a road.

8.7 THE MAN-MACHINE INTERFACE

CARIN's main function is navigating the driver to the required destination. To enter the destination and issue commands, a dialogue between the user and CARIN is necessary. Such a dialogue can occur only when the car is stationary. Communication is menu-driven, for instance, using a keyboard and display.

The driver is guided to his or her destination through speech-synthesized commands, supported by stylized pictures on the flat panel display. The speech module generates directions in natural language ("first turn left," or "straight ahead"). The pictures on the display are stylized intersections, which are presented "heading up."

The time at which an instruction is given depends on:

- the current speed of the car;
- the type of road;
- the duration of the route instruction;
- the reaction time of the driver to the instruction; and
- the time needed for execution of the instruction.

If actions are expected on successive adjacent intersections, a concatenation of route instructions will be given. If the driver ignores directions, the system plans an alternative route and presents the driver with new instructions.

Traffic safety is a major issue in the design of CARIN's human-machine interface. CARIN contributes positively to traffic safety (less search behavior, avoidance of congestion and unsafe routes, shorter journeys), but new guidance instructions are also given to the driver, which could be prejudicial to safety if the system is not properly designed. Speech instructions are preferred because the driver needs to watch the road. If a direction is not understood or forgotten, the display can be consulted. The pictures are designed for minimum distraction and short interpretation time.

8.8 SOME CONTROL SOFTWARE ASPECTS

The CARIN hardware is based on a multiprocessor architecture for the on-board computer, and decentralized intelligence in the peripheral units. Communication between them is logically handled by *virtual interfaces*. Several processes may run in parallel in the on-board computer. For example, a new route may be planned while the current position of the vehicle is constantly being determined and the CD-I disc is being read. This requires the operating system to have a multitask capability.

The operating system used is CD-RTOS (*compact disc–real-time operating system*). Its features are:

- multitasking with real-time response;
- versatile modular design;
- ROM-capable;
- device-independent and interrupt driven;
- able to support a variety of arithmetic and input-output (I/O) coprocessors;
- able to handle multilevel tree-structured directories;
- able to support both byte-addressable random access files and real-time files; and
- OS9/68K (trademark of Microware) compatible.

The organization of CD-RTOS is shown in Figure 8.12; it is composed of four major blocks:

1. libraries;
2. CD-RTOS kernel—a customized version of the OS/9 68K kernel;
3. managers (on virtual device level; hardware independent);
4. drivers (interface between virtual level and actual hardware).

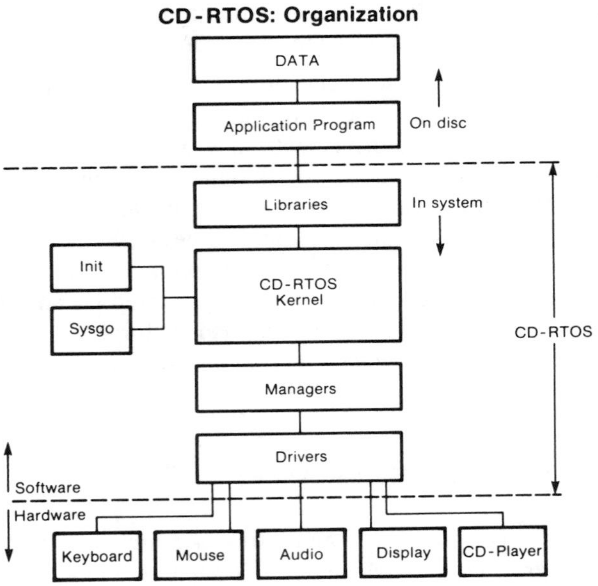

Figure 8.12 A block diagram of the organization of the CD-RTOS operating system.

Flexibility and compatibility are increased by combining the electronic road map and the CARIN application software on the same disc. Upon initialization of CARIN, the application is loaded by a bootstrap program from the CD-I disc. Map updates and feature enhancements are synchronized in this way.

8.9 SUMMARY AND CONCLUSIONS

CARIN, which has been described in this chapter, comprises a vehicle-locating unit, a compact disc player that can read both music discs and discs with digital map databases, and on-board computer and devices for communication with the driver. In the Philips laboratories, the feasibility of this concept has been demonstrated, and prototypes are being designed and built.

The main benefits of the system include efficient route planning, which results in time and cost savings, and safe route guidance, using a voice synthesizer supported by a schematic display of junctions. CARIN can utilize a data link through the car radio and the RDS to receive warnings of hazards and congestion.

The compact disc is used to store digital map databases because it has a large storage capacity, random access capability, and high data integrity. Automobile CD players can be designed to play back digital database discs as well as digital audio music discs. For navigation and interactive use, dedicated database structures are required for the storage of geographic data. Information such as telephone listings and tourist information can also be stored.

Experience shows that international standards are essential in such technical areas. CD-I, a logical evolution of CD-DA, provides a complete standard combining audio, video, and text recording with powerful data processing facilities. Because CD-I permits the storage of the geographic database together with related application programs on a single medium, the discs can be used on both home and car systems, provided that they are designed to the standard's specifications. This compatibility should be beneficial to the acceptance of CD-I in the marketplace.

ACKNOWLEDGEMENTS

This chapter is based on two articles recently published by the CARIN team: "The Car Information and Navigation System CARIN and the use of the Compact Disc Interactive," by M.L.G. Thoone, L.M.H.E. Driessen, C.A.C.M. Hermus, and K. van der Valk, *SAE Technical Paper 870139,* International Congress and Exposition, Detroit, Michigan, February 23–27, 1987 ©1987 Society of Automotive Engineers; and "CARIN, A Car Information and Navigation System," by M.L.G. Thoone, *Philips Technical Review,* Vol. 43, No. 11-12, December 1987. We are grateful to the Society of Automotive Engineers and N.V. Philips Gloeilampenfabrieken for permission to use these articles.

REFERENCES

Benning, H.J.G.M., "Digital maps on compact discs," SAE International Congress and Exposition, Technical Paper No. 860125, 1986.

Blanchard, W.F. "The plain Navigator's Guide to GPS," *Navigation News,* The Royal Institute of Navigation, Vol. 1, Issue 4, July 1986, pp. 28–32.

Broome, F.R., "TIGER preliminary design and structure overview: the core of the geographic support system for 1990," *Annual Meeting of the Association of American Geographers,* 1984.

Burrough, P.A., *Principles of Geographic Information Systems,* Chapter 2, "Data structures for thematic maps," Oxford University Press, London, 1986.

Carasso, M.G., J.B.H. Peek, and J.P. Sinjou, "The compact disc digital audio system," *Philips Tech. Rev.,* Vol. 40, 1982, pp. 151–155.

Cooke, D.F., "Map storage on CD-ROM," *Byte Magazine,* May 1986.

Friedl, H., "Guidance of vehicle traffic flows—ALI," *Radio Elektron.,* Schaum 51, 1975, pp. 266–269.

Heemskerk, J.P.J., and K.A. Schouhamer Immink, "Compact Disc: system aspects and modulation," *Philips Tech. Rev.* Vol. 40, 1982, pp. 157–164.

Jeffery, D.J., "Ways and means for improving Driver Route Guidance," TRRL Report No. 1016, Transport and Road Research Laboratory, UK, 1981.

Jeffery, D.J., "Options for the provision of improved driver information systems; the role of microelectronics and information technology," *Proc. IEE Colloquium on Vehicular Route Guidance,* London, 1985, pp. 1–3.

Lambert, S. and S. Ropiequet, *CD-ROM, The New Papyrus—The Current and Future State of the Art,* Microsoft Press, Redmond, WA, 1986.

Philips "An introduction to compact disc interactive," New Media Information Center, Philips International BV, Eindhoven, The Netherlands, October, 1986.

Thoone, M.L.G. and R.M.A.M. Breukers, "Application of the compact disc in car information and navigation systems," *SAE Int. Cong. and Exposition,* Tech. Paper 840156, Detroit, MI, 1984.

Thoone, M.L.G., L.M.H.E. Driessen, C.A.C.M. Hermus, and K. van der Valk, "The car information and navigation system CARIN and the use of compact disc interactive (CD-I)," *SAE Int. Cong. and Exposition,* Tech. Paper 870139, Detroit, MI, 1987.

Zimdahl, W., "Guidelines and some developments for a new modular driver information system," *Proc. IEEE Vehicular Technol. Conf.,* Pittsburgh, PA, 1984, pp. 178–182.

PART IV
INFORMATION TECHNOLOGY IN VEHICLES

Chapter 9
PROMETHEUS and DRIVE—European Initiatives

I. CATLING
IAN CATLING CONSULTING

9.1 INTRODUCTION

PROMETHEUS and DRIVE are major European research and development programs with the same objectives—the use of advanced technology in systems leading to more efficient, safer, and less environmentally damaging traffic. Both are multidisciplinary programs involving many participants from industry, traffic authorities, universities, and research organizations.

There is a difference in emphasis, however, which follows partly because PROMETHEUS is an initiative of the automotive industry and DRIVE is a program of the European Community (see Table 9.1 for the members of the EC). While PROMETHEUS is far from being totally vehicle-based, there is a natural concentration on developments within the vehicle and in its immediate environment; DRIVE is mainly concerned with the infrastructure requirements. There are also differences of time scale—PROMETHEUS is an eight-year program, whereas DRIVE has been defined, at least initially, over a three-year period (a more substantial program, DRIVE II, is planned to follow this initial period). Nevertheless, the two programs are working very closely together, and they have many formal and informal links.

The concept of an integrated system of road traffic is common to both PROMETHEUS and DRIVE. The objective of integration is to ensure that different applications (advanced traffic control, route guidance, driver information, driver support, fleet management, automatic debiting systems, and others) share a common infrastructure and common on-board components as far as possible. Both programs clearly recognize that this integration can only be achieved by the close cooperation

Table 9.1
Member States of the European Community (EC) and the European Free Trade Association (EFTA)

EFTA:	EC:
Austria	Belgium
Finland	Denmark
Iceland	France
Norway	Federal Republic of Germany
Sweden	Greece
Switzerland	Republic of Ireland
	Italy
	Luxembourg
	The Netherlands
	Portugal
	Spain
	United Kingdom

of governments, the scientific community, and industry, the latter comprising automotive manufacturers, electronic systems and component suppliers, and telecommunication service providers.

Fundamental to the development of the *Integrated Road Transport Environment* (IRTE) is the need for a widely available two-way communication system. To maximize the potential for *Road Transport Informatics* (RTI), significant amounts of real-time and other data must be transmitted to and from mobile vehicles. The quality of these data will depend on the direct collection of information from equipped vehicles as they use the road network. Both PROMETHEUS and DRIVE have a strong emphasis on communication.

There is also a recognition in both programs of the need to act quickly to achieve the integrated approach; many components of the integrated traffic system (e.g., road pricing, dynamic route guidance, advanced traffic control, and improved driver information systems) are under active and rapid development, and many of the systems will be introduced during the 1990s. Both PROMETHEUS and DRIVE have already had a substantial effect in bringing together the many organizations involved in these individual subsystem components, putting the overall objective of an integrated system into perspective.

9.2 EUREKA

PROMETHEUS is one of the larger projects contained within the EUREKA framework. EUREKA was established in 1985 when nineteen European countries, together with the Commission of the EEC, agreed upon a framework for international technological cooperation between firms and research institutions. The aim of EUREKA is to encourage the development of products, processes, or services involving new technologies to strengthen the competitiveness of European industry.

EUREKA projects are funded by the individual governments involved, so there is no central source as there is in commission-funded programs such as DRIVE. (Funding for both programs is on a partial basis, where industry contributes directly, and normally at least 50% of total costs.) The EUREKA agreement, however, means that the governments view applications for support of collaborative research within EUREKA in a particularly favorable light.

There are currently about two hundred EUREKA projects [EUREKA, 1986], each of which involves participants from at least two European countries. Project costs range from around one million ECUs (European Currency Units) to well over 100 million ECUs. PROMETHEUS is one of the larger projects, and the total budget for the eight years could reach 200 million ECUs.

9.3 PROMETHEUS

PROMETHEUS is an acronym meaning "*Programme for a European Traffic with Highest Efficiency and Unprecedented Safety.*" It began in 1986 when the major European motor manufacturers agreed to use the combined effort of their research departments to work toward common approaches to the traffic of the future. The companies involved are BMW, Daimler-Benz, M.A.N., Porsche, Volkswagen, Matra, PSA (Peugeot), Renault, Saab, Volvo, Fiat, Jaguar, Rolls-Royce, and Steyr-Daimler-Puch.

In addition to the automotive manufacturers, some forty research establishments are actively involved in the basic research aspects of PROMETHEUS. The program was widened in 1988 to include the participation of equipment suppliers and component manufacturers. PROMETHEUS now has more than three hundred scientists and engineers working on analysis, development, and evaluation. The program, therefore, is a substantial one, and its management is a difficult and demanding task. The program is run by the PROMETHEUS Steering Committee with administrative support by the PROMETHEUS Office in Stuttgart, which is jointly funded by the participating companies.

9.3.1 PROMETHEUS OVERVIEW

The objective of PROMETHEUS in the longer term is the concept of the driver being informed and supported by an intelligent automatic vehicle "copilot." It will monitor conditions aboard and outside the vehicle; the copilot will give the driver advice on the best route, based on actual traffic conditions, and will provide other relevant information to help on the journey. The copilot will sense the slipperiness of the road surface, the curvature of the road, and, as with antilock braking, will be able to intervene to help the driver, not only by braking more safely, but also by avoiding wheel spin and actively controlling the suspension. By communicating directly with other vehicles, the copilot will alert the driver to immediate hazardous conditions, sense potential collisions, and, in extreme circumstances, be able to react more quickly than the driver can to avoid them. Ultimately (sometime during the twenty-first century) the copilot will be able to offer the driver the option to join a convoy or "road train" in which the controls are automatically handled.

PROMETHEUS is therefore designed to deal with both immediate developments, such as route guidance systems, and the longer term possibilities, such as systems enabling longitudinal and lateral control of vehicle movements. PROMETHEUS deals with all aspects of the development of such systems: the technologies, such as communication, sensors, and control systems; the strategies for control; and the framework in which the traffic of the future will need to adapt to take best advantage of these developments.

The time scale for all the developments envisaged within the program will be well over twenty years. PROMETHEUS seeks to provide the foundation on which to base the important decisions for investment in both new, advanced products and infrastructure.

The program includes the analysis of benefits from the type of system envisaged. PROMETHEUS also includes demonstrator systems and field tests. Although individual subsystems will be developed by different teams, there is a strong emphasis on a systems approach, using descriptive scenario techniques to work toward an overall integrated system. The scope of PROMETHEUS includes not only car traffic, but also commercial vehicles and (to some extent) public transport and the interests of vulnerable road users (pedestrians and cyclists). There is a strong emphasis on the potential for using new technologies to produce a safer system of traffic, with the objective of causing a significant reduction in the number of severe road traffic accidents.

The designers of PROMETHEUS recognized from its start that to make the best use of new technological developments in traffic applications, the program itself would need to influence and be involved in those basic developments. The program is therefore structured to include not only the industrial research needed to develop applications, but also the basic scientific research necessary to provide the foundation for the applications development. Cooperation between universities, research institutes, and industry has been firmly established in a number of fields, including artificial intelligence, microelectronics, communication, information technology, and traffic engineering.

The international aspect of collaborative work is as important as the interdisciplinary cooperation. Traffic control has not historically been one of the areas in which different countries have worked closely together, but with the advent of onboard communication systems, there will be an increasing emphasis on the need for standardization. As a starting point, the UK and West German governments established a joint working party in 1987, which produced a draft standard for the road-vehicle communication link. This standard will be used in route guidance systems, paving the way for PROMETHEUS and DRIVE to achieve international standardization. The working group is being expanded to include other European countries, and the draft standard is providing the basis for developments in both programs.

9.3.2 Program Timetable

PROMETHEUS was set up to run in three phases: the *definition* phase, the *launching* phase, and the main *research and development* phase.

During the definition phase, reviews of current technological developments were combined with requirement analyses to identify the main areas for research during the later phases of the project.

The launching phase occurred during the second year of the project, and was completed at the end of 1988. During this phase, detailed work programs of many subprojects were developed, harmonized, and synchronized. These activities were at two levels. The first involved coordination between companies and institutes within each country. The second established the methods for international cooperation. Work in the main research and development phase is now in progress, concentrating on more than a dozen different industrial subprojects and four principal areas of basic research.

9.3.3 Program Structure

There are three primary areas of industrial research, called *improved driver information and active driver support* (PRO-CAR), *cooperative driving* (PRO-NET), and *traffic and fleet management* (PRO-ROAD). Each area forms a coherent subprogram of research within the overall PROMETHEUS framework. There are four subprograms concerned with basic research—three of these, PRO-CHIP (custom hardware for intelligent processing), PRO-COM (communication), and PRO-ART (artificial intelligence) deal with the technologies and methodologies necessary to implement the applications being developed within the industrial subprograms; the fourth, PRO-GEN (traffic engineering), seeks to develop and evaluate the traffic scenarios in which PROMETHEUS developments will be implemented.

Within the three industrial subprograms, there are eleven "thematic projects," listed in Table 9.2. Each thematic project is concerned with one area of technology relevant to the overall development of PROMETHEUS systems. It is under the control of a single "European Lead Researcher" (ELR), who is responsible for coordinating all the PROMETHEUS research relevant to his or her subject area.

Parallel to the thematic project structure, there are a number of demonstrator projects which seek to bring together the work of thematic projects into integrated

Table 9.2
The Eleven Thematic Projects and their Lead Companies

PRO-CAR	
1. Sensors and signal processing	Saab
2. Actuators and vehicle operation	FIAT
3. General architecture	Peugeot
4. Man-Machine Interface	Saab
5. Vehicle safety and dependability	BMW
PRO-NET	
6. PRO-NET system engineering	VW
7. Communications	Renault
8. Emergency warnings	Saab, VW, and Daimler-Benz
PRO-ROAD	
9. Information processing and data acquisition	Volvo
10. Infrastructure-based systems	Daimler-Benz
11. On-board elements	Renault

subsystems, to demonstrate PROMETHEUS functions. Table 9.3 lists the initial demonstrator projects, which were developed during an initial stage in which much of the work was done by individual companies. The second development stage involves much closer collaboration between different companies and organizations, and is leading to the emergence of a better defined set of common approaches and solutions. These will then be demonstrated in further projects called *Common European Demonstrators*.

Figure 9.1 shows the overall structure of PROMETHEUS and the relationships of the seven subprograms.

9.3.4 Improved Driver Information and Active Driver Support (PRO-CAR)

This subprogram is primarily concerned with the development of subsystems wholly contained within the vehicle. PRO-CAR comprises five thematic projects (see Table 9.2).

Sensors and Signal Processing

The objective of this thematic project is to develop new concepts and knowledge bases for on-board sensors and multisignal processing systems. These will be needed

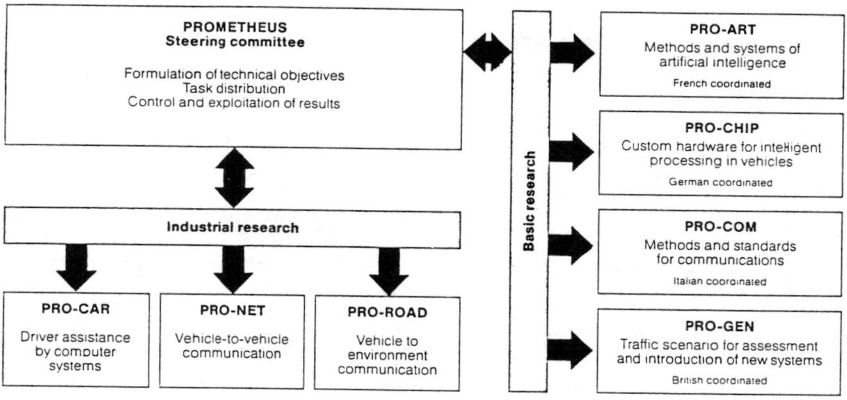

Figure 9.1 The overall structure of PROMETHEUS.

to support other PRO-CAR autonomous systems, as well as for PRO-ROAD and PRO-NET functions.

There are five application areas, as shown in Figure 9.2:

- *Environmental sensors* are concerned with radar for anticollision systems, a laser system for distance supervision, the "CAR VIEW" vision enhancement system, and sensors for obstacle detection.
- *Road surface sensors* aim to monitor friction and adhesions. The scope of the work extends to the prediction of adhesion.
- *On-board vehicle status sensors* are continually being developed as new models are introduced by the automotive manufacturers. PROMETHEUS is particularly concerned with improved speed sensing and angular velocity sensors. (Other types of sensors, and the application of electronics to engine and transmission controls, are covered in detail in Chapter 10, "The Intelligent Car.")
- Fundamental to the success of all technology-based systems being developed in PROMETHEUS and DRIVE is the level of user acceptance, and the effect on driver behavior. To support the advances being made in the study of driver behavior, the fourth area of *sensor technology development* in PROMETHEUS is concerned with monitoring driver status. Work in PRO-CAR will therefore include the development of driver condition sensors, such as the analysis of steering wheel movements, to give a normal *steering* signature. Significant deviations from the normal signature might be used to detect overtiredness, for example.
- The fifth work area in this thematic project is *signal processing*. Topics covered include multisensor processing, image processing, knowledge representation, and decision finding.

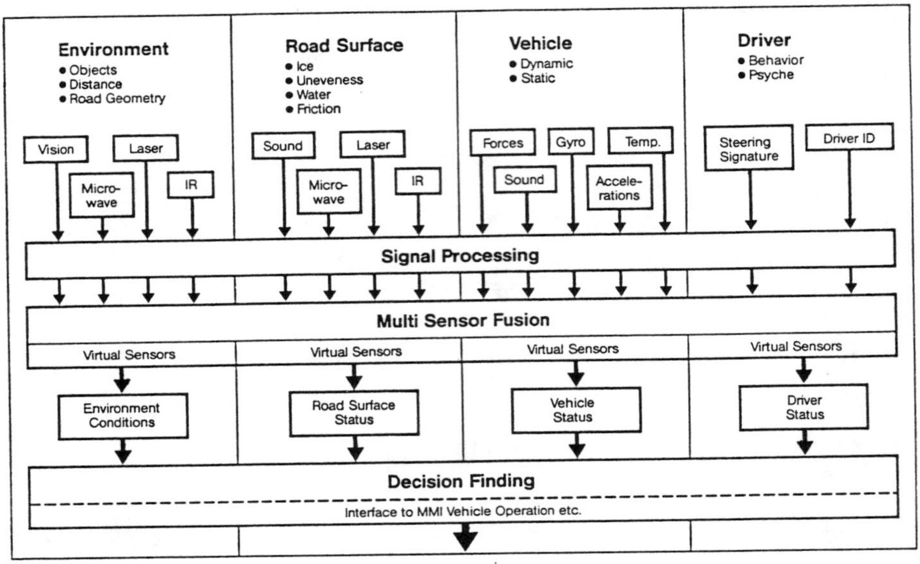

Figure 9.2 Sensors and signal processing.

Actuating Systems and Vehicle Operation

The objective of this thematic project is the development of active systems for PROMETHEUS "copilot" functions for longitudinal and lateral vehicle control, including the difficult question of how such advanced concepts might be integrated with the conventional driving task.

There are three main work areas. The first concerns vehicle control, and includes the development of actuating systems, application methods and strategies for copilot assistance. The use of "*drive-by-wire*" systems for collision avoidance and longitudinal and lateral control will be investigated, together with actuating systems for cruise control-throttle, brakes, and steering, including variable ratio steering and rear wheel steering.

The second work area is the development of artificial intelligence (AI) algorithms and mechanisms for providing feedback from the vehicle control system to the driver. This would include investigations of the "drive-by-wire" concept, as well as analysis of the situations and circumstances in which the intervention of copilot control functions might be applicable.

The third work area is *vehicle dynamic conditioning,* or active suspension control. This will include the development of models, strategies and devices for real-time adaptive response to driving conditions through active chassis control and positive yaw angle control.

General Architecture

This thematic project is intended for the development of compatible on-board control and communication systems.

The project will include the development of a standard ruggedized computer, common software, and interfaces to a *Vehicle Area Network* (VAN). The VAN will provide a bus-based data link between sensors, actuators, displays, controller units, on-board computers, video units, and communication links with external equipment.

This project will attempt to achieve the harmonization of existing multiplexed buses for common gateways to a VAN. It is also concerned with proposals for standardization of hydraulic systems aboard the vehicle.

Man-Machine Interface

The fourth thematic project (Figure 9.3) is concerned with an integrated and coherent design for the man-machine interface necessary to handle the new types of information and assistance which will be introduced to help the driver. There are three main project areas:

- The first work area concerns the presentation of information to the driver, at the right time and in the right way, and includes selection of information, timing of presentation, and the optimal choice of media.

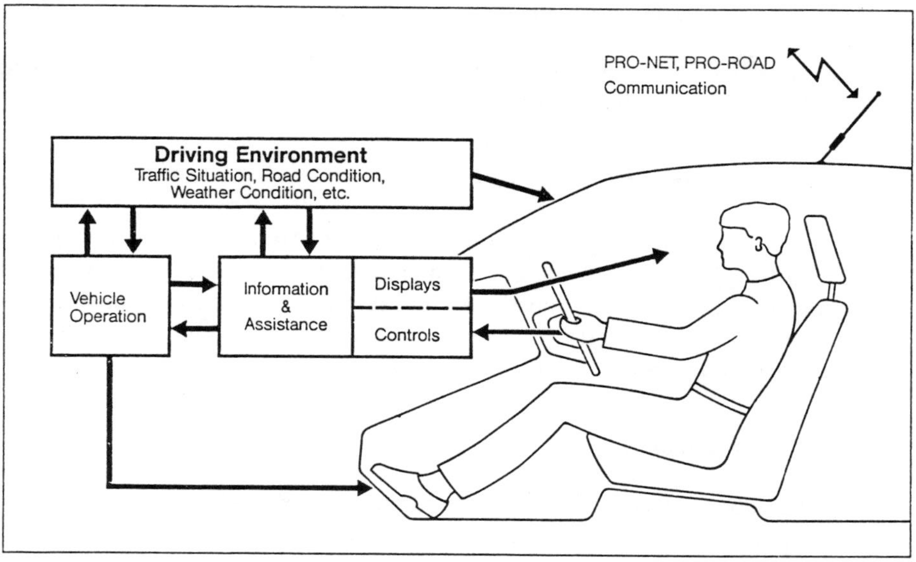

Figure 9.3 The man-machine interface.

- The second project area concerns assistance given to the driver, and includes the difficult problem of task allocation between PROMETHEUS systems and the driver, the timing of system actions, and the integration of assistance functions with the needs of the driver.
- The third area is the basic study of the human-machine interface, comprising driver task and workload analysis, capacity, and modeling. The study includes evaluation methods and tools, the objective being to produce guidelines and specifications for good design of the human-machine interface.

Vehicle Safety and Dependability

This thematic project is intended to ensure overall safety and reliability standards for all PROMETHEUS systems by using common standards and failure management strategies for both hardware and software developments. The term *dependability* includes reliability, maintainability, repairability, and testability. There are four main application areas:

- The first area is overall system reliability and safety standards, based on a functional analysis of driver assistance systems.
- The second project area is fault-tolerant systems, including the study of distributed electrical system architectures, communication networks, fail-safe computers, and actuating systems.
- The third area is failure management in driver assistance systems, including monitoring, diagnostic, and predictive algorithms and systems.
- The fourth project area is software reliability, including the use of common specification, design, and documentation standards.

9.3.5 Cooperative Driving (PRO-NET)

The second subprogram of industrial research is PRO-NET, which covers cooperative driving based on intervehicle communication. PRO-NET is probably the area with the longest time for the introduction of any systems; nevertheless, the ultimately achievable benefits, particularly in terms of safety, are very significant. There are three thematic project areas, listed in Table 9.2.

PRO-NET System Engineering

The sixth thematic project deals with the basic requirements for the PRO-NET concept of a dynamic local network of intercommunicating vehicles. There are four application areas:

- The first area deals with the development of cooperative strategies for exchange of information between vehicles, in particular for the high priority applications of distance keeping and coordinated braking, and for providing improved vehicle position information for lane change and overtaking maneuvers.

- The second area concerns vulnerable road users and the possibility of including transponder techniques for bicyclists and pedestrians into the PRO-NET scenario.
- The third area, in conjunction with PRO-GEN, will use simulation studies of traffic scenarios to assess the benefits of cooperative strategies. This will be achieved partly by the use of modeling techniques, and partly by direct simulation, using facilities such as the Daimler-Benz simulator in Berlin [Daimler-Benz, undated].
- The fourth area, in particular cooperation with PRO-COM and the electronics and supply industries, deals with field trials to provide practical experience of PRO-NET concepts. The *High-Net* project (described in Section 9.3.7) is an example of the type of development included in this area.

Communications

The seventh thematic project is concerned with the communication methods both for PRO-NET (i.e., intervehicle), and for PRO-ROAD (i.e., vehicle-to-roadside).

The main project areas are the definition, testing and standardization of communication systems, working closely in conjunction with PRO-COM and other thematic projects.

Particular subprojects will deal with emergency warning systems and the special requirements for convoy driving applications.

Emergency Warning Systems

This function is designed to use both PRO-ROAD and PRO-NET communication systems to provide faster and more efficient response to incidents (see also Chapter 5 for traffic information broadcasting systems).

The PRO-NET concept will be used to develop and establish prototype systems for the direct exchange of hazard warning information between vehicles. Existing communication systems for emergency services will be studied, with an assessment of the potential for improvement using PROMETHEUS systems. Other project areas are field testing and assessments of benefits and costs.

9.3.6 Traffic and Fleet Management (PRO-ROAD)

The third subprogram of industrial research deals with systems dependent on communication between vehicles and roadside equipment, which are, in turn, connected back to a control center. There are three thematic project areas (see Table 9.2).

Information Processing and Data Acquisition

The ninth thematic project deals with control center processing requirements and data needs. The project is particularly concerned with route guidance and driver information systems, and the integration of such systems with traffic management and control systems.

This thematic project is also concerned with the collection of traffic data and information from sources other than the IRTE itself. (Chapter 11 describes an "intelligent knowledge-based system," which extracts data on traffic incidents from police traffic incident logs, and prepares a message for broadcasting to motorists, although the system is not part of PROMETHEUS.)

Infrastructure-Based Systems

The tenth thematic project has the closest links with DRIVE, because it is designed to develop an integrated approach toward systems based on vehicle-to-roadside communication. One project area will concern the integration of improved information and traffic control systems.

The main active project area allocated to this thematic project is work on the development of standards for digital road maps. PROMETHEUS has taken a lead in setting up a European *Benchmark* project to bring together a task force containing the main organizations involved in digital mapping. The Benchmark project has now been incorporated into the DRIVE program.

On-Board Elements

The final thematic project takes an integrated view of all the on-board requirements of different PRO-ROAD systems. The project is particularly concerned with the requirements for route guidance systems. There are three specific project areas:

- The first area concerns the improvement and development of map matching algorithms to use map data to correct the errors inherent in any on-board dead reckoning system.
- The second specific project area is the development of improved route finding algorithms, particularly for autonomous on-board route guidance systems.
- The third specific area will cover an assessment of different types of vehicle location systems, including satellite based systems.

Chapter 6 deals with route guidance in more detail.

9.3.7 Demonstrator Projects

Within the overall framework of the eleven thematic project areas, a development program of demonstrator projects has been defined (see Table 9.3). These are in-

Table 9.3
The Initial PROMETHEUS Demonstrator Projects

Friction monitoring	Porsche
Cooperative driving	VW
Electronic architecture	Peugeot
Integrated MMI	Saab
Heading control	BMW
High-Net	Daimler-Benz
Intervehicle communication	FIAT
ARTHUR	Daimler-Benz
MOMS	Volvo
Satellite communication	M.A.N.
CARMINAT	Renault
AUTOGUIDE	Jaguar
Longitudinal control	Porsche

tended to show the technical feasibility of individual applications, and to enable the investigation of their relevance to future traffic systems. Initially, some fifteen demonstrator projects have been developed to demonstrate specific features. During the next stages of the project, these will be integrated into the common European demonstrators.

Each of the initial fifteen demonstrator projects addresses some of the specific functions within one or more of the thematic project areas, and is being implemented in at least a prototype form during the PROMETHEUS time scale. Most of the projects are developments wholly within PROMETHEUS; however, some relate to separate projects of direct relevance to PROMETHEUS, for which there is a specific interface with PROMETHEUS developments. Each of the initial demonstrator projects is briefly described below.

Friction Monitoring

This project aims to improve safety by monitoring friction between tires and the road surface, thereby recognizing that an adequate safety margin is being maintained. It will use multisensor concepts and on-line signal processing to detect road surface conditions, tire properties, and *dynamic driving condition* indicators.

Cooperative Driving

One of the most ambitious concepts within PROMETHEUS, but also one with very high potential benefits, is the use of PRO-NET communication concepts, which will enable vehicles automatically to maintain their relative positions on the road. Such systems will have major effects in reducing fuel consumption and emissions, but most significantly, they will dramatically increase the capacity of crowded roads.

The technology needed to accomplish such objectives includes advances in sensors, actuators and control processors. The demonstrator project is investigating infrared lasers for distance control, image processing for both longitudinal and lateral control, ultrasonic sensors for lateral control, and gyro, accelerometer and brake sensors for vehicle stabilization. Actuator systems are needed for steering, brakes and throttle. Processing functions are needed for controlling the relative positions of vehicles within the convoy and for organizing entry to and exit from the convoy.

Safety and the evaluation of failure modes will be critical to the development of this type of advanced concept.

Electronic Architecture

The objective of this project is to bring together advances in PRO-CAR sensors, the man-machine interface, PRO-ROAD navigation, information and communication systems, PRO-NET communication possibilities, and advanced concepts for electronic communication aboard a vehicle.

The different electronic modules will be interconnected through a number of multiplexed buses, using an enhanced VAN protocol, and the software implementation of VDX (*Vehicle Distributed Executive*). A prototype ruggedized PC AT and communication interface cards will be developed, including interfaces to existing proprietary systems manufactured by Philips, Bosch, and Peugeot.

Integrated Man-Machine Interface

The objective of this project is to develop an integrated driver-vehicle interface for many of the advanced functions within PROMETHEUS. Technologies to be applied include multilingual speech synthesis, head-up visual displays, simple graphic displays incorporated into standard dashboard systems, and voice recognition. Voice recognition in the noisy environment of the vehicle is particularly difficult (see also Chapter 3). The demonstrator project will also include the development of a first generation information management system.

Heading Control

The objective of this project is to monitor the vehicle's lateral position, and to provide feedback to the driver to assist him or her in maintaining correct heading, together with obstacle detection and collision zone monitoring.

Image processing will be used to monitor the vehicle's position, and *haptic* (i.e., using the driver's sense of feel through the steering wheel) methods will be used to alert the driver to any wandering from the desired course. This demonstrator project will provide the first step toward "forgiving and teach-back systems," which will use artificial intelligence techniques to give predictive and supportive assistance to the driver. The project will begin to investigate the scope of partial compensation for limitations of driver skills and capabilities.

High-Net

This project (see Figure 9.4) will demonstrate PRO-NET techniques in an autonomous safety network for collision avoidance. High-Net is based on the continuous broadcast by each equipped vehicle of its position and status to neighboring vehicles. Each network member monitors its own safety zone. No active control systems are included within this project, but the driver will be informed of potential conflicts and dangerous situations.

The project will use advanced vehicle location techniques based on other projects within PROMETHEUS, e.g., combined autonomous and infrastructure-supported systems such as ALI-SCOUT [Siemens, 1986] and AUTOGUIDE [Catling and Belcher, 1989], both covered in Chapter 6, and in this section AUTOGUIDE

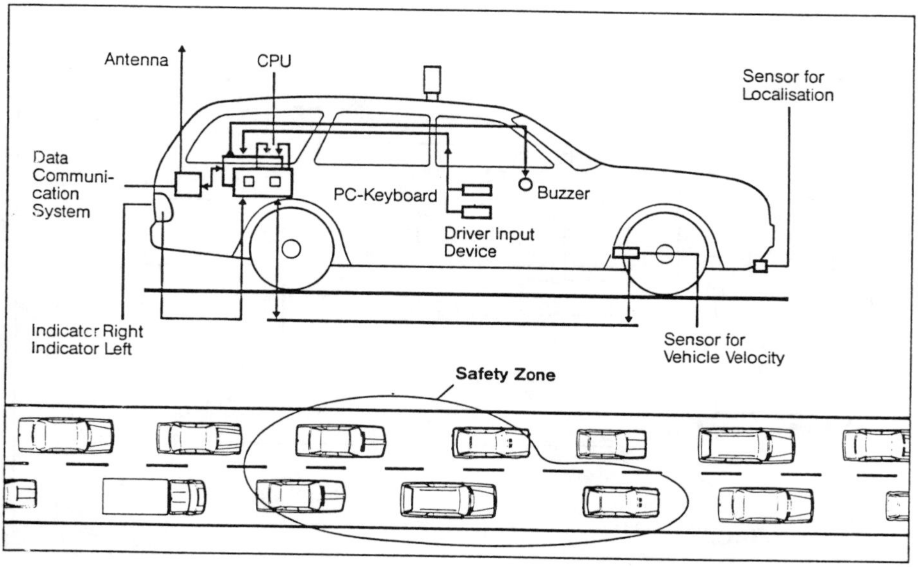

Figure 9.4 The High-Net project.

is treated together with advanced microwave communication systems for intervehicle communication.

Vehicle-to-Vehicle Communication

This project will also use PRO-NET techniques, but is primarily concerned with the exchange of information between vehicles traveling in opposite directions, to give advanced warning of adverse weather conditions, traffic congestion or other incidents.

In the demonstrator project, researchers will use VHF-FM transceivers operating at around 145 MHz with a ten-channel *Carrier Sense Multiple Access* (CSMA) protocol at 1200 baud. Information will be transmitted based on data received from speed sensors, wiper motor, fog lamp, and other on-board equipment, as well as possibly via infrastructure-supported systems for driver information.

ARTHUR (Automatic Radiocommunication System for Traffic Emergency Situations on Highways and Urban Roads)

ARTHUR (see Figure 9.5) is an electronic emergency radio system which is triggered either automatically or manually in the event of accident or breakdown, to warn other traffic in the area. Unlike other emergency radio systems developed to

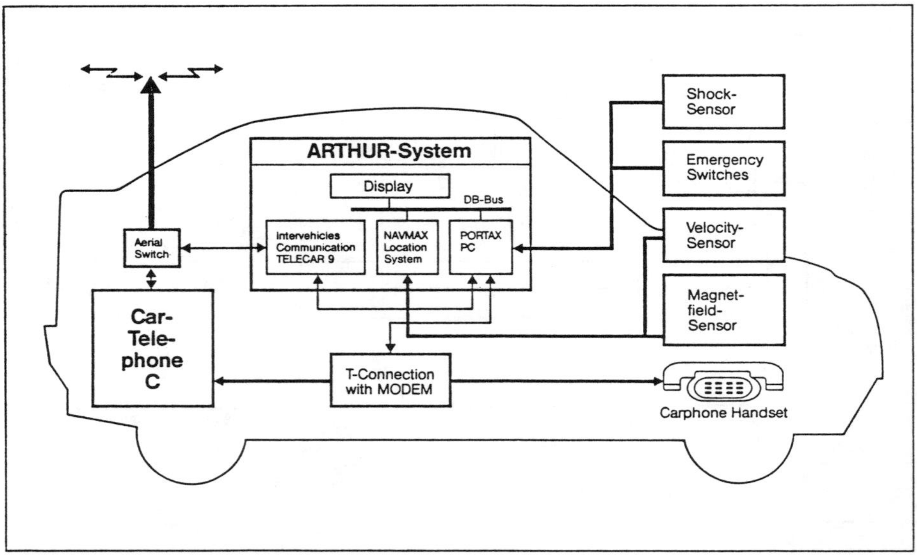

Figure 9.5 ARTHUR.

date, ARTHUR uses PRO-NET techniques to pass information directly and immediately to other vehicles. The project will investigate the use of cellular radio systems (see also Chapter 3) in order to communicate directly with emergency services.

MOMS—Mobile Messaging System

The objective of this project is to integrate the *Radio Data System Traffic Message Channel* (RDS-TMC) information system (see Chapter 5), mobile data systems (specifically MOBITEX, the Swedish digital data radio system) and mobile telephones to demonstrate a range of possibilities for RTI systems based on intervehicle communication.

In MOMS, the RDS system will be directly linked to a road and weather monitoring station able to transmit local traffic information and data on current and forecast weather conditions. The MOBITEX mobile system will be used to demonstrate a system of hotel reservations using mobile terminals in vehicles. The project will include some investigation of the potential for the new Pan European mobile telephone standard (GSM; see Chapter 3), which is expected to operate throughout Europe by the mid-1990s, and this part of the project may be closely linked to the DRIVE SOCRATES project investigating the full potential for using GSM for traffic information systems (see Section 9.4.3 and Figure 9.8).

Satellite Communiciations

This project will demonstrate the potential for satellites to provide the communication medium between driver, vehicle, and control center. The geostationary satellite MARECS-82 (see Chapter 4) will be used for digital data transfer in the L-band (frequency range 1.5–1.6 GHz) at a transmission rate of 300 baud.

CARMINAT

This demonstrator project within PROMETHEUS will establish links with the separate EUREKA project, CARMINAT (see Chapter 6), which in turn integrates developments in vehicle navigation systems and on-board vehicle information and human-machine interface systems. CARIN is the autonomous navigation system being developed by Philips, which will use compact disc for extensive on-board map storage (see Chapter 8 for a full description of CARIN). CARMINAT will use route-finding algorithms to present the driver with clear instructions to follow for the best route to his or her destination. ATLAS is an on-board system for integrating the exchange of information and control between the driver and electronic systems aboard the vehicle. CARMINAT will integrate these systems, and include the use of RDS to provide real-time traffic information for enhancing the route-finding capabilities of the CARIN system.

AUTOGUIDE

ALI-SCOUT [Siemens, 1986] and AUTOGUIDE [Catling and Belcher, 1989] are dynamic route-guidance systems which represent one of the most important developments in RTI systems; the UK Department of Transport's guidelines for the *London Pilot System* [Department of Transport, 1989] anticipate that a full commercial system for London could be in place by 1993. The *London AUTOGUIDE Demonstration Scheme* [Belcher and Catling, 1987], which has been operational since April 1988 (see Figure 9.6), includes two vehicles purchased and specially equipped within the PROMETHEUS program. We expect that the implementation of the next stage toward route guidance in the UK, namely the major London pilot system planned to be operational in 1992, will be implemented with the cooperation and participation of PROMETHEUS partners.

The main criterion that the British government has required of potential commercial operators of AUTOGUIDE is close adherence to the draft standard for road-to-vehicle communication that was developed by a joint Anglo-German working party, now expanded to include other countries as members. The standard allows for bi-directional communication between vehicles and beacons typically mounted at traffic signal installations.

The primary purpose for such an infrastructure is to provide dynamic route guidance to vehicles, so that recommended routes vary with traffic conditions almost in real time. However, the widespread installation of such an infrastructure would

Figure 9.6 The London AUTOGUIDE Demonstration System installed in a vehicle.

provide the basis for many of the RTI applications envisaged in both PROMETHEUS and DRIVE. Close cooperation during the AUTOGUIDE pilot installation, and its subsequent expansion to full commercial operation in London and all of the UK, is therefore important. More details of AUTOGUIDE, and of the technically compatible ALI-SCOUT system, which is the subject of the major field trial project LISB in Berlin [von Tomkewitsch, 1987], can be found in Chapter 6.

Longitudinal Control

This project will be specifically designed to develop a copilot system capable of electronically controlling the drive train and braking system throughout the vehicle's speed range. The project will include sensors to determine the distance to the vehicle ahead, and will also rely on information received via PRO-NET and PRO-ROAD systems. The longitudinal control is closely connected to the cooperative driving project.

9.3.8 Basic Research

In addition to the three main areas of industrial research (led jointly by the automotive manufacturers and the electronics and supply industries), PROMETHEUS includes four further areas of research, which are led by academics, research institutes, and highway and traffic authorities. Three of these constitute the basic research on technologies; the fourth is PRO-GEN (see Section 9.3.9).

PRO-ART

This subprogram deals with the development of artificial intelligence techniques specifically for application within vehicles and traffic systems of the future. Sandeval [1989] describes some 30 subprojects which will contribute to the overall PRO-ART goals. These concentrate on the development of methods for the following two capabilities in the vehicle:

- a *maneuvering capability* (i.e., the vehicle's ability to regulate its trajectory and speed), within the framework of an overall plan, taking the positions and movements of other vehicles into account, and cooperating with them as appropriate; and
- a *dialogue capability* (i.e., the ability of the vehicle as an integral system to maintain a satisfactory dialogue with the driver-user). The need for an AI contribution to such a dialogue capability arises because of the accumulation of facilities in the vehicle which serve a longer time frame, going beyond the driver's direct control of the vehicle.

PRO-CHIP

PRO-CHIP has as its objective the development of integrated microelectronics systems, which will be required by the applications being developed in the industrial research programs. The research program (described in [PROMETHEUS PRO-CHIP, 1989]), distinguishes five main areas of research in PRO-CHIP:

- technology support;
- sensing;
- intelligent processing;
- communicating; and
- actuating and indicating.

PRO-COM

This area deals with research into communication systems. PRO-COM is designed to take a fundamental look at the requirements of communication both within and between vehicles, and between vehicles and control centers.

The PRO-COM work plan (described in [PROMETHEUS PRO-COM, 1989]) distinguishes between the requirements of PRO-CAR, PRO-ROAD, and PRO-NET. It deals with the following communication aspects:

- protocol selection strategies;
- protocol design techniques;
- communication transmission techniques;
- communication protocols;
- a combined PRO-NET–PRO-ROAD network;
- communication management; and
- information handling.

9.3.9 PRO-GEN

PRO-GEN represents mainly the highway authorities of the European countries participating in PROMETHEUS, and has the difficult objective of redesigning the traffic system of the future; PRO-GEN will do this by evaluating the many different PROMETHEUS subsystem developments, and integrating the successful ones into a coherent framework.

Following an initial review and requirement analysis, the main PRO-GEN work program (PROMETHEUS PRO-GEN, 1989) was originally drafted in 1987 to cover the system design aspects of the evolution of an integrated system of road traffic. The program covered seven main areas:

- Traffic system framework—this work deals with the overall framework of the traffic system of the future, its likely constraints and requirements, and levels of traffic demand.
- System design—this work area is intended to impose a top-down approach to research in individual industrial projects, so that the developments which result are suited to an integrated traffic system of the future.
- Behavioral aspects—the main task in this group is to describe and predict how the PROMETHEUS user, and groups of users and nonusers in society, will react and behave in relation to various future characteristics and functions of PROMETHEUS systems, and thus influence developments taking behavioral aspects into account.
- System simulation—by the mathematical representation of key traffic variables, models can be designed that represent a dynamic environment for the study of the effects of PROMETHEUS functions.
- Evaluation—this is concerned with the global evaluation of proposed systems and functions, using the results of simulations, and the overall assessment of subsystem developments against the PROMETHEUS objectives of increased traffic efficiency, improved safety, and reduced environmental consequences.
- System introduction—this deals with the administrative, legislative, institutional and financial aspects of the introduction of systems developed within PROMETHEUS.
- Safety—because of the key place of traffic safety in the objectives of PROMETHEUS, a special safety group was established to monitor the effects of PROMETHEUS developments in relation to safety, and to guide industrial research to maximize the potential for safety improvements.

The work program was drawn up in association with the industrial and basic research programs within the PROMETHEUS project. Recognizing the need to evaluate the traffic system of the future and taking into account the requirements of society as a whole, it includes the objectives of highway and traffic authorities as well as those of the driver.

Because the DRIVE program so closely matches that of PROMETHEUS in many respects, but particularly overlaps with the work program of PRO-GEN, it was decided that within PROMETHEUS PRO-GEN should take responsibility for coordinating bids for DRIVE projects. This was done with remarkable success, as described in Section 9.4.

Following the start of work within DRIVE, designers of the PROMETHEUS program, but in particular PRO-GEN, reassessed its work program to remain both complementary to DRIVE and supportive of PROMETHEUS. This has been done successfully. As DRIVE progresses, some of the activities to be covered in PRO-ROAD, concentrating on the development of infrastructure-supported systems, will be well covered in DRIVE. Cooperative driving and active driver support may depend more on intervehicle and intravehicle systems, and therefore fit more naturally within PROMETHEUS. The revised PRO-GEN work program recognizes this slight change of emphasis within PROMETHEUS.

9.4 DRIVE

The DRIVE (*Dedicated Road Infrastructure for Vehicle Safety in Europe*) program is an initiative of the Commission of the European Community with three main objectives:

- to improve road safety;
- to improve road transport efficiency; and
- to reduce environmental pollution.

These objectives are to be achieved through the application of RTI, both to road vehicles and to the roadside infrastructure. The approach is to work toward the development of an IRTE, in which different RTI applications are brought together with common components, infrastructure, and strategies.

The program was approved by the European Council of Ministers for a three-year period, during which funds of about 60 million ECU are being made available to support precompetitive research. Research funding is on a 50% basis, whereby input by the organizations involved is matched by the Commission's contribution.

9.4.1 Development of the DRIVE Program

In 1985, the Commission initiated a pilot study to assess the many RTI developments in Europe, recognizing the potential role that the European Community could play in harnessing the different approaches being taken, and helping to move toward standards and integrated systems. This was particularly so for the provision of roadside infrastructure equipment, for which pan-European compatibility would be important.

The results of the pilot study were presented at a seminar [DRIVE, 1986] in Brussels in September 1986, followed by a series of workshops designed to gather expert opinion from all of Europe.

These resulted in a proposal by the Commission to the Council of Ministers in July 1987. The proposal included an outline workplan for a three-year program of research.

A National Representatives Committee was established, consisting of government officials from the member states (see Table 9.1 for a list of member states), with a parallel working group containing traffic, transport and communication systems experts from industry, research establishments, and administrations throughout the EC. The working group assisted the commission in preparing the complete workplan, which was made available in its final draft form [CEC, 1988] in April 1988, following a wide consultation of all relevant European organizations (about 65 in all). A Management Committee was established following the formal adoption of the program by the Council of Ministers. The Management Committee approved the workplan at its meeting of June 1988.

9.4.2 The DRIVE Workplan

The workplan is organized into five sections, as shown in Figure 9.7.

Management and Systems Approach

The management tasks are designed to ensure that a systems approach is taken which can integrate the "top-down" analysis of objectives with the "bottom-up" development of industrial prototypes. DRIVE seeks to bring together the traditionally independent fields of information technology and telecommunication with traffic engineering, to produce an IRTE.

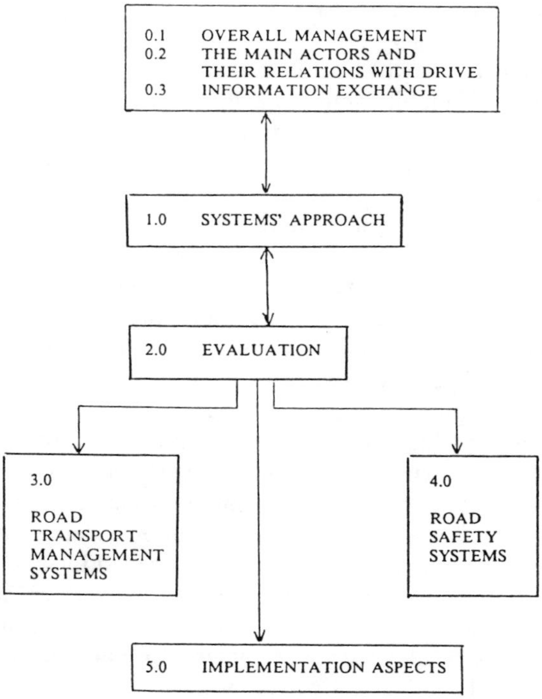

Figure 9.7 The DRIVE workplan.

Evaluation

The purpose of evaluation is to provide a basis for decision-making, so DRIVE contains a structured evaluation framework. The framework will play three roles in the evolution of new systems. It will:

(1) Help to indicate how existing subsystem designs should be modified, as part of an iterative design process;
(2) Identify which subsystems should be considered as viable candidates for implementation; and
(3) Help in steering the amount of resources allocated at a later stage to the development of each potential subsystem.

Computer-based modeling techniques form an important part of the analytical process. These will comprise a number of distinct submodels, but the overall modeling procedure can be thought of as a single "reference model" within an evaluation framework.

Road Transport Management Systems

The foundation of the IRTE will be a communication system enabling information to be exchanged between mobile vehicles and traffic control centers. Developments in route guidance systems, which will form a major part of significantly more efficient traffic systems in the future, are already underway in West Germany and the UK (see Section 9.3.7 and Chapter 6).

In addition to incorporating these developments in infrastructure-supported route guidance systems, the IRTE will be concerned with driver information systems in general, including such current developments as the RDS-TMC (see Chapter 5).

IRTE's scope is much wider, however, and includes advanced traffic control methods, including incident detection, congestion and traffic monitoring systems, and the development of integrated strategies for advanced traffic control. The workplan includes the development of road, weather and pollution monitoring systems. The IRTE will support parking management, control and information systems, and automatic debiting systems (which may or may not be part of an integrated demand management strategy). Road pricing is becoming an increasingly important issue in Europe. The technical success of the Hong Kong pilot system [Catling and Harbord, 1986] paved the way for a number of European investigations; The Netherlands has made the political decision to implement a widespread road pricing system by 1992 [Ministry of Transport and Public Works, 1989]. Ramp metering and other access control systems will also be covered (see Chapter 7 for more details of ramp metering and other aspects of Urban Traffic Control systems).

DRIVE is not just concerned with private traffic, however. Public transportation, passenger information systems and vehicle scheduling and control systems should also be integrated within an overall system, as should commercial vehicle fleet management systems.

The integration of all these systems will eventually enable such services as trip planning systems, which may be available in homes to allow preplanning of journeys, allowing passengers to decide on both the mode and time of travel.

Within the DRIVE program, as in PROMETHEUS, there are also more advanced concepts such as position keeping and convoy control systems, which could radically change driving conditions of the future and significantly increase highway capacities.

Road Safety Systems

A significant part of the program is dedicated to improved road safety. The primary technical objective is to make significant advances in collision avoidance assistance and the improvement of the human-machine interface to reduce accidents.

Implementation Aspects

The key to an integrated pan-European IRTE is standardization. The DRIVE workplan therefore addresses the development of standards for vehicle and roadside equipment, but most importantly for the communications systems which, when standardized, will allow the development of separate but compatible RTI systems in different countries in Europe. The workplan also addresses legislative, administrative and financial aspects of the implementation of RTI systems within an overall IRTE.

9.4.3 Start of the Drive Program

Following approval by the European Council of Ministers, a formal invitation to submit proposals was issued on July 2, 1988 for projects to be carried out under the rules established by the European Community's "framework" research and development program. Proposals were required to be submitted by consortia representing at least two European countries, and containing at least one industrial partner. They were to address specific areas of the DRIVE workplan, and to cover research and precompetitive development during the DRIVE time frame of 36 months.

Support is normally given at 50% of the full cost of the research, so an important principle was that the industrial and commercial interests in particular should be strong enough to justify the research proposed.

By the closing date in October 1988, nearly 200 proposals had been received for about 5000 man-years of work at a proposed cost of over 500 million ECU (1 ECU = \$1 US, approximately). Behind these proposals were some 650 manufacturers, universities, research institutes, as well as user and other organizations.

The number of proposals was sufficiently large for effective competition to take place. At the end of October 1988, a two-week evaluation period took place, during which more than 60 experts from throughout the Community assessed the proposals on their technical merits. This resulted in recommendations for the retention of about 30% of the proposals, amounting to about 11000 man-months of work, plus about 500 from EFTA participants (i.e., non-European Community; see Table 9.1).

Following a negotiation period, nearly 60 projects were successfully contracted, and most of them began at the start of 1989, covering the vast majority of the workplan. These projects are listed in Table 9.4 and described in more detail in CEC [1989]. Together they involve some 450 participants. Some minor parts of the workplan were not addressed by these projects, so a small proportion of the program's budget was retained for a supplementary call for proposals in mid-1989.

For the purposes of managing the overall program, the projects have been divided into seven main groups, basically in accordance with the original workplan outlined above.

Evaluation and Modeling

Fourteen projects are concerned with modeling the effects of RTI systems in Europe, and with evaluating new systems against the three main objectives of the DRIVE program. Together they provide a balance between innovatory and established procedures. They include behavioral modeling, traffic simulation, emission prediction, demand modeling, and guidelines for field trials and their evaluation.

Behavioral Aspects and Traffic Safety

Fourteen projects deal with the following topics: accident data analysis, vulnerable road users, behavioral evaluation, human-machine interface, collision avoidance, data recording (including one project, *DRACO,* developing the equivalent for road traffic of the aircraft accident black box), effects and implementation.

Traffic Control

Seventeen projects cover the following subject areas: demand management, parking management and information systems, urban and interurban traffic signal control systems, incident detection and congestion prevention, tidal flow systems, tunnel control systems, AI applications, and road and weather monitoring systems (see also Chapter 7).

Route Guidance, Vehicle Location, Maps and In-Vehicle Information Systems

This group of eight projects includes: the integration of route guidance with UTC, linked closely with the development of ALI-SCOUT [Siemens, 1986] in West Germany

Table 9.4
DRIVE Projects and their Prime Contractors

V1001: Public transport scheduling CGA-HBS ALCATEL F	V1002: Short range microwave links CGA-HBS ALCATEL F
V1003: Requirements and systems specif. MIZAR AUTOMAZIONE SPA I	V1004: Dev. of vehicle mounted device TUV RHEINLAND D
V1005: PREDICT CASTLE ROCK CONSULTANTS UK	V1006: Factors in elderly people's abilities KING'S COLLEGE UK
V1007: SOCRATES IAN CATLING CONSULTANCY UK WITH TATE ASSOCIATES LTD. UK	1008: Strategies for integrated demand CASTLE ROCK CONSULTANTS UK
V1009: Vehicle location systems FIAR I	V1010: PANDORA AUTOMOBILE ASSOCIATION UK
V1011: Integration of route guidance SIEMENS AG D	V1012: Road safety management system UNIV LEUVEN B
V1013: Comparative eval. radiating cables INRETS-CRESTA F	V1014: IMAURO CENTRE DE RECHERCHES ROUTIERES B
V1015: Artificial intelligence systems INRETS F	V1016: Infor. syst. for road user safety device TFK/VTI TRANSPORTFORSCHUNG D
V1017: Changes in driver behaviour HUSAT RESEARCH CENTRE UK	V1018: Total traffic management environment IAN CATLING CONSULTANCY UK
V1019: CASSIOPE Computer aided System CETE-MEDITERRANNEE F	V1029: Tidal flow systems HEUSCH/BOESEFELDT GMBH D
V1021: European digital road map DAIMLER-BENZ AG D	V1022: Real time urban traffic control system GTM ENTREPOSE F
V1023: EUROTOPP OXFORD UNIVERSITY (TSU) UK	V1024: Driver Information Systems HEUSCH/BOESEFELDT GMBH D
V1025: EURONETT OXFORD UNIVERSITY (TSU) UK	V1026: Integr. of computer vision techniques ETRA E
V1027: EUROFRET TRADEMCO CONSULTANTS GR	V1028: TUNICS SAIT ELECTRONICS B
V1029: Standards for RDS-TMC throughout Europe CASTLE ROCK CONSULTANTS UK	V1030: Microwave communications UNIVERSITY OF NEWCASTLE UK
V1031: System for vulnerable road users UNIVERSITY OF LEEDS UK	V1032: STRADA CETE-MEDITERRANEE F

Table 9.4 (cont'd)

V1033: Automatic policing information systems RIJKSUNIVERSITEIT GRONINGEN NL	V1034: Road inform. and management systems BELCOTEC B
V1035: Motorway traffic flow and control INRETS F	V1036: Evaluation process for road informatics TECHNISCHE UNIVERSITAET MUNCHEN D
V1037: Definition of standards INRETS F	V1038: DACAR BAKKENIST MANAGEMENT CONS. NL
V1039: Application of artificial intelligence FORSCHUNGSZENTRUM INFORMATIK D	V1040: Identification of hazards HUSAT RESEARCH CENTRE UK
V1041: Generic intelligent driver support systems (GIDS) UNIVERSITY OF GRONINGEN, TRC NL	V1042: Accident data collection and analysis TECHNISCHE UNIVERSITAT MUNCHEN D
V1043: Drive integrated telecommunications BRITISH TELECOMMUNICATIONS PLC UK	V1044: Freight log efforts for European traffic DORNIER GMBH D
V1045: Parking management, control and information NATIONAL TECHN. UNIV. OF ATHENS GR	V1046: FRIDA NATIONAL TECHN. UNIV. OF ATHENS GR
V1047: OD information vs traffic control CENTRO STUDI SUI SISTEMI DI TRASPORTO I	V1048: Advanced Control Strategies CENTRO STUDI SUI SISTEMI DI TRASPORTO I
V1049: Field trials ZELT F	V1050: Driving and accident co-ordinating QUEEN MARY COLLEGE UK
V1051: Procedure for safety submissions TUV RHEINLAND D	V1052: ICARUS UNIVERSITY OF SOUTHAMPTON UK
V1053: Model. of emissions & consumpt. in urban areas INRETS F	V1054: System and scen. simul. for test. RTI syst. UNIVERSITAT POL. DE CATALUNYA E
V1055: AI techniques for traffic control AUTOMA I	V1056: Incident detection congestion system TRRL UK
1057: Drive System Engineering and Consensus Formation Office (SECFO) DAIMLER-BENZ AG D	1058: ROAD VEHICLE RESEARCH Road conditions and weather monitoring TNO NL
1059/2: Strategies for preventing congestion WOOTTON JEFFREYS CONS. LTD UK	

and AUTOGUIDE [Catling and Belcher, 1989] in the UK; driver information systems; digital mapping and road database projects, including the European Benchmark project, proposed and led by PROMETHEUS participants; standards for RDS-TMC; satellite location systems; and a potentially important new project, SOCRATES (*System of Cellular Radio for Traffic Efficiency and Safety*).

This project (see Figure 9.8) is investigating the potential for a system based on the forthcoming GSM pan-European cellular radio standard, and is particularly designed to develop a route guidance system which can integrate with autonomous systems such as CARIN and EVA (see Chapter 3 for more information on cellular radio, Chapter 6 for route guidance, and Chapter 8 for a detailed description of CARIN).

Public Transportation and Freight Management

Four projects are designed for the integration of RTI systems specifically for goods vehicles and public transportation within the IRTE. Two deal with specific issues for public transport operation: scheduling, network planning, database management, real-time strategies and control, traffic effects, and driver and passenger information systems.

Two projects are designed for investigating user and market requirements for freight transport. One concentrates on the development of strategies and will analyze the areas in which RTI can be used to improve road freight operations; the other project focuses on the integration of fleet management systems, and will generate technical specifications.

Telecommunication

Besides the SOCRATES project, which includes cellular radio, there are eight other projects investigating the many possible communication systems and media which might form part of the IRTE infrastructure. Three deal with an overall approach to communications, and the others are concerned with specific transmission media. These include one project developing microwave links specifically for traffic monitoring and road pricing.

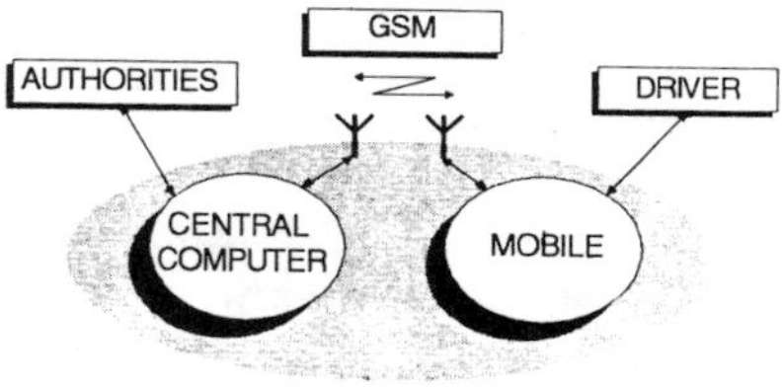

Figure 9.8 The SOCRATES project.

This group of projects is of fundamental importance to the development of DRIVE and of the IRTE. The basic hypothesis of the IRTE is that a common communication infrastructure should provide for all the RTI applications within it:

route guidance;

advanced traffic control;

parking management and information systems;

fleet management;

public transportation management and information systems;

hazard warning;

driver information;

tourist information;

data for traffic management and traffic planning; and

trip planning.

One of the questions to which DRIVE designers seek an answer is the nature of a common communication infrastructure: three projects are investigating the use of microwave frequencies, which are currently being implemented in a number of automatic toll collection systems around the world (the first fully operational system, as opposed to experimental evaluations, was the system in Alesund [JIHT, 1988], which uses the Philips PREMID identification system).

However, the use of infrared beacons is already becoming widely established with the LISB and AUTOGUIDE projects in West Germany and the UK. Infrared is not subject to the same restrictive frequency allocation requirements as microwave, but the debate as to which, if either, should provide the single IRTE communication medium has not yet been resolved.

An alternative solution is the use of the GSM system as proposed by SOCRATES designers. Although there are a number of outstanding issues, for example over its use for automatic debiting, the great advantage of the GSM system is that the infrastructure will be installed over all of Europe, and will not need separate funding for RTI applications.

The DRIVE telecommunication project designers will seek to resolve these issues by identifying the strengths and weaknesses of the different contenders.

Systems Approach and Concensus Formation

There are two projects which are specifically designed to ensure that the overall objective of a single IRTE is kept firmly in view.

TARDIS (*Traffic and Roads: DRIVE Integrated Systems*) aims to specify the functional requirements of the IRTE. Early success was achieved with it by providing the forum for agreement on common requirements for automatic debiting systems [DRIVE, 1989]. The groundwork has been laid for a common system which will be used for both road pricing in the Netherlands and toll collection on French, Spanish, and Italian motorways.

SECFO is the *System Engineering and Consensus Formation Office* proposed by PROMETHEUS designers, which seeks to establish that DRIVE developments do indeed fulfill the requirements of the many groups concerned with the development of RTI systems (see Figure 9.9).

9.5 THE FUTURE

PROMETHEUS and DRIVE are concerned with possible futures. Within their program structures, they are each designed to identify scenarios for road transportation during the next twenty or more years, and to lay the groundwork for the implementation of new systems in a coordinated, rather than haphazard, way.

Some of the ideas are very close: TRAVELPILOT, the autonomous electronic map display, which is the European version of ETAK [Bosch, 1989], was actively marketed during 1989; AUTOGUIDE systems should be commercially available during the early 1990s. Automatic debiting systems are already operational in Norway, and will be implemented in the form of road pricing in the Netherlands by 1992.

Beacon-based systems potentially offer the means of achieving many RTI applications in a straightforward way. Dynamic route guidance will allow the widespread installation of beacons, which will soon become multifunctional. The actual frequencies to be used might prove difficult to resolve, but there are strong arguments for a single communication infrastructure. However, there are commercial pressures for each of the different types of possible beacon system; the level of compatibility between different systems will, in particular, be a measure of DRIVE's success.

An alternative to beacon-based systems is that being used by designers of SOCRATES, who are investigating the use of the forthcoming pan-European cellular

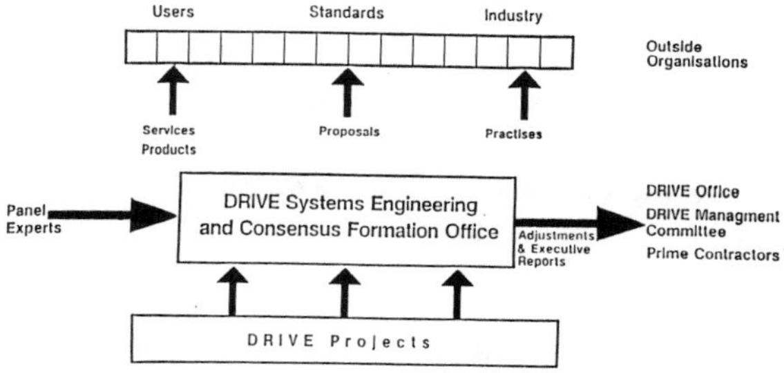

Figure 9.9 The SECFO project.

radio system, GSM, for RTI applications. GSM will not be widely available until the mid-1990s, by which time beacons may be fairly widely, but by no means universally, installed. Again, DRIVE may have an important role in achieving the installation of RTI systems in a complementary way to each other, rather than being directly competitive.

In the longer term, the potential of intervehicle communication, on which PROMETHEUS is concentrating perhaps more than DRIVE, is enormous. The implementation aspects are likely to be profound for any system capable of sharing the vehicle control task with the driver. Nevertheless, the results to date are encouraging.

In studies of future traffic scenarios, in fact, there is a fair degree of unanimity. Route guidance will become commonplace by the end of the century. Many of the associated RTI applications will have been implemented within the same sort of time scale. The totally automatic car is not likely within the foreseeable future, whereas some of the semiautomatic concepts such as the intelligent cruise control are rapidly gaining credence. An analysis of possible timescales for implementation was carried out in the ARISE project (described in [Sviden 1989]).

9.6 CONCLUSION

The time is right in Europe for the rapid development and introduction of many systems based on new technologies. These have the potential to make the whole road traffic system significantly safer, more efficient, and less environmentally damaging. PROMETHEUS and DRIVE designers recognize that, to achieve the maximum effectiveness, many different strands need to be coordinated.

PROMETHEUS has now become well established as a world-leading research program that is the envy of many outside Europe. The PROMETHEUS workplan is broad, addressing all aspects of RTI applications, although the plan tends to emphasize the potential of cooperative driving techniques.

DRIVE is complementary to PROMETHEUS in that it concentrates on the infrastructure potential from RTI systems. DRIVE provides a framework for the rapid development of RTI systems within the overall context of the IRTE.

With the introduction of LISB in Berlin and the London AUTOGUIDE pilot system, the first steps toward a dedicated road infrastructure have already been taken. CARIN, EVA, and TRAVELPILOT are autonomous on-board systems all expected to be on the market during the early 1990s. RDS trial systems are well on the way to the effective utilization of the extra broadcasting capacity offered by the Traffic Message Channel. SOCRATES opens the possibility of a widely available two-way communication link dedicated to the exchange of traffic information.

All of these projects are entirely within, or have very close links to, either PROMETHEUS or DRIVE. Designers of the two programs have maintained close liaison throughout the development of the work programs, and they are likely to continue to do so to bring about radically improved traffic systems of the future as soon as possible.

REFERENCES

Belcher, P., and I. Catling, "Electronic route guidance by Autoguide: the London Demonstration," *Traffic Engineering and Control,* Vol. 28, No. 11, 1987, pp. 586–592.

Bosch, "Bosch Travelpilot Vehicle Navigation System," Bosch, 1989.

Catling, I., and P. Belcher, "Autoguide—electronic route guidance for London and the UK," *IEE Conference "Road Traffic Monitoring,"* Publication No. 299, February 1989, pp. 182–190.

Catling, I., and B. J. Harbord, "Electronic road pricing in Hong Kong—the technology," *Traffic Engineering and Control,* October 1986.

CEC, Commission of the European Communities, "DRIVE Workplan," DRI 100, Draft of April 26, 1988.

CEC, Commission of the European Communities, DGXIII, "The DRIVE programme in 1989," DRI 200, March 1989.

Daimler-Benz, Daimler-Benz simulator [undated].

Department of Transport, "Autoguide—guidelines for pilot Autoguide system proposals," Issued by the United Kingdom Department of Transport in connection with Part II of the Road Traffic (Driver Licensing and Information Systems) Bill, January 1989.

DRIVE, *DRIVE Seminar Proc.,* September 1986.

DRIVE, "Automatic debiting systems," TARDIS Inception Report No. 5, April 1989.

EUREKA, Eureka Project List, Eureka Secretariat, Brussels, 1986.

JIHT, "First "no-stop" road toll station," *Journal of the Institution of Highways and Transportation,* February 1988.

Ministry of Transport and Public Works, the Netherlands, "Road pricing project—programme of requirements," 1989.

PROMETHEUS PRO-CHIP, PROMETHEUS PRO-CHIP programme description, May 1989.

PROMETHEUS PRO-COM, PROMETHEUS overview of PRO-COM research, May 1989.

PROMETHEUS PRO-GEN, PROMETHEUS PRO-GEN work programme, June 1989.

Sandeval, E., Overview of National Plans for PRO-ART Research, in *PROMETHEUS Research Newsletter* No. 3, PROMETHEUS Office, Stuttgart, April 1989.

Siemens, Ali-Scout system description, 1986.

Sviden, O., "Scenarios," Linkoping Studies in Management and Economics, Dissertation No. 19, 1989.

von Tomkewitsch, R., "LISB: large-scale test 'Navigation and Information System Berlin,'" *Proc. ITTT Seminar,* P302, PTRC, 1987.

Chapter 10
The Intelligent Car

M. H. Westbrook
Ford Motor Company Limited

The rapid development of electronic technology in the last twenty years has brought it to a point where it is practical to use electronic control systems, with suitable associated sensors and actuators, to directly replace a number of the existing mechanical or electromechanical systems in the motor vehicle.

This is being done where there is a need for the improved accuracy, speed and reliability available from electronics, and also to provide new functions and features where these are necessary either to meet legal requirements, to provide new competitive product features, or to improve the performance or operation of the vehicle.

This chapter describes the applications of electronics that have already taken place, and will also look at the automotive electronic systems expected to be introduced over the next few years, and which will result in what we could justifiably call *the Intelligent Car*.

10.1 INTRODUCTION

The use of electronics in vehicles started with the first installation of car radios in the early 1930s, but it was not until the advent of the low cost transistor in the 1950s that the electronic engineer's thoughts were seriously turned toward applying electronics in any other area of the vehicle.

The performances of high-speed car engines had always been limited by the loss of spark energy in the ignition system at high engine speeds, due to the limited time available to charge the coil with the short contact period available in the cam-contact breaker ignition system. The system also suffered from a relatively short life and a continuing deterioration in timing accuracy due to wear between the cam and cam follower, and the buildup of the deposits of arcing on the contact surfaces. In fact, if these two effects had not been in opposite directions and partially canceled each other, the deterioration in performance with time would have been much more rapid than it actually was.

Automotive designers realized during the 1950s that if the coil could be controllably charged using a power transistor (suitable Darlington power transistors were then becoming available) and triggered by a noncontact inductive or photoelectric method that had purely a timing function rather than a timing and current switching function, the system would operate without these major disadvantages. So, *breakerless* electronic ignition was born, and first used successfully in the racing and rally cars of the 1960s.

This new ignition system, however, was not adopted for use in mass produced cars for many years for a reason that is critical when any application of new technology is considered for the motor vehicle—the electronic ignition cost significantly more than the contact breaker system it replaced!

One must appreciate the extremely critical nature of costs in motor vehicles to understand why it has taken so long for electronics to become a significant factor in automotive design. Electronic breakerless ignition thus was in existence as an available improvement, but was not adopted outside the specialist car market until the increasing environmental pressures in the US and, in particular, in California, on the "smog" issue forced an agonizing appraisal of what could be done to reduce the levels of carbon monoxide (CO) and hydrocarbons (HC) emitted from the exhausts of US vehicles. This appraisal led to introduction of the first emission control regulations in California in 1966.

One of the actions quickly established was that better control of ignition timing had played an important part in making an improvement, and said improvement could be maintained throughout the life of the car (and certainly over the required 40,000 mile test distance) if breakerless ignition were used. Designers also discovered that accurate control of the timing advance characteristic of the ignition system, which relates engine speed and load (represented by the manifold vacuum pressure level) to ignition advance angle, was critical in obtaining the lowest emission levels.

This initial introduction of electronics to ignition control triggered much more detailed studies of how engine performance, particularly with respect to exhaust emissions, could be improved by the use of electronics. These studies resulted in the development of electronic fuel injection by Bendix in the US, Bosch in West Germany, and Lucas and Associated Engineering in the UK.

By 1967, Bosch and Lucas were in production with fuel control systems for Europe designed primarily to improve performance rather than reduce emissions. In the US, however, in spite of the early work by Bendix, electronic fuel control was not introduced in production cars until 1975, when the increasing severity of the exhaust emission regulations made the use of catalysts in the exhaust system necessary. With this catalyst, the need for precise electronic methods of controlling fueling was unavoidable. The reason for this reluctance was, of course, primarily cost, but also concerns about the reliability of electronic systems in the very difficult environment of the motor vehicle.

At the time, designers did not fully appreciate how difficult these environmental conditions were, particularly under the hood of the vehicle, where electromagnetic interference is both radiated and conducted from the ignition system and from any switching of inductive loads. Interference is also generated by external sources such as broadcast and radar transmitters, where we now know field strengths of as high as 200 V/m may exist on roads close to such installations.

In addition to this severe electromagnetic environment, substantial mechanical accelerations up to 20*g* have to be protected against, as well as extreme temperature fluctuations from −30° C to +120° C for components under the hood. In fact, these environmental conditions, in general, are significantly more severe than those encountered in aircraft and spacecraft, and comparable to those found in many military applications. In the automotive case, these extreme conditions must be withstood by low-cost electronics that is economically possible; the very sophisticated protection used in military installations cannot be employed.

However, with cooperation between the electronics industry, the automotive components industry and the vehicle manufacturers, these problems have been progressively better understood. As a result, low-cost systems are more effectively engineered, although their development has proved to be a classic illustration of the old adage: "*good engineering is doing for a penny what any fool can do for a pound.*"

So electronic ignition and electronic fuel injection became established very widely in the US in the late 1970s as the only effective method, in combination with exhaust system catalysts, of meeting the exhaust emission regulations, which by now had added oxides of nitrogen (NOx) to the previously controlled HC and CO emissions.

During the 1970s, electronic engineers had been looking at other possible vehicle systems to which electronics could be applied. The first two to be successfully developed and used in production were the replacement of the electromagnetic relay voltage control system in the alternator, and the replacement of the mechanically driven tachometer by an electronic system which integrated the low voltage ignition trigger pulses.

The electronic alternator voltage control system development brought home to many people, in both the electronics and automotive industries, how severe the environmental challenge in the vehicle was. Many early systems failed catastrophically due to the large voltage spikes (+200 V) generated during switching operations in the alternator circuits, and not until the development and use of protective surge diodes in the circuit was this problem overcome and the present high reliability (over ten times more reliable than the old electromagnetic control system) attained.

There was less of a reliability problem in the case of the electronic tachometer, primarily because it is essentially a system that samples and integrates the low voltage ignition pulses to the distributor so that large voltage spikes from any source are correspondingly attenuated. Also, the tachometer merely provides a useful function, rather than the essential function of the alternator regulator, so that a momentary misreading on a tachometer has no serious effect. This division between the essential functional system (the ignition system) and the useful but nonessential system (the driver's instruments) is important in establishing the reliability required from an automotive electronic system.

In the 1970s and 1980s, many other areas of the vehicle have been considered for the application of electronics, and those for which systems have been developed are shown in Figure 10.1 and will be described in this chapter. In addition to the engine control systems already referred to, those systems include electronic control of transmission ratio and its combination with engine control to provide integrated powertrain control; the replacement of the conventional wiring harness by multiplex wiring; the use of electronic driver instrumentation in place of the familiar mechanical and electromechanical dial and pointer instruments; and developments in the electronic control of the vehicle power supply system. Most importantly, if all these systems are to be effectively maintained, computerized fault diagnostic facilities must be available within and external to the vehicle. Some longer term developments, including the electronic control of suspension characteristics and the possible future use of collision avoidance vehicle radar, will also be discussed.

The driver of the intelligent car of the 1990s will also rely heavily on receiving a wide range of information about traffic conditions, routing, and vehicle location from outside the vehicle, in addition to the entertainment services he receives today.

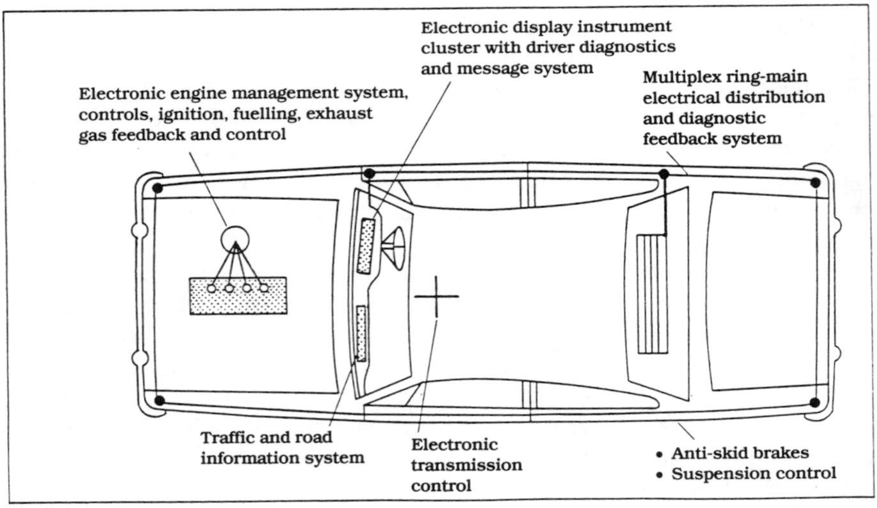

Figure 10.1 Application areas for electronics in the motor vehicle.

The technology being developed to make that possible has been covered in detail in Chapters 5 to 9, and will not be covered in this chapter. But this need for external awareness of the driver and his vehicle should never be forgotten, and illustrates the interactive nature of the internal and external vehicle-traffic-road electronic systems.

10.2 ENGINE CONTROLS

10.2.1 Breakerless Ignition

As we have already mentioned, the early development of breakerless ignition was the first step in the use of electronics for a critical function in the vehicle; however, that only replaced the contact breaker points and cam with an electronic trigger and power transistor to charge and discharge the ignition coil at the correct time in the engine cycle. The really critical factor in controlling the ignition of a gasoline engine is the time in the cycle at which the spark plug is fired and ignition starts. There is an optimum angle or timing before the piston reaches top dead center on each cylinder, at which ignition should commence to obtain the best performance from that engine. This optimum angle is dependent on engine load and speed, and can be represented by a three-dimensional surface giving an individual ignition timing for each load and speed condition. In a conventional distributor, the required changes to timing are produced by physically rotating the distributor with a centrifugal weight system for speed control, and a vacuum diaphragm-lever system (connected via a tube to the inlet manifold vacuum) for load control. This then produces a specific advance spark timing for the ignition for each load and speed condition.

Because of the mechanical limitations of this system, however, it is only possible to produce a relatively simple three-dimensional surface form, whereas, if the engine is put on a test bed under load and optimized for each load and speed condition, a much more complex three-dimensional surface relationship is produced (see Westbrook [1988], Figure 2). The only way to obtain this complex relationship in a practical ignition system is to use electronic control of the advance spark timing. This is done by storing the three-dimensional map as a look-up table in digital form in an electronic memory. Then, using the ignition trigger signal frequency, which gives a measurement of engine speed, combined with the signal from a suitable pressure transducer measuring vacuum pressure in the inlet manifold, the coordinate information required to select the optimum ignition advance from the memory is obtained. This technique was incorporated into the early "engine control" systems used in the US cars of the late 1960s and early 1970s to meet the increasingly severe emission regulations then being introduced, and appeared in Europe in special "economy" versions of cars which normally would have only a simple breakerless electronic ignition system.

10.2.2 Electronic Fuel Injection

The parallel development of electronic fuel injection did not appear in production cars in the US until 1975, when the manufacturers were forced by the increased severity of exhaust emission standards to add this improved control of fueling to the existing ignition control. In Europe, however, Bosch and Lucas were in production with electronic injection by 1967, although its usage was confined to expensive performance cars at a considerable cost premium. The general principle of most electronic fuel injection systems is illustrated in the diagram of Figure 10.2, which shows the widely used Bosch L-Jetronic system.

In this system, an electromagnetic injector, in which a small pintle valve is opened by applying voltage to a solenoid, is placed as near as possible to the inlet valve for each cylinder of the engine, so that if fuel is supplied to the pintle valve at suitable constant pressure, when the valve opens, a fine spray of fuel will emerge at high velocity in the direction of the inlet valve. This fuel will then be sucked into the cylinder, together with the appropriate amount of air, which has been induced into the inlet manifold under the control of the throttle flap valve. The period for

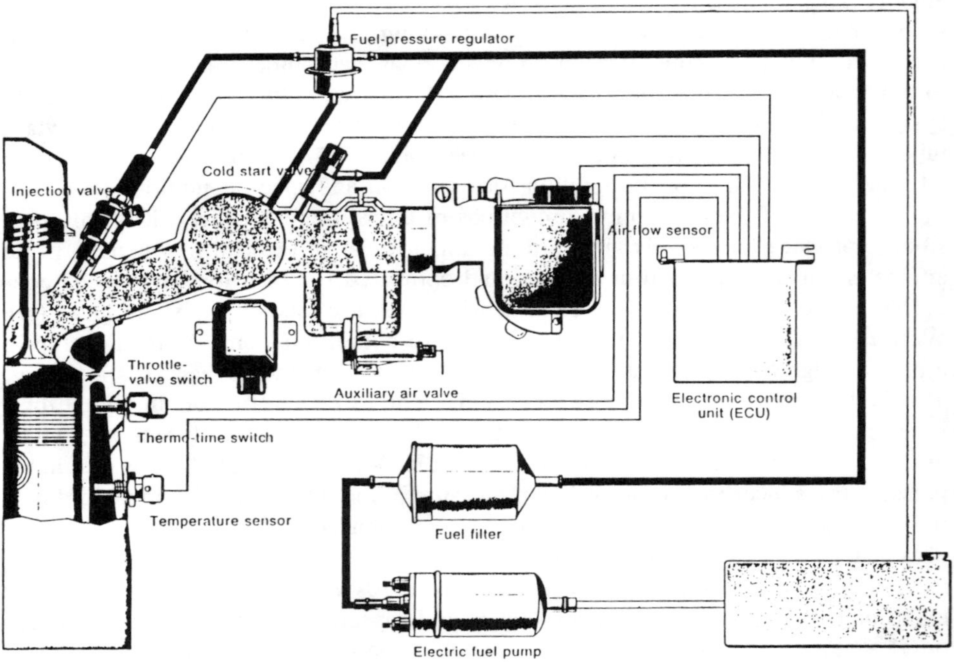

Figure 10.2 Schematic of Bosch L-Jetronic fuel control system.

which this injector valve is open then determines the total amount of fuel fed to that particular cylinder.

In this relatively simple system, the critical parameters necessary to control the time period of injector opening prove to be the same as those required to control ignition timing—engine speed and load. It is therefore possible to establish a three-dimensional surface relating injector opening time to engine speed and load as shown in Figure 10.3 and to use devices similar to those used for the control of the electronic advance ignition system for making speed and load measurements.

In the same way as for ignition, the optimum injector opening time can be established for every speed-load combination, a process known as *engine mapping,* and the resultant matrix information stored in electronic memory to be called up as required.

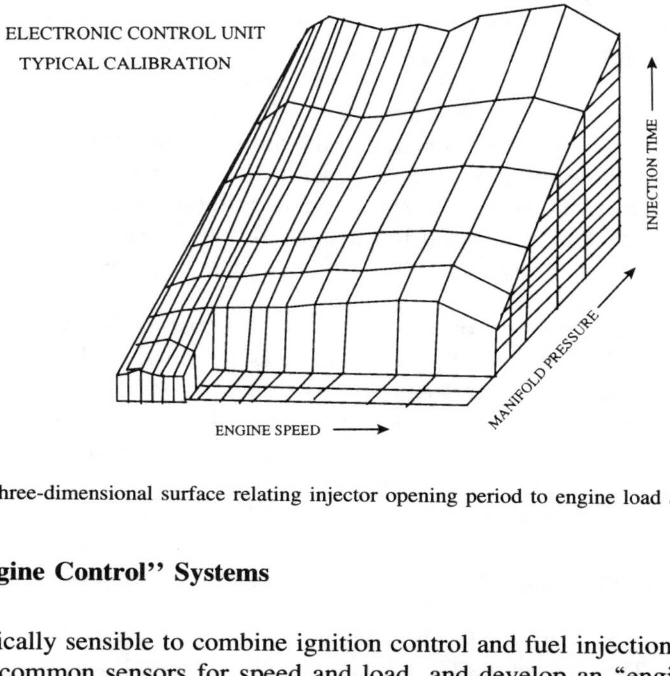

Figure 10.3 Three-dimensional surface relating injector opening period to engine load and speed.

10.2.3 "Engine Control" Systems

It is economically sensible to combine ignition control and fuel injection on a single engine with common sensors for speed and load, and develop an "engine control" system. This has been done, and such systems are available and provide significant improvements over conventional distributor-carburetor methods of controlling ignition and fueling.

Unfortunately, however, on more detailed investigation, we see that the single measurement of inlet manifold vacuum as an indication of load is inadequate to specify accurately the required engine fueling, particularly under transient conditions. What is really required is a measurement of the mass of air entering each cylinder, or failing that, the combined mass of air entering the manifold for all cylinders.

Sensors have been developed to make this measurement, the most widely known and used being the Bosch *air vanemeter,* used in the Bosch L-Jetronic fuel control system shown in Figure 10.2. This uses a simple pivoting vane in the inlet manifold air stream to move a potentiometer according to air flow velocity. An alternative device, which has a much faster response time and does not obstruct the inlet air flow is the *hot wire anemometer,* also developed by Bosch, which has the advantage of measuring true mass air flow directly.

Given the accurate measurement of mass air flow, it is then possible to control the amount of fuel being injected, and therefore the air-fuel ratio being supplied to the engine. However, the control engineer is still faced with two major problems; the first is that the system is open loop, and relies entirely on maintaining the accuracy of the engine calibration made during engine mapping, a condition which cannot be achieved in practice because this calibration will change with fuel quality, humidity, and various types of engine wear (corrections for barometric pressure and ambient temperature are normally included in the system). The second is that a measurement is being made which relates to the particular condition when the intake valve of a particular cylinder is open, and this information is used for the open loop control of the combustion process in different cylinders operating later in the engine cycle. The same cylinder does not operate again until three (or more) other cylinders have fired, and by this time, under transient conditions, the engine will be operating differently. We therefore assume that each cylinder performs in exactly the same way, a condition we know is not true.

The answer to the first of these problems is to use feedback rather than open loop control. A decision then has to be made as to which of the conditions—performance, economy, or exhaust emissions—should be the one for prime feedback optimization, and should therefore be the major feedback parameter. When this has been decided, it is necessary to find a method of making a suitable measurement which describes the condition accurately.

In countries such as the US (and now in Europe as well), where severe restrictions on exhaust emissions exist, the control of exhaust emissions is clearly the prime requirement. With current engine technology, the only way to meet these regulations is to combine the engine control system with a "three-way catalyst" in the exhaust system to reduce the levels of the critical pollutants: CO, HC, and NOx (see Figure 10.4). This determines the fueling condition required, because for its correct operation, such a catalyst requires that the air-fuel ratio fed to the engine should always be as close as possible to the optimum stoichiometric level of 14.7:1. This is achievable by the use of a feedback control system in which the air fuel ratio (λ) is sensed by means of an exhaust gas oxygen sensor (a λ sensor) in the engine exhaust manifold.

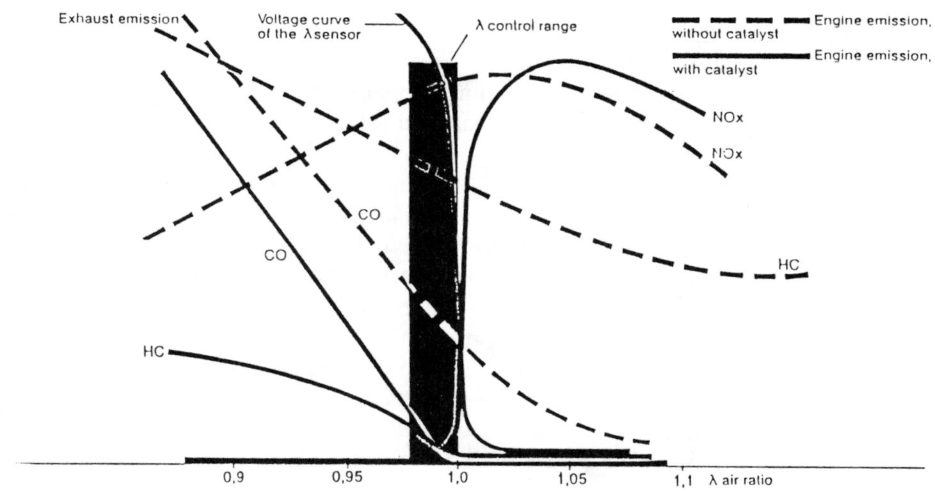

Figure 10.4 Reduction of exhaust pollutants by use of a three-way catalyst.

This sensor (see Figure 10.5) makes use of the fact that the migration of oxygen ions across a suitable membrane or filter from one gas to another is only dependent on the partial pressure of oxygen in the two gases. At the stoichiometric air-fuel ratio, the partial pressure of oxygen in the engine exhaust gas equals that of ambient air, so that if suitable electrodes are attached to each side of the ceramic filter used in the sensor, a voltage is generated by the ion migration when the air-fuel ratio is below or above stoichiometry, with a rapid voltage change occurring over the transition between those two conditions (see Figure 10.4). This voltage transition is ideal as a feedback signal to control the amount of fuel injected, and therefore keep the air fuel ratio at stoichiometry.

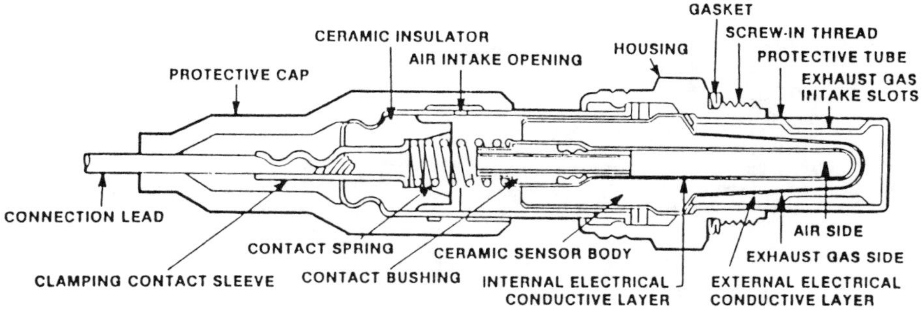

Figure 10.5 Stoichiometric exhaust gas oxygen (EGO) sensor.

Unfortunately, the device requires a number of engine cycles to respond accurately, so that sudden changes in engine speed or load have to be compensated by open loop adjustments based on a three-dimensional map similar to that for ignition timing, but relating transient fueling to input manifold pressure or mass air flow and engine speed.

Developments in emission controlled engines, particularly in Europe, were until recently expected to be toward the use of *lean-burn* technology to obtain low levels of emissions, rather than by the use of the expensive three-way catalyst with stoichiometric engine and with consequent poor fuel economy and performance. Lean-burn operation requires effective control of the engine at air fuel ratios between 14:1 and 22:1, which give much better economy while still maintaining adequate drivability. If this is to be done by feedback methods, then the availability of a lean-burn exhaust oxygen sensor becomes very important, and prototype sensors are now available using a technique shown in Figure 10.6 and known as *oxygen pumping,* in which the oxygen partial pressure within the sensor is positively controlled by applying a voltage across a filter which forces oxygen ions to migrate to or from a small pumping cell. The increased partial pressure of oxygen in this cell is then compared (using a conventional oxygen sensing cell) with the increased partial pressure of oxygen in the exhaust gas stream from a lean-burn engine, and the resultant voltage change produced at balance being as in a conventional exhaust oxygen sensor.

The current in the pumping cell required to produce this balance condition is then a measure of the increase above the stoichiometric level of the air-fuel ratio being supplied to the engine. A lean-burn oxygen sensor of this type still unfortunately suffers from a response time which is long compared to the rate of change of engine conditions, and the associated control system therefore continues to require open loop compensation for transient conditions.

An ideal system would be able to measure some meaningful property of the combustion process in the cylinder quickly enough to be able to control the engine operation accurately on a cycle-to-cycle basis. Methods of doing this could involve

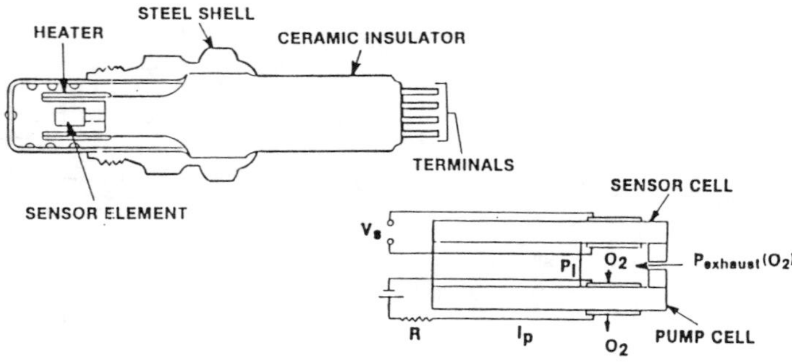

Figure 10.6 Lean-burn EGO sensor.

either the direct sensing of combustion chamber pressure, which would then be used as a measure of both work done [Rhodes and Werson, 1983] and (by timing the pressure peak) of combustion efficiency; or alternatively by the detection of the position and timing of the flame front arrival in a selected part of the cylinder, using an ionization flame-front sensor [May, 1986]. These still would be "after the event" measurements made when the combustion process was almost complete, and could be used only to control subsequent engine cycles. This problem is compounded in a multicylinder engine by the fact that, as we have already said, each cylinder performs slightly differently, and the next combustion in the same cylinder will not come around in a four-cylinder engine until four engine cycles later, so that a decision has to be made whether to accept imperfect fast feedback on the next combustion in a different cylinder, or an imperfect slower feedback to the same cylinder on its next combustion. The ideal method would be to measure some parameter such as inlet gas velocity, which is thought to determine the subsequent combustion event quality, and use this to feed forward through a very fast control system to define and control the ignition timing and the fuel injected for that particular combustion event. However, there is still a long way to go before that becomes a practical possibility.

The electronics necessary to provide any of these more complex control methods is typified in Europe by the Bosch "Motronic" system (Figure 10.7), or in the US by the Ford Electronic Engine Control Module Mk IV (EECIV) (Figure 10.8). Each of these systems provides full three-dimensional map control of ignition and fueling, and can be adapted to either open-loop or feedback control using any available feedback sensor, although currently the exhaust gas oxygen sensor is normally used.

In summary therefore, engine controls can cover a wide range of functions from the simple control of idle speed and fuel cut off during deceleration (as in the BL Maestro [Meyer *et al.*, 1983]) to the full complexity of an optimized ignition, fuel injection and feedback emission control system on a large US car. The continually changing legal and customer requirements in the main markets of the US, Japan and Europe seem destined to generate more and more complexity in the control system unless some international agreement on exhaust emission standards is reached. Recent agreement in Europe (1989) to implement US level emission controls has made it impossible to meet the required emission levels with "lean-burn" engines, and it will now be necessary to go to full three-way catalyst stoichiometric operation on virtually all gasoline engined vehicles by 1992.

10.3 TRANSMISSION CONTROL

Moving to the transmission, which is fitted between the engine and the road wheels, we must understand that its function is that of a power-matching device between

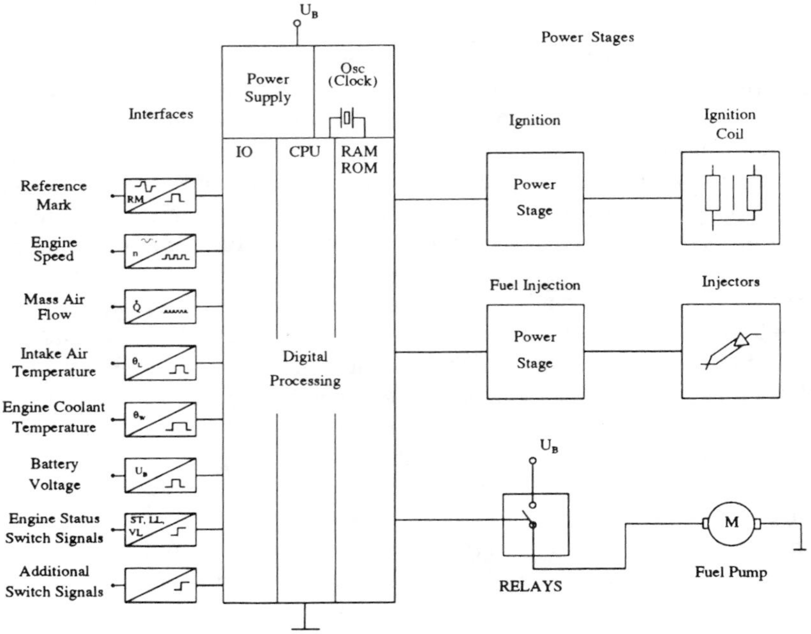

Figure 10.7 Bosch "Motronic" engine control system.

power source (the engine) and load (the steering wheel-road system). With a conventional manual transmission, the driver acts as the feedback loop, sensing speed and load and adjusting the transmission ratio within the mechanical limitations to what he or she perceives as the best operating conditions for what he or she is trying to achieve. One of the main feedback parameters is, however, engine speed in the form of the audible pitch and noise level of the engine, which unfortunately is a rather bad representation of engine power output. Therefore, gear changing, although it may optimize subjective acceleration and drivability, does not give anything like optimum operation for economy and performance. In fact, if we study the road load curves for a typical engine (Figure 10.9), we see that optimum economy is obtained by keeping the engine at the lowest possible speed as long as possible during acceleration, changing the gear ratio to give increased vehicle speed, and only increasing engine speed to produce more power output when a wide open throttle condition is reached. Good acceleration performance will require some modification to this strategy.

Operating the transmission in this way requires the use of either an automatic stepped transmission with electronic control (with smooth changes and sufficient steps

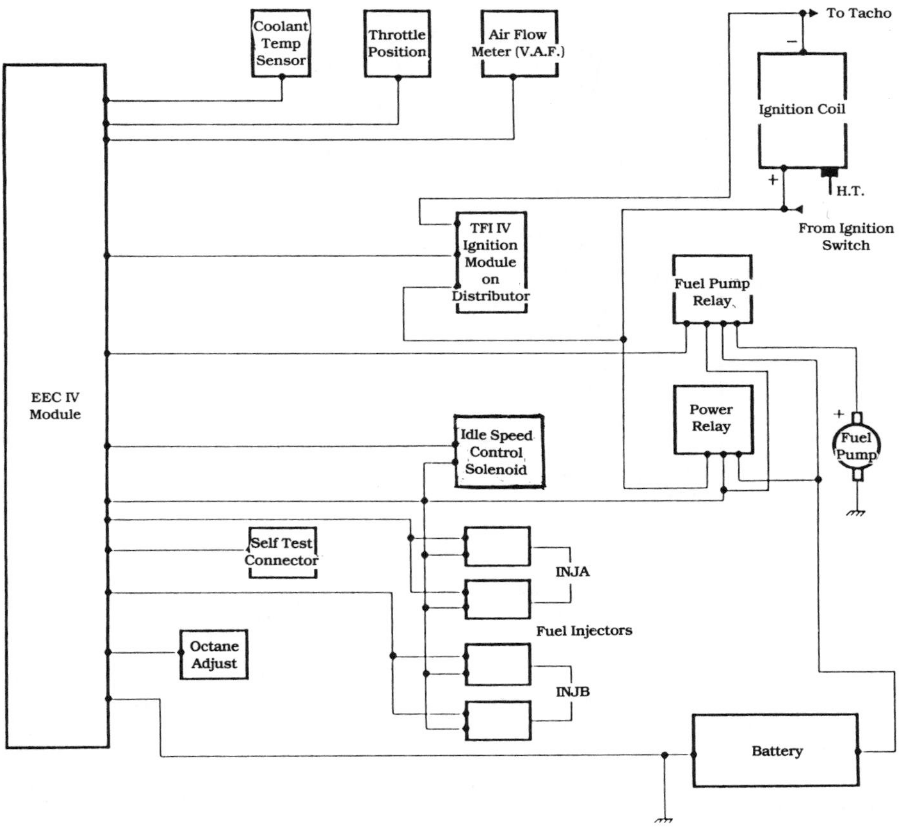

Figure 10.8 Ford electronic engine control unit (EEC IV).

(six) to give an adequate range of gear ratios) or an electronically controlled *continuously variable transmission* (CVT) with an adequately wide ratio.

Both these approaches have been tried, but only the CVT has been developed to production level. A number of experimental electronically-controlled stepped transmissions have, however, been built, the most recent being that described in Richardson *et al.*, [1983]. In this system, a conventional three-speed automatic gearbox was modified by the addition of an epicyclic gear set, to give a total of five stepped speed ratios. The control system is shown in Figure 10.10, the critical items controlled being the clutch engagement, the engine throttle and the rotary hydraulic valve (which controls the gear shift). Engine speed and road speed are fed into the microprocessor, and together with information for accelerator pedal angle and

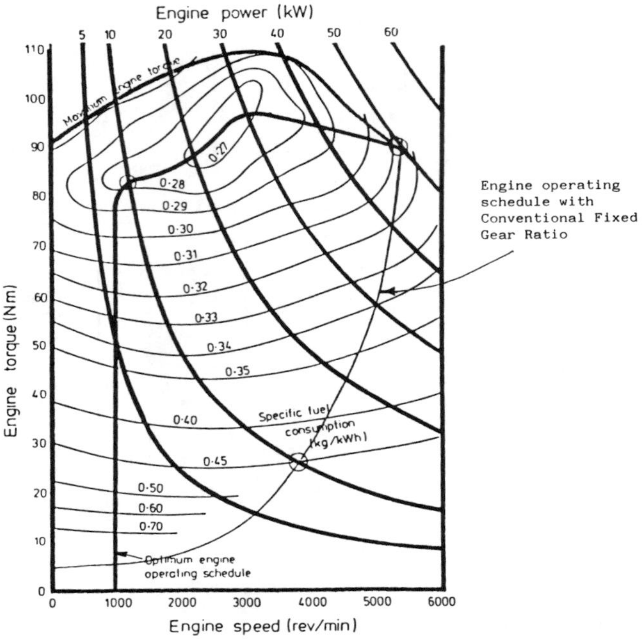

Figure 10.9 Road Load curves relating engine speed, torque, power output and specific fuel consumption under vehicle operating conditions, and showing engine operating schedules with conventional and optimized transmission gearing.

manifold vacuum, compared to the optimum engine and road speeds recorded in memory for those particular operating conditions. The control system then changes gear via the clutch and shift controls to optimize the gear ratio appropriately.

To obtain smooth gear changes under power ("hot" shifts), it is necessary to carefully control the clutch engagement and disengagement, and at the same time back off the throttle control to avoid power surges. Maintaining good "hot" shifting in production is one of the major problems in realizing a mass produced version of this system. Because of this, the major effort to date has gone into CVT and its controls, where by definition no "hot" shifting is required.

Two types of CVT have been developed: the Perbury, which makes use of discs with toroidal hollows in their faces separated by rollers with spherical rolling surfaces, the rollers being constrained in a cage system so that they can tilt and provide a variable ratio drive between the discs; and the Van Doorne pulley and belt system, in which the power transmitting and power receiving pulleys are varied in diameter in opposing senses, so that a belt going around both of them remains at constant length, but the rotation speed ratio between their shafts changes.

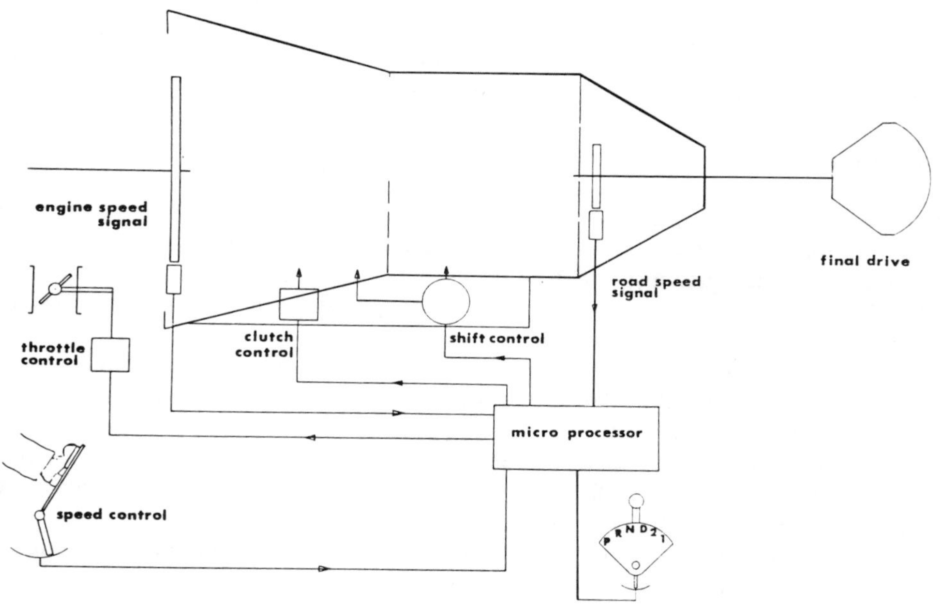

Figure 10.10 Block diagram of microprocessor controlled stepped automatic transmission.

Generally, the control requirements of these two systems are similar, although the Perbury can reverse direction if taken through its "geared neutral" position and this, together with the potentially very high torque produced at or near the geared neutral condition, must be taken care of in the control strategy by the use of a drive by wire throttle control to back off engine power when required [Ironside and Stubbs, 1981]. Apart from these special considerations, transmission ratio is determined by looking at input and output speed and power demand as signaled by accelerator pedal position and rate of movement, and then optimizing transmission ratio for maximum economy or performance as specified by the torque-speed curve.

Using these techniques of transmission control and following the best fuel economy operating curve for the transmission shown in Figure 10.9, it would be possible to improve economy by up to 30% over a conventional automatic gearbox, and about 15–20% over a manual gearbox, if the efficiency of the transmission were as high as a manual gearbox (85–95%).

Unfortunately, the belt-type CVT efficiency is much lower than this, and dependent on speed and power transmitted, which means that the overall effect of fitting an electronically controlled belt-type CVT produces an economy equivalent to a well-operated five-speed manual gearbox.

In the case of the Perbury, because the power is transmitted by metal to metal contact rather than through a flexible belt, efficiency is much higher, and this accounts for the continuing interest in developing a commercially viable transmission with electronic controls based on this principle.

10.4 POWER TRAIN CONTROL

We have looked at engine control and transmission control as separate systems, but it is clear that the maximum operational advantage is obtained by integrating the two systems into a comprehensive *power train control system* which contains both the power source (the engine) and the output matching device (the transmission) [Main, 1985].

In both systems, some of the transducers (sensors and actuators) have a common function. In particular, the sensing of engine speed inlet manifold vacuum and the direct electronic actuation of the throttle are vital to the optimum operation of both systems. Also, changes to the engine control system change engine performance, and therefore modify the optimum gear ratio required. The two systems are therefore, by their very nature, interactive.

In such a power train control system, this interactive operation of the engine and transmission can be effectively optimized to provide best economy, best performance or best emissions, and these conditions can be arranged to be selected by the driver or built in by the manufacturer; for example, in those countries where stringent exhaust emission regulations exist, it would probably be necessary to optimize for best emissions and not give the driver the option to switch to best economy or best performance.

Only this type of integrated interactive power train system offers the possibility of meeting the latest exhaust emission regulations by using lean-burn technology, and therefore offering a relatively lower cost power train system with reasonable operational economy.

If it is then further developed to make use of the fast feedback techniques of in-cylinder combustion described earlier, and the control system is made electronically adaptive so that it can maintain a modified operational mode which has been "learnt," then we are well on the way to realizing the "intelligent" car which can adapt itself to whatever its operating conditions are, and to whatever is demanded of it by the user.

The diagram of Figure 10.11 shows the layout of a typical integrated power train control system.

10.5 TRANSDUCERS

In describing the various methods of control used in the vehicle power train, the role of sensors and actuators (transducers) is a vital one.

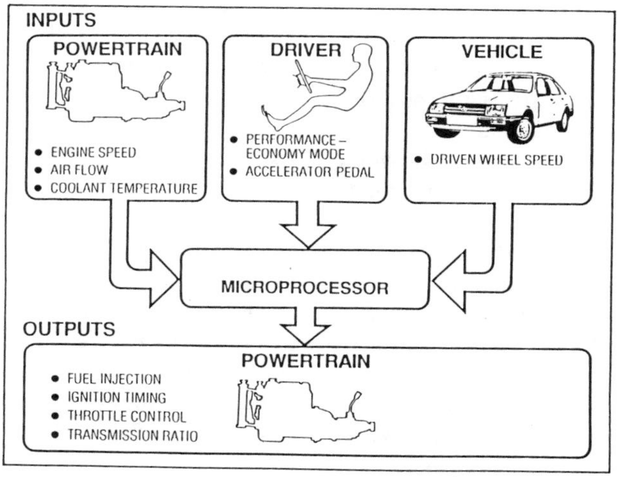

Figure 10.11 Integrated power train control system.

In a typical complex engine control system, about 65% of the total cost is in the transducers, compared to only 20% in the electronics, so that the scope for significant economies by the use of low cost transducers, or by devising control systems which require fewer transducers, is very considerable.

The really critical sensors are those used for measuring engine timing, inlet manifold air mass flow, manifold vacuum pressure, exhaust gas oxygen level, transmission control valve position, transmission input and output speed, and throttle and accelerator position. The critical actuators are those for controlling fuel injection, exhaust gas control valves, throttle position, transmission control valve position, and transmission clutch lock-up.

A detailed description of the critical transducers is given in [Westbrook, 1985, 1988]. The important point to note however, is that future control systems for vehicle power trains will be designed to reduce the number of transducers wherever possible by the use of feedback techniques with adaptive "learning" control systems. For example, the combustion feedback systems described earlier in this chapter can potentially be used in an adaptive control system and remove the need for sensors for most of the relatively slowly changing parameters, such as air temperature, engine water temperature, and atmospheric pressure, as well as optimizing power train performance for changes in fuel quality or mechanical wear in the moving parts.

Transducers, however, will continue to be critical devices in the realization of effective control systems, and much needs to be done to develop more accurate, more reliable and lower cost devices.

10.6 DIAGNOSTICS

Another vital area of development is that of providing an effective means of sensing malfunction of any part of the power train control system and signaling this to the driver in the form of a warning, while also, if necessary, readjusting the system to permit continued operation of the vehicle, even if at a reduced level.

Also highly desirable is that such a diagnostic system, if correctly interrogated, be capable of indicating to those required to service the vehicle the nature of the fault and the identity of the faulty component. Servicing can then be rapidly effected by changing the component concerned.

Such a system has been introduced on the new 2.0 litre EFI Sierra with the use of full electronic control of ignition and fueling using the Ford EEC IV (Electronic Engine Control IV) microprocessor-based system (see Figure 10.8). This has begun a new era of diagnostics. Apart from the operational programs stored in the electronic memory, which are used by the microprocessor to control ignition and fueling from stored three-dimensional tables, the EEC IV also has a self-test program in its permanent memory. When activated by the technician using a special plug-in test module, this checks the memory integrity and processing capability of the system, and verifies that various sensors and actuators are connected and operating properly. Intermittent faults can also be stored and retrieved later. These facilities, combined with the automatic capability of running in a fixed ignition timing, fixed fueling, "limp-home" mode in the event of a malfunction on the road, give high reliability and high service capability for rapid system checking—essential features if these more complex control systems are to be widely accepted.

10.7 ELECTRONIC INSTRUMENTATION

Electronic displays for vehicles, using various display technologies, have been progressively developing over recent years as possible replacements for existing mechanical and electromechanical instrument cluster displays. The first use of electronic display techniques in a production vehicle was by Aston Martin in 1976, where *light-emitting diodes* (LEDs) were used, and this was followed in 1980 by Ford in the US, where vacuum fluorescent speedometer and fuel gauge displays were offered in the Lincoln and Mercury vehicles for that year. Recently, the Austin-Rover Maestro and the Renault have used vacuum fluorescent and *liquid crystal displays* (LCDs) respectively, in some high series models for a range of instrument cluster functions.

A considerable number of display technologies have been considered for use in the vehicle environment, and much development has been done on some of them to meet the stringent requirements, not only for appearance, color and brightness, but also for low cost and high reliability. The display technologies which have been investigated over the last few years are: LEDs, gas discharge, electroluminescence, vacuum fluorescence, LCDs, and *cathode ray tubes* (CRTs).

LEDs have been developed to a point where high brightness, large digit displays are available, although these are restricted in color to the red-green end of the spectrum, and are relatively expensive. For a single digital display, such as a digital clock or car radio frequency indicator, this technology can be a suitable display method. However, LEDs are unsuitable for the large complex displays necessary for a complete instrument cluster in a mass-produced car, because such a display would require a large number of devices at high cost with high power consumption.

As with all the other self-lit display technologies, LEDs also suffer from the condition known as "washout" under conditions of high ambient lighting (for example, when direct sunlight falls upon the display, all the display segments, both lit and unlit, become invisible to the viewer, so the information content of the instrument is lost).

Gas discharge technology, using an individual gas discharge cell for each point of light or *pixel* (picture element) required, has been used for experimental automotive displays, and provides the capability of easy matrix multiplexing where alphanumeric or bar graph displays are required. The displays, however, are complex, particularly if a range of colors is required, because these must be obtained by using different phosphors in conjunction with the groups of gas discharge cells to be used. The discharge cells also require operating potentials in the 200 V region for satisfactory operation, and do not compare with some of the other display technologies for brightness.

Direct (dc) *and alternating current* (ac) *electroluminescence* have both been used as display technologies for large automotive displays. The technology makes use of a sandwich construction, with an electron emitting layer and a phosphor layer, which can be activated by electron bombardment under the control of a suitable potential on the order of 200 V applied to suitable electrodes across the part of the layer to be illuminated.

Colors are restricted to the red-green end of the spectrum due to the relatively low electron energies available, and this also restricts the brightness of the display elements, so that washout in sunlight is a severe problem. There is also a tendency for the brightness of those display segments used most to decrease, thereby giving a variable brightness between segments after a period in service.

Vacuum fluorescent technology was used in the first mass production electronic instrument cluster, and its use continues to increase. In this technology, a phosphor is applied to the anode of a flat structure vacuum tube and viewed through a fine wire-heated cathode and grid assembly. When the anode is connected to a suitable potential in the 100–200 volt region, the electron flow to the anode, with its phosphor coating, gives a high brightness (up to 1000 ft lamberts) display which can be in a range of colors depending on the phosphor used.

The BL Maestro uses this technology very successfully in its electronic instrument cluster (see Figure 10.12), as does the Ford Sierra for its trip computer and vehicle condition monitor display. However, as with all the other emissive displays, washout in bright sunlight remains a problem.

The flat structure required of the glass vacuum tubes used in vacuum fluorescent displays necessitates a split of a normally sized vehicle instrument cluster into two or three sections, partly because of the difficulty of maintaining adequate physical strength to avoid implosion in the flat glass tubes used, and partly because of the bulk and weight of the tubes.

Because the phosphor deposits on the anodes are generally relatively reflective (although usually loaded with a blackening coloring agent) a very dense neutral filter must be used in front of these tubes to reduce reflective washout effects. Densities of up to 90% are used with a corresponding reduction in wanted light from the display as well as unwanted reflected light.

LCDs are now being widely used in production vehicles after a long period of development to meet the extreme environmental conditions existing in vehicles.

The practical displays developed to date have used the twisted nematic type of liquid crystal, in which the change in orientation of the long liquid crystal molecules caused by the application of an electric field is made visible by the application of a polarizing film to both the front and rear of the cell, as in watch displays. However, in many cases, the mode in which transparent figures or letters appear against dark surroundings is used (by rotating one polarizer by 90°), instead of the watch display mode of dark figures against a transparent surround.

This mode makes possible an effective combination of partial rear illumination and partial reflected ambient illumination via a semireflective layer on the rear of the cell, and this is usually combined with color filters to give suitable colors to various parts of the display. An automotive instrument panel-size display operating in this mode is shown in Figure 10.13.

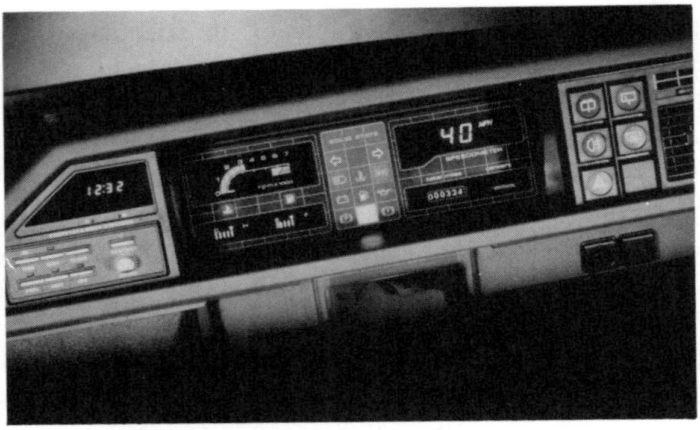

Figure 10.12 BL Maestro vacuum fluorescent electronic instrument cluster.

Figure 10.13 LCD instrument cluster.

This combination of rear illumination and reflected ambient illumination overcomes the major problem of emissive displays, that of washout in bright sunlight, because under these conditions the reflected ambient illumination predominates, and the display operates almost entirely in the mode in which contrast actually improves as ambient illumination increases, but such is not the case with the emissive displays.

Another advantage of the LCD is the relative ease with which the rear glass of the display, which normally carries the printed electrodes representing the symbols to be illuminated, can be expanded in size so that printed connectors to each segment of those symbols can be formed into a circuit on which surface-mounted decoding semiconductor chips can be mounted directly on the glass. This dramatically reduces the number of connectors required to be made externally to the display, and directly improves the reliability of the system, because in most of these systems, reliability under the high vibration and corrosion conditions of vehicle operation is directly related to the number of system connections required. This method of construction of LCDs is known as the *chips on glass* technique.

The main disadvantage of LCDs is their susceptibility to extreme temperatures. At the high temperature end, the display cannot be controlled and goes blank above about 80° C, and at the low temperature end, switching from one display configuration to another becomes slower and slower until the switching rate becomes unacceptable. In recent years, this lower limit has been moved down considerably by the use of new liquid crystal materials, and is now acceptable down to −20° C, so that for almost all operational conditions inside the passenger compartment, operation is entirely satisfactory. As proof of this, a number of manufacturers (e.g., Renault, Mitsubishi) have introduced LCDs in their mass produced vehicles.

Some work has been done on what are known as *dye phase change* LCDs in which a colored dye is added to the liquid crystal material, and the dye molecules attach themselves to the long liquid crystal molecules and take on their orientation, with the result that displays of acceptable contrast can be produced without the use of polarizers, because the dye molecules act to cut off or pass the light, depending on their orientation. However, it is still difficult to obtain adequate contrast for automotive use with this method, and there is some susceptibility to eventual fading after exposure to ultraviolet light.

LCDs offer potential for multiplex or matrix methods of addressing the characters in alphanumeric displays, in which individual pixels are activated by applying the correct voltages to row and column addressing leads on a matrix, so that the voltage available at the liquid crystal pixel at the row and column junction is just sufficient to cause switching to take place. Matrixed displays of this type do, however, suffer from restrictions on viewing angle and temperature range.

These problems, however, can be overcome by the use of individual transistor switches associated with each display pixel. In this technology, amorphous silicon (Si) thin-film driver transistors are deposited on the LCD glass substrate in direct contact with the liquid crystal material. Switching can then be done electronically by the transistors, which directly apply the correct voltage for each LCD matrix, and no reductions of viewing angle or temperature range occur.

This technology, which is under intensive development for flat screen television displays, will undoubtedly provide the majority of the automotive displays of the future, including those for navigation systems such as AUTOGUIDE (see Chapter 6), and for displays of frequency and channel name in RDS (see Chapter 5).

CRT displays have one major advantage over all other display technologies: the ability to show large amounts of totally different information in the same display space at different times. This means that in an emergency the normal display, which might consist of speedometer, fuel gauge and temperature gauge characters, could be replaced by a single warning display, which could be flashed to give greater "attention-getting" capability.

CRT displays also have full color capability, and can be run at high brightness if high voltages are used, as in the latest aircraft systems. Their disadvantages are those related to the potential problems in an accident, caused by their high voltage requirements and risk of implosion, combined with the problems of washout in direct sunlight, which can only be overcome by the use of very high tube voltages with their attendant effect on reducing tube life.

For intermittent, noncritical use, as in a navigational system map display, there seems to be a place in the car for the CRT, but it seems unlikely to be used to provide the permanent instrumentation system.

10.7.1 Digital Displays

One of the issues that has been addressed in considerable detail within the automotive industry is the viability of a digital display for vehicle speed.

Studies which have been done [Baines *et al.*, 1981] indicate that the digital display of speed is read more accurately and is subjectively preferred by drivers, particularly those who use glasses, because of the greater ease and speed of reading which is possible. However, often there is some initial concern by drivers before they have used such displays and this, together with the fact that, to date, electronic displays have been more expensive than electromechanical, has meant that there are relatively few examples of digital electronic displays in European cars, although the Austin-Rover Maestro and Renault are exceptions. These displays have, however, proved very popular in the US and Japan, and are widely used in vehicles in those countries.

10.7.2 Trip Computers

Most automobile manufacturers now offer a *trip computer* on their top-of-the-line vehicles. With this device, measurements of vehicle speed, fuel flow rate, and fuel tank level are made and combined in a microprocessor with a time signal to derive information on both the instantaneous and trip fuel rates of use (mi/hr), miles to go before the tank is empty (this makes use of an algorithm to modify the result displayed according to the rate at which fuel is being used (i.e., according to driving style and speed)), and to provide speed limit and trip mileage information, as well as a basic clock and stopwatch functions.

10.7.3 Speech Synthesis and Recognition

Communicating vital information to the driver is the fundamental objective of any vehicle instrumentation system, but communication does not have to be exclusively by the conventional visual means of dials, digital displays and warning lights. Other methods may be employed, of which the most obvious is the use of sound.

Buzzers and gongs have been used for warning purposes for some time, but recently the technology of voice synthesis has reached a stage where it is practical for use in a vehicle to convey specific messages.

The technology involves actually duplicating electronically the operation of the human vocal cords, throat, mouth, and tongue system to permit the synthesis of artificial speech. Systems have been marketed by Austin-Rover and Renault, but public acceptance has not been good. This emphasizes the importance of drivers not being overloaded with information and the use of aural means of communication being very carefully controlled. However, speech synthesis can be of considerable value if correctly used. An appropriate application is in navigation systems to add specific short-term information, such as required lane position, to visual data.

Much work is currently being done on electronic speech recognition. In the noisy environment of the car, this is particularly difficult, but a successful system could be very useful for future driver controls.

10.7.4 Head-Up Displays

Head-up displays, in which information is projected on the windshield in front of the driver, of course, are well known in the aircraft industry.

Technically possible, although relatively expensive, is to provide this feature in cars, but, before such a system can be introduced, great care will be necessary to investigate driver reaction, particularly on wet, dark nights, to having a changing display of speed continually in his or her line of sight even if the brightness and focal distance are appropriately controlled. An important fact to remember in this connection is the wide range of capabilities and visual acuity of car drivers compared to the highly trained, physically fit professionalism of aircraft pilots.

10.8 MULTIPLEX WIRING

One of the major areas of unreliability in any car is electrical connections and switches. Even when assembled correctly during the installation of the car electrical wiring loom, connections are always liable to subsequent mechanical damage and corrosion in service, yielding high contact resistance and resultant rapid deterioration. Switches are also prone to most of these problems, together with arcing problems where high currents are to be broken, and they are required to be physically large to accommodate multiple pole high current connections. Also, the ever increasing complexity of vehicle wiring systems makes some method of getting away from the need for individual wiring harnesses for each vehicle variant essential; for example, the Ford Escort alone requires more than 30 separate wiring harnesses for different variants.

The multiplex ring main offers an alternative approach, which overcomes many of these problems and offers an additional "free" capability for providing driver diagnostic information on the condition of the various electrical systems and components on the vehicle.

Multiplex wiring operates by using a single, large gauge ring circuit to distribute electrical current to all parts of the vehicle (see Figure 10.14), then, at each takeoff point where current is required (i.e., for rear lamps, stop lamps, direction flashers, heated backlight), is an integrated circuit control module, which includes an electronic switch or relay capable of controlling the current to the component, plus a sensing circuit that can detect and act on control information in the form of pulses passed either down the bus bar itself or via a separate control wire, or even down an optical light guide. The control switch on the dashboard hence incorporates

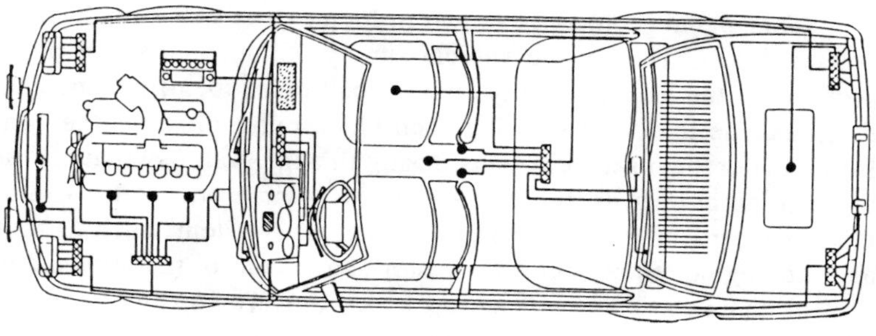

Figure 10.14 Multiplex wiring system (low series vehicle).

a low current or even a touch type of switch, together with a suitable code pulse generator in integrated circuit form. The control switch can be as physically small as required by the car interior designer.

A driver diagnostic display with a further code detector can also be incorporated, which can detect how each component sensing circuit responds to an interrogating pulse, and therefore whether the component is in a normal state, short-circuited, or open-circuited; thus, for example, a diagnostic display of bulb condition can be easily provided without additional wiring or sensors.

Systems have been developed by Bosch, Lucas and Philips and also by vehicle wiring manufacturers such as Salplex (GEC/Marconi), Reinshagen and VDO. Vehicle manufacturers such as Ford have also developed experimental systems [Robins *et al.*, 1983], and the Ford system will be described in some detail because much work has gone into making it suitable for the difficult electromagnetic environment of the vehicle.

The system, shown in Figure 10.15, makes use of a single signal wire in addition to the power cable, and all the control and diagnostic information is transmitted between the control modules and the dashboard using a TDM signal sent over this wire.

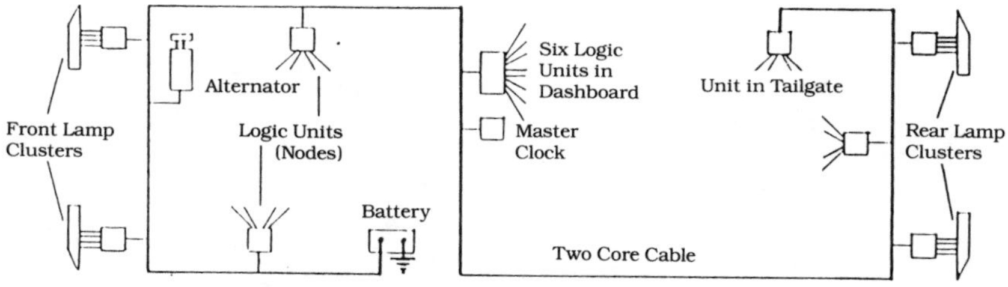

Figure 10.15 Ford multiplex wiring system (low series vehicle).

This signal protocol is shown in Figure 10.16. In this diagram, level A is zero volts and level D is at the vehicle supply voltage. Levels B and C are fixed proportions of the vehicle supply. Each clock pulse identifies one channel of information, and the interval following each clock pulse provides the control information. If the level remains at zero volts during this interval, "switch off" is being signaled, but holding at level B throughout signals "switch on." Pulse width modulation provides a means of transmitting analog signals for items such as the fuel tank sender and the engine temperature thermistor. Each control module contains four switched control channels, each with an up to ten ampere capability, plus four channels on which diagnostic or other information can be transmitted back to the dashboard.

Power switching is currently by special low saturation voltage drop bipolar transistors with the vital capability of dropping less than 0.4 V at 10 A current, a critical requirement for electronic switching in a 12 V automotive power system. However, the latest MOSFET (metal-oxide semiconductor transistor) switching devices, with their much lower current driving requirements, seem certain to replace bipolar devices in future systems. By the use of a very low frequency pulse repetition frequency for the signal protocol (about 5 kHz), it has been possible to minimize both electromagnetic radiation from the system and the system's susceptibility to external radiation. This last item is extremely important because we now know that vehicles can be exposed to field strengths up to 200 V/m close to powerful broadcast or radar transmitters; if the frequency concerned happens to be such that the lengths of wire in the car are at or near resonance, large voltages can be induced.

The frequency chosen is such that the worst-case switching delay of a lamp is just noticeable, and this takes into account that the system is so arranged that the complete control protocol of 128 pulses must be transmitted twice and received correctly for switching action to take place. This feature, together with a parity check after each group of 128 pulses to confirm that exactly the correct number of pulses has been received, has been found to give excellent protection against electromagnetic interference of all types.

Other systems using different communication protocols, which are adapted to the transfer of more complex packets of information, in particular sensor information, have also been devised [Johnson *et al.*, 1985].

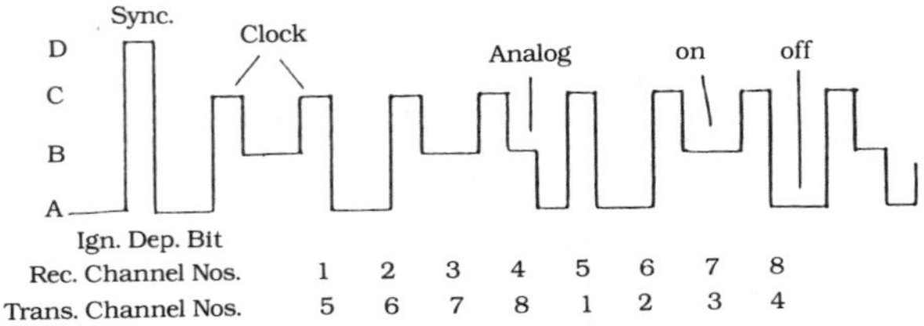

Figure 10.16 Pulse coded TDM switching protocol.

The system which seems likely to become a standard in the US uses what is known as a *contention* protocol. In this, any remote logic unit source can generate a digital word describing the information or instruction to be transferred at any time, but each source has a given priority, and if two messages coincide in time, whichever message has the highest priority is allowed to proceed to its destination, and the lower priority message source must try again. In this system, it is only necessary to send the control instruction once and have it acknowledged to switch on or off a particular load, or send a particular piece of information. This makes possible, at the expense of significantly increased electronic complexity, more effective use of the available bandwidth of the multiplex communication link.

The advent of low-loss, low-cost optical fiber connectors in the late 1990s is expected to make the use of optical fiber multiplex systems possible, and it seems likely that this will require a further change to the system protocol.

10.9 OTHER SYSTEMS

To complete the picture of the automotive electronics of today and tomorrow, we must look at some of the other vehicle systems to which electronics is being applied, but which perhaps are more specialized and more related to driver convenience, comfort, and safety than the major applications so far described. Those to be described are:

- anti-lock brakes;
- automatic temperature and air conditioning control;
- speed control;
- driver diagnostics; and
- headlamp leveling.

10.9.1 Antilock Brakes

Antilock brakes are an example of a system which has been available for some years, but which, until recently, has only found application in a few specialized and expensive vehicles. The system does, if properly designed, provide a substantial improvement in braking performance, particularly under slippery road conditions. Operation is by detecting the wheel slip condition under braking action by the driver, and controlling this to not become greater than a predetermined amount by rapid on-off application of braking to the brake discs or drums, the period of brake "on" to brake "off" determining the effective total braking effort. The diagram of Figure 10.17 illustrates the action.

Early systems, particularly those used to meet a legal requirement on trucks in the United States in the late 1970s, suffered severely from electromagnetic interference

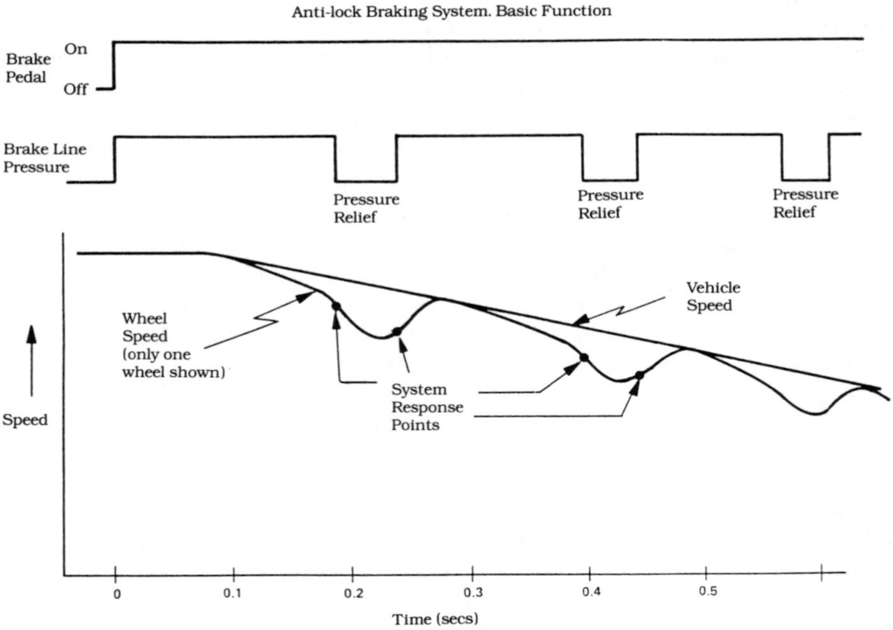

Figure 10.17 Antilock braking system control sequence.

to the extent where there were stories told of antilock systems in trucks near airports being turned on by high power radio transmissions from aircraft! As a result of these problems, the US legal requirement for truck antilock brakes was removed.

In more recent years, by the use of redundancy in the system and protection against electromagnetic interference, very reliable systems have become available and been fitted to some luxury cars. The major step forward (in April 1985) was the fitting of antilock brakes as standard to the full Ford Scorpio Granada range of vehicles, initiating a trend which is continuing with other manufacturers to introduce such safety improving systems on a much wider range of vehicles. There are also now strong indications that reliable low-cost antilock systems are becoming available, which seem certain to ensure their much wider use.

10.9.2 Automatic Temperature and Air Conditioning Control

Drivers today expect the temperature within the vehicle to be maintained in a comfortable condition without continual adjustment of the heating and, where fitted, air conditioning controls. To achieve this effectively, automatic thermostatic control of the heating and air conditioning is essential. To obtain full control, there needs to be control of both the temperature and velocity of the incoming air, and this requires motor control of the deflector flaps within the heating and air conditioning system, as well as the capability of automatic clutching in the air conditioning pump drive.

The more sophisticated systems provide this control via a microprocessor and a number of temperature sensors within the passenger compartment.

10.9.3 Speed Control

In the US, to have a device fitted to the car which automatically controls the throttle has been common for some years, to maintain the vehicle at a set speed irrespective of the load or the type of terrain. This speed control system is widely used on long trips on the US freeway system to reduce fatigue and to avoid exceeding strict speed limits. The system most commonly used has the throttle controlled by a manifold vacuum servo, which, in turn, is controlled by measuring vehicle speed by means of a gearbox-driven digital sensor and comparing this with the required speed set manually in a control unit by the driver pressing a "set" button when the vehicle reaches the required speed.

The control can be overridden by the deliberate application of the brake or accelerator, but will resume the original speed setting when these are released and a "resume" button pressed. Speed control systems of this type are being more widely used in Europe, although it seems unlikely that the high usage rate seen in the US will ever be reached because of different road and economic conditions.

In the sophisticated integrated power train control systems described earlier in this chapter, in which electronic control of the throttle is required to implement interactive controls, speed control can relatively easily be incorporated as an additional feature without the expense and package problems of adding a vacuum servo throttle control.

If it becomes possible to develop an effective system of vehicle headway control—whether this uses radar, infrared, or laser obstacle detection—the capability of electronic throttle control to precisely control vehicle speed will be critical if some of the more advanced ideas suggested in the PROMETHEUS project for traffic control are to be realized. (See Chapter 9 for more details on PROMETHEUS.)

10.9.4 Vehicle Diagnostics

A very useful display for the driver to have in front of him or her would comprise warnings on the state of the systems essential to the safe and convenient operation of the vehicle. These are generally considered to be:

- low fluid levels in the fuel tank, radiator, screen washer reservoir, brake fluid reservoir and oil pan;

- integrity of all driving and external lamp bulbs;
- state of wear of all brake pads;
- the external temperature close to the road surface to permit warning of freezing road conditions; and
- the pressure level of all four tires.

Low fluid levels have been measured by using different methods depending on type of fluid concerned. The warning of low fuel in the tank is normally derived directly by triggering a warning at a particular level of current or voltage as obtained from the conventional float and potentiometer fuel sender, which has been used in vehicles for many years to provide the fuel gauge control signal.

The disadvantages of the system are entirely related to the inaccuracy of the float and potentiometer sensor used in a tank which has progressively become shallower and of more complex cross section as car design has changed over the years. There is, therefore, the need for a better way of measuring fuel tank fluid level and a number of low cost alternative methods are currently being investigated.

In the radiator, screen washer and brake fluid reservoir, a simple float and micro switch are used to give a warning when the liquid has dropped to a particular level; these devices are normally very reliable and of low cost, so that there is little incentive to look for alternatives.

The oil level in the oil pan, however, presents a much more difficult problem because of the high temperature, considerable turbulence and change of level and condition during operation. Because of the viscosity of oil and the way it is pumped around the engine and mixed with air by engine action, an oil-air emulsion of relatively low density is formed in the oil pan during operation. At the same time, a considerable quantity of the oil is pumped up into the galleries and bearings in the engine, and depending on its temperature and viscosity, a significant quantity will be missing from the oil pan reservoir during operation. The result of these various effects is that to measure the level of engine oil during operation is extremely difficult, and most systems currently in use only make a static measurement before the engine is started and give a warning, which is then retained at that time. The measurement method usually used is a thermistor operating in a "hot wire" mode: the thermistor is heated by applying a current and its change in resistance between being immersed in oil and in air is used as the warning signal. Additionally, it is convenient if the measuring device is small enough to form part of a standard size oil level dipstick, and this can be achieved by using a thermistor in this way. Of considerable interest to the automotive industry would still be to find a low-cost way of continuously measuring true effective oil level during operation because, with the present static measurement before start-up, a complete loss of oil during operation would not result in a warning being given.

Bulb integrity has increasingly become a desirable warning to provide in view of its potential contribution to improved safety. The method normally used is to measure the current in each bulb circuit by means of a small series resistor, and to use the presence or absence of current within certain defined limits to provide the warning of bulb failure to the driver. The bulbs normally monitored are those used in headlamps, side lamps, rear lamps and stop lamps. The display arrangements for this information are often technically the most interesting part of the system. Typical of the type of display used is the vacuum fluorescent vehicle outline display used in the Ford Sierra. This also displays "door ajar" warnings and a low road temperature warning.

Brake pad wear is a clear safety feature. Measurement is by the use of a wire embedded in the brake lining at a certain depth, so that when the lining has worn to that depth, the wire makes contact with the metal of the brake shoe and a warning signal is given. The arrangement is very successful in avoiding the excessive wear and the associated loss of braking power which used to occur.

Tire pressure warning has not so far been available as a driver diagnostic feature. This is not because measuring tire pressure is difficult, but rather because of the difficulty of finding a low-cost way of transferring the signal from the tire pressure sensor and the rotating wheel to the main part of the vehicle to operate a suitable warning device. Two methods have been developed which have some prospect of being adopted for mass produced vehicles. The first of these makes use of the absorption characteristics of a tuned circuit when inductively coupled to an oscillator. The level of tire pressure at which warning is required is sensed by a simple sealed diaphragm switch mounted in the well of the wheel inside the tire. This switch circuit is shown in Figure 10.18; the associated inductance being placed as close as possible to the inner wall of the tire, where it can couple once every revolution with a similar inductance mounted as close as possible to the tire on the vehicle structure.

This second inductive circuit is energized by an oscillator at the resonant frequency of the tuned circuit in the tire, and is also coupled to an amplifier to detect the loss in signal caused by the absorption of resonance of the tire-tuned circuit. When the tire pressure drops to the critical level, the sealed diaphragm switch closes,

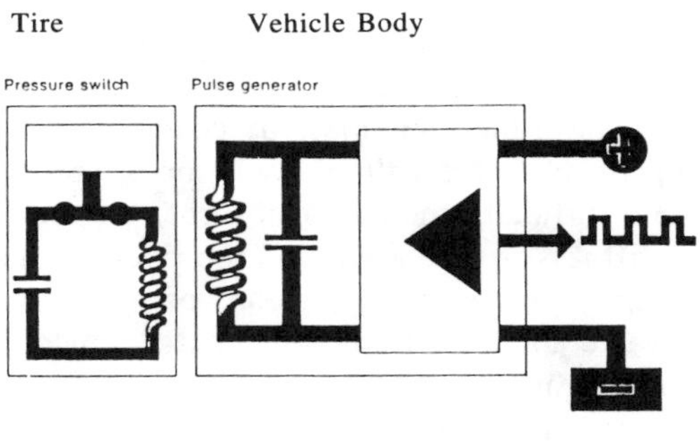

Figure 10.18 Tire pressure measurement using inductive coupling.

shorting the tuned circuit to ground and removing the absorption effect. This change in signal of the inductance on the vehicle body is detected, amplified, and transmitted as a warning signal to the driver. The warning operates in a fail-safe mode, because loss of the tire signal will give a low tire pressure warning.

The second of the tire pressure warning systems, which is commercially available as an optional "add-on" unit, makes use of a small radio transmitter in the well of the tire, controlled by the pressure switch, to transmit a signal to an adjacent receiver on the vehicle.

With both of these systems, a complete sensor-transmitter-receiver system is required for each wheel, so that the cost of a complete vehicle system can be high. Primarily for this reason neither of these systems has been widely adopted.

10.10 FUTURE DEVELOPMENTS

Two important areas in which electronics is expected to be applied to vehicles in the future are the control of suspension rate and deflection, and vehicle collision warning radar.

10.10.1 Suspension Control

In the case of suspension control, the objective is to provide a suspension system which will give a suitable ride for all conditions, so that normally a soft ride would be provided at low speeds, with the ride becoming stiffer at high speeds to reduce vehicle roll and pitch. The suspension would also become stiffer under heavy load conditions and during cornering.

This can be achieved by control of the shock absorber stiffness, a method known as *adaptive damping*. A suitable method is shown in the block diagram of Figure 10.19. In this system, sensors continuously detect the wheel-to-body height and feed this information to a central control unit, together with information on vehicle speed.

The objective in the particular system shown is to maintain the dampers at a "soft" setting for as long as possible, and only go to a "hard" setting when additional damping is required to control the body mass by changing the aperture in the bleed valve between the upper and lower half of each shock absorber. The smaller the valve opening, the more difficulty the shock absorber has in being compressed; as a result, the effective stiffness is increased, and the ride becomes harder. The usual method of control currently proposed is by means of solenoids, which can provide two or three different valve apertures to provide "soft," "medium," and "hard" settings. Some development is also underway to provide continuous aperture size control by means of a proportional solenoid.

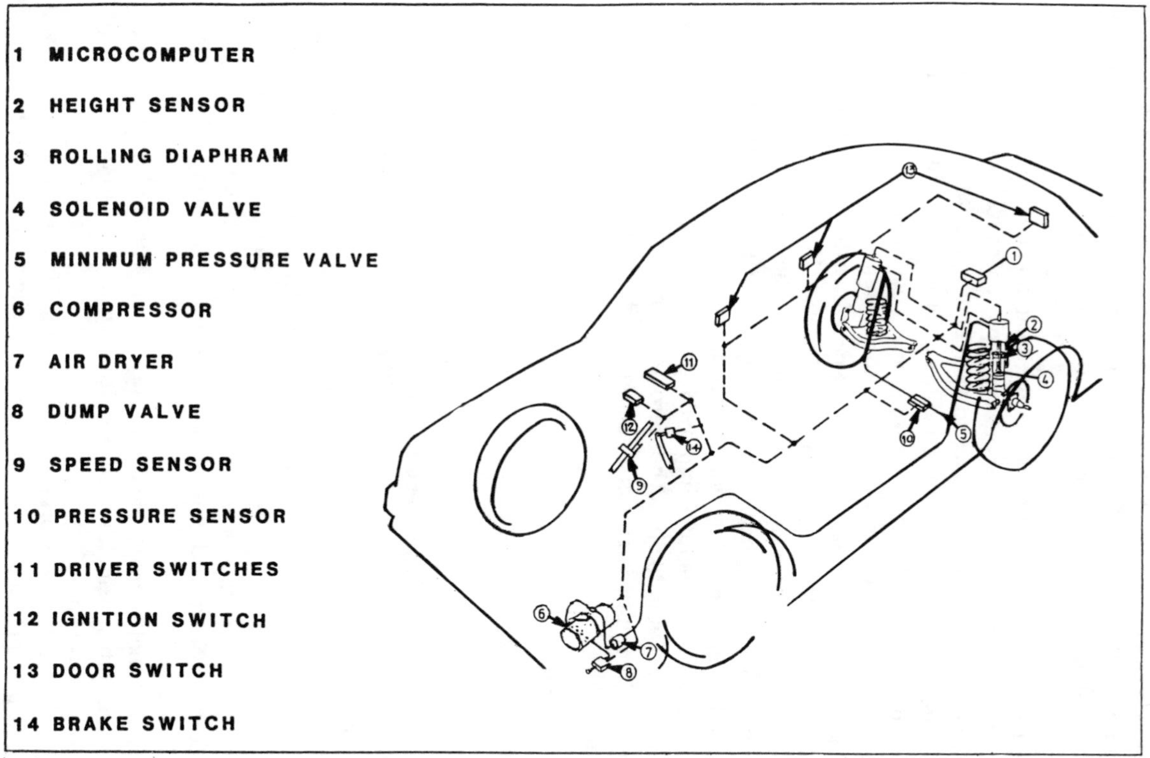

Figure 10.19 Adaptive damping and leveling system.

The adaptive damping system therefore can control the dynamic operation of the shock absorbers, but is unable to compensate for changes in vehicle attitude and height. The slowly acting control required for this is also incorporated in the system shown in Figure 10.19. The rear height is adjusted by controlling the air pressure in an air-spring system according to the input from the height sensors and door opener switches.

A further step towards the ultimate is the system known as *active suspension,* in which the springs and shock absorbers are replaced by fast operating hydraulic jacks. This permits the control by valves to each end of the jack, of attitude, height and damping rate under all operating conditions. However, considerable power is required to control these jacks under bumpy road conditions, when large fast suspension movements are required if the car is to remain level.

Lotus has developed a system of this type which has been fitted to a Lotus Esprit and also used on two occasions on "Formula 1" racing cars. The control system is similar to that for adaptive damping with the addition of accelerometers on each wheel measuring vertical acceleration. These suspension control systems give the possibility of improving ride, comfort, and safety, because handling and stability in cornering are generally improved, but the more sophisticated and costly active suspension seems unlikely to be introduced in mass-produced cars for many years, because of its high power consumption and the increase in fuel consumption that will result.

10.10.2 Vehicle Collision Warning Radar

Work has been done for many years on the application of vehicle mounted radar to the detection of other vehicles in the direction of travel, and the provision of a warning to the driver of the equipped vehicle if the obstacle has been approached too closely and quickly.

With the use of Gunn diodes and simple antennas, low-cost radar systems can be made for use in vehicles, and there is no difficulty applying them to detect the presence and approach rate of an obstructing vehicle. The problem is that the radar will also detect vehicles on sidewalks, which are not obstructing the road, bridge pillars, and other assorted road objects, with the result that many false warnings are generated.

Another potential problem that will only occur with widespread usage is the effect of meeting other vehicles with operating collision warning radars, possibly even on the same frequency.

Despite these problems, working systems which perform remarkably well have been developed in Japan and West Germany [Takehana *et al.,* 1981] using narrow beam patterns of the order of $\pm 20°$ and accurately controlled range sensitivity.

However, to obtain fully acceptable performance, the incidence of false warnings must be extremely small, otherwise real warnings will be ignored. This desirable situation has not yet been achieved, and how it can be with present techniques is difficult to see. Therefore, we are unlikely to see collision avoidance radar in production vehicles for a number of years.

10.11 COST AND RELIABILITY

A major challenge in developing the new technology for use in mass produced vehicles is that of achieving low cost and high reliability when using that technology in the very difficult vehicle environment.

A fact often not appreciated is that, although the electronic systems must survive levels of mechanical vibration up to 20 *g*, a temperature operating range of −30° C to as high as +120° C in the engine compartment—and, in some positions on the vehicle, salt water splash conditions—the really severe problem is that of *electromagnetic interference* (EMI).

10.11.1 Electromagnetic Interference

The electromagnetic interference in a vehicle has two different components. One is the interference generated by the various electrical and electronic systems within the vehicle, and which is under some control by the vehicle and system designer; the other is externally generated interference from high-power broadcasting and radar transmitters. The first of these normally consists of either radiated or conducted interference spikes from the ignition system, large voltage spikes caused by the switching of inductive loads, or the connection or disconnection of the alternator or battery.

Some of these interference spikes can be as high as 200 V, with durations of a few milliseconds, and were the cause of the breakdown problems with the early electronic alternator regulators. With the electronic system designs now currently used, the power input to the modules concerned is adequately protected and failure from this cause is rare. The only point at issue is whether it would be better to provide a separate suppressor and regulated power supply to the electronic modules rather than provide suppression and regulation at each module, as is presently done.

In the case of the second component of electromagnetic interference, from outside the vehicle, the situation is much more difficult. As has been established, on roads close to the antenna of a high-powered broadcasting or radar transmitter, field strengths up to 200 V/m may be measured. Although the metal of the vehicle may help to reduce this field within the vehicle structure, the effect is uncertain, and protection against this level of high frequency interference is required to be absolutely certain of correct operation under these conditions.

The problem becomes particularly acute when the interfering signal wavelength is on the same order as the length of the wiring in the car connecting the electronic modules, when resonance can considerably enhance the effect. The frequencies concerned here are generally between 50 and 200 MHz, although other frequencies, rather unpredictably, also give problems. Low-pass filtering of the electronic module power supplies is normally the first countermeasure, with special screening and filtering of particularly vulnerable parts of the system, or, in the case of some digital control protocols, the use of redundancy and parity checking. Where safety is critical, as in antilock braking, hardware redundancy may be used to protect against both disruption by interference and hardware failure.

Multiplex wiring is particularly vulnerable to this type of interference, and may result eventually in the use of optical fibers instead of wire connections. Interconnections between microprocessor-based control units are also vulnerable, but much work on this has improved their reliability dramatically.

In those few systems which have passed through the initial stages and have become mature technology, such as the electronic breakerless ignition system and the electronic alternator regulator, reliability has been shown to be better by more than an order of magnitude over the reliability of the electromagnetic systems they replaced. This augurs well for the future reliability of the more sophisticated automotive electronic systems to come.

10.11.2 Self-Diagnostics

Future reliability will also be improved by the extension of the use of the self-diagnostic system for engine control described earlier in this chapter to the other electrical and electronic systems, and to the diagnosis of incipient mechanical failure conditions. The technique, therefore, for effective self-diagnostics will follow that precedent, and use the microprocessor to run a special diagnostic program which can self-test in sequence the main control systems and all the sensors and actuators used, and then give a fault signal, which is displayed as a number on a suitable readout, either built in to the vehicle or as a plug-in test module, to identify the fault component, which can subsequently be replaced.

A slightly different concept needs to be applied to the diagnosis of mechanical functions where performance deterioration or noise vibration increase in some form is usually the first sign of oncoming failure.

Here, the major method of diagnosis must be by sensing changes in mechanical performance and noise-vibration levels from those of the original system (the system signature). This is particularly applicable to the engine and transmission system, where it is possible to look at torque and noise-vibration characteristics (signatures) during use, and compare these with the signatures obtained when the power train was new. The development of suitable low cost sensors and accurate computer methods for such signature analysis still requires considerable work, but will eventually transform the ability to detect incipient deterioration and failure in mechanical systems.

Assuming that these techniques of failure detection are suitably developed, the motorist of the future should be able to enjoy a great improvement in the effectiveness of diagnostic procedures at dealers, and a fast turnround of his vehicle. However, there will still be vehicles around for many years which do not have these sophisticated systems, and there will also be faults which these diagnostic systems cannot economically detect, so there is one further step that must be taken to improve dramatically the general diagnostic capability of dealers for all classes of vehicle and all classes of fault. This step is to use a central computer store of diagnostic expertise; an *expert system,* which can be interrogated from each dealer by means of a question and answer sequence down a computer link, which may be an extension of that already available for ordering parts and making warranty claims. This will permit a relatively inexpert mechanic to describe the symptoms of the fault in response to suitable questioning from the computer, and also eventually to enter information from particular sensors on the vehicle, to permit a fast and accurate diagnostic procedure, which, in the great majority of cases, will very quickly provide a lead to the faulty component.

10.11.3 Cost

There is much still to be done to reduce the cost of automotive electronic systems. In general, the cost of the actual electronics is relatively low, but the costs of the sensors and actuators required for the more complex control systems required for engine, transmission, suspension and diagnostics is still much too high for the accuracy and reliability provided. This, therefore, must be a major area of development over the next decade, and one in which the UK could have a major share.

However, where electronic systems are replacing existing electrical, mechanical, or electromechanical systems, for example, in electronic instruments or multiplex wiring, the cost of the electronic system is expected to become progressively more competitive with the system it replaces, thereby making the economic arguments for using electronics as great as the technical arguments.

10.12 CONCLUSIONS

In this chapter, we have tried to give some breadth as well as depth to the subject of automotive electronics, and to describe the systems which, when combined in the vehicle, will yield the "Intelligent Car" of the title.

This integration of currently separate systems will take place through the 1990s, and will result in the increase in electronic content of vehicles from the current US figure of nearly $800 per car, to the $1350 per car (in 1992) projected in the 1984 University of Michigan Delphi III forecast.

In meeting the stringent cost and reliability requirements of the automotive industry, we expect that there will be a major spin-off into other industrial and consumer markets, with the resultant proliferation of low cost, high reliability controls in areas other than automotive between now and the end of the century.

We can look forward to this period of rapid expansion of intelligent computing and control technology both inside the vehicle and outside it, with increasing confidence that it will result in cars which are more fuel efficient, safer under both normal and emergency conditions, have a reduced effect on the environment, and ultimately are enjoyable to drive.

ACKNOWLEDGEMENT

The author wishes to thank the Institution of Electrical Engineers for permission to reproduce much of the material used in the IEE review, "Automotive Electronics," published in the *IEE Proceedings* in June 1986.

REFERENCES

Baines, P.A., *et al.*, "Ergonomics in Automotive Electronics," *Proceedings of 3rd International Conference on Automotive Electronics,* IMechE 1981, pp. 33–38.

Ironside, J.M. and P.W.R. Stubbs, (1981) "Microcomputer Control of an Automotive Perbury Transmission," *3rd International Conference on Automotive Electronics,* IMechE, 1981, pp. 283–292.

Johnson, W.J., *et al.* "Systems Considerations for Incorporating Vehicle Data Networks (Multiplex) into Automobiles," *5th International Conference on Automotive Electronics,* IMechE Publication 1985—12, pp. 209–218.

Main, J.J., "New Developments in Power Train Control," *Proceedings of the 5th International Conference on Automotive Electronics,* IMechE 1985, pp. 135–142.

May, M.G., "Flame Arrival Sensing Fast Response Double Closed Loop Engine Management," SAE Paper 840441, 1984.

Meyer, E.W., *et al.*, "An Electronic Fuel Control Carburettor System," *4th International Conference on Automotive Electronics,* IEE Publication 229, 1983, pp. 67–75.

Rhodes, D.M., and M.J. Werson, (1983) "The Patrol Engine—A Case for Closed Loop Control," *4th International Conference on Automotive Electronics,* IEE Publication 229, 1983, pp. 62–66.

Richardson, R.M., *et al.*, "A 5-Speed Microprocessor Controlled Economy Transmission," *4th International Conference on Automotive Electronics,* IEE Publication 229, 1983, pp. 32–38.

Robins, R.F., *et al.*, "A Car Multiplex Wiring System with Self-Coding Control Modules," *4th International Conference on Automotive Electronics,* IEE Publication 229, 1983, pp. 139–145.

Takehana, T., *et al.*, "Automotive Radar using MM Wave," *3rd International Conference on Automotive Electronics,* IMechE 1981, pp. 123–128.

Westbrook, M.H., "Sensors for Automotive Application," Institute of Physics, *J. of Physics E: Sci. Instrum.*, Vol 18, 1985 pp. 751–758.

Westbrook, M.H., "Automotive Transducers—An Overview," *IEE Proceedings* Vol. 135, Pt D, No. 5, September 1988, pp. 339–347.

General References on Current Technology

Proceedings of the 4th International Conference on Automotive Electronics, IEE 1983.
Proceedings of the 5th International Conference on Automotive Electronics, IMechE 1985.
Proceedings of the 6th International Conference on Automotive Electronics, IEE 1987.
Proceedings of the 7th International Conference on Automotive Electronics, IMechE, 1989.

PART V
THE FUTURE

Chapter 11
Mobile Information Systems—the Future

J. Walker
Racal Research Ltd.

11.1 INTRODUCTION

Chapter 1 introduced the subject of *mobile information systems* (MIS), and indicated how *information technology* was beginning to be applied in the mobile environment. It also alluded to some common themes and overlaps in interest and application. Chapters 2 through 10 have covered particular areas of RTI in more detail. We will now consider again the common themes, see where the research is leading and what problems remain to be solved. First, however, we will look at the financial aspects of mobile information systems, and the markets for some of the products and systems covered in the book.

11.2 MARKETS FOR MOBILE INFORMATION SYSTEMS

Forecasting future market sizes is interesting and important from the business point of view, but doing so is extremely difficult. Forecasting markets for mature products is not easy, but to forecast for products that do not yet exist or are not well known to the general public is an order of magnitude more difficult and imprecise. It is also something of a circular process because progress in a technology depends on investment in it, which is determined by the potential market. But investment will not occur unless a market looks big enough! Nonetheless, estimating potential demand is important in order to assess the scale of investment that is appropriate.

11.2.1 Paging Markets

Table 11.1 indicates the number of pager users in various countries at the present time, and the "penetration" (i.e., the percentage of the working population of the

Table 11.1
Pager Populations in 1988

Country	*Penetration (%)*	*Number of Pager Users (thousands)*
United States	6.6	8000
Japan	4.8	3500
Europe	0.86	1500 + 1700
United Kingdom	2.4	628 + 238
Norway	2.4	—
West Germany	0.7	180 + 558
France	0.32	—
Italy	0.05	—
Australia	—	7
New Zealand	—	7

NOTE:
Where two figures are given, the first refers to public and the second to private paging systems.
Sources: Manuel [1989], Tyler [1988], and 1989 press reports of the Frost and Sullivan study, "Radio Paging Market in Europe." *See also* Stewart [1987], and Chapter 2 herein.

country with a pager). Penetration in the US and Japan is higher, which suggests that higher sales can be expected in Europe, particularly in France and Italy.

"The radio paging market in Europe," a recent (1988) market research report by Frost and Sullivan, predicts that the number of pager users in Europe will double between 1988 and 1993, from 3.2 million to 6.6 million. Public paging service revenues will increase during the same period, from $291 million to $1.2 billion. By 1993, France will become Europe's second-largest market in the public services area after the UK, which currently has 40% of Europe's pager users.

The European Commission expects that the number of pagers in Europe will increase to 13 million in 2000; spending on paging equipment will triple to $430 million per annum in 1995; revenue to operators will increase from $770 million in 1988 to $5.9 billion in 2000. Pagers will become a consumer item. However, only 5% of users will need a pan-European pager.

11.2.2 Cellular Radio

Cellular radio companies are in the fortunate position of seeing their most optimistic forecasts in the growth of cellular radio continually turning out to be underestimated.

In some countries, penetration has already reached the several percent level (see Table 11.2) in less than 10 years of operation.

According to Carrington [1987], there were 1.5 million users of mobile radio services in the UK in 1987, split as follows: cellular radio, 200,000, PMR, 400,000, wide-area paging, 500,000, cordless telephones 400,000; this total will increase to 3.5 million in 1992. By the year 2000, there will be 10 million users of mobile radio in the UK, and over 50 million in Europe as a whole.

Focusing on cellular telephones, the number of cellular telephone users worldwide is currently well over three million (see Table 11.3). If the Norwegian penetration is achieved in the rest of Western Europe, the total European market will be in excess of ten million cellular telephones. The world market could be on the order of 30 million, of which 12 to 15 million could be in the GSM system [Wilson, 1989a].

Guppy [1988] has produced detailed forecasts for the GSM and analog systems in Europe. He expects only a minority of users to need the facility of pan-European roaming, particularly because the subscriber equipment will initially be more expensive than analog versions. However, congestion on the analog networks will mean that some subscribers are prepared to pay extra even if they remain in one country. Guppy expects the highest demand for GSM to be in Belgium, West Germany, and the UK because of this saturation. He also expects hand-held portables to capture

Table 11.2
Cellular Radio Penetration in 1989

Country	*Cellular System*	*Telephones per 1000 Inhabitants*
Norway	NMT	35.7
Sweden	NMT	28.0
Iceland	NMT	26.8
Finland	NMT	21.1
Denmark	NMT	19.2
United States	AMPS	8.6
United Kingdom	TACS	8.4
Austria	NMT	4.7
Switzerland	NMT	4.7
Singapore	AMPS	3.6
France	Radiocomm 2000	1.8

Sources: Bergqvist [1989]; Green [1989]

Table 11.3
Worldwide Cellular Markets

	NMT-450	*AMPS*	*TACS*	*NMT-900*	*NAMTS*
Number of users worldwide	600 k	2 M	500 k	150 k	?

Sources: Bergqvist [1989]; Wickham [1988]

more than 50% of the market in the long term. His predictions are summarized in Table 11.4.

Large markets are also forecast for the proposed "telepoint" systems—some 15 to 20 million subscribers ultimately in Europe [Wilson, 1989b]. Ferranti has predicted three million in the UK.

11.2.3 Traffic Information, Route Guidance, and UTC

These topics, which have been covered in separate chapters, are combined here because they overlap, and also because forecasting markets for traffic information and route guidance systems is difficult insofar as there are few in use, and therefore little on which to base the forecasts. However, there is considerable concern about traffic congestion in developed countries, with some sectors of the public wanting

Table 11.4
European Cellular Radio Demand, 1991–1996

Number of Users (thousands)	*1991*	*1992*	*1993*	*1994*	*1995*	*1996*
GSM Subscribers	24	278	811	1588	2569	3688
Analog Subscribers	3470	4037	4445	4705	4818	4873
TOTAL	3494	4315	5256	6293	7387	8561

Source: Guppy [1988]

more roads to reduce congestion, and other sectors opposing roads because of the cost or the environmental damage caused. The application of information technology will not remove the need for new roads, but more effective use of existing roads will help. So, there will be considerable pressure to use traffic information and route guidance systems.

For example, as we saw in Chapter 5, 80% of car radios sold in West Germany incorporated an ARI decoder, and a survey of UK motorists suggested that up to 30% of drivers in London would install AUTOGUIDE units in their vehicles. According to Hampton *et al.* [1988], 25% of Toyota Crowns sold in Japan include a route guidance system, despite the extra cost of $2700. This indicates that there is a significant market for both traffic information and route guidance, although there is little published information on the size of these markets.

The "market research" that has been done concentrates on quantifying the achievable benefits rather than the number of products to be sold. For example, Chapter 6 indicates that a "historic" route guidance system (i.e., one which does not take congestion into account) can result in savings in the UK of $1.5 billion per year, increasing to perhaps $7 billion if "dynamic" information on congestion is included; $1.7 billion of that will be in London—hence, we have the AUTOGUIDE proposal.

Similar figures would apply to other developed countries. Hazard and congestion warnings alone could save $50 million per annum in the UK. UTC systems also face an expanding marketplace. They often pay for themselves in less than one year; a system installed in New York was priced at $16 million, with annual operating costs of $3.75 million, but saved $25 million in the first year of operation (see Section 7.7.2).

11.2.4 The Intelligent Car

Forecasts are easier in this field because the market is mature, and the production figures for automobiles are known. The total worldwide vehicle production in 1990 will be 48 million vehicles, with a potential electronics market of $42 billion, compared with $27 billion in 1989, according to Matsushita; Hitachi forecasts a 12% annual growth rate to $23 billion in 1990, with digital vehicle instruments growing 40% to $420 million [Hartley, 1987].

The electronics content of passenger cars is growing 19% per annum in a market worth $2.4 billion in 1988, and $4.8 billion by 1992, according to the UK Electronic Component Industries Federation; electronics will account for 25% of the manufacturing cost of the average automobile in the late 1990s, compared with 7% in the late 1980s [Goodall, 1988].

The semiconductor device content of European cars has increased from about $10 per vehicle in 1980 to $50 in 1989, according to Texas Instruments. The figure for electronic equipment in US cars is $800 (see Chapter 10); Motorola predicts a world market of $10 billion for automotive chips by 1993; according to General Motors, electronics could account for 30% of the cost of a new car by the year 2000, yielding a world-wide market of $60 billion [Hampton *et al.*, 1988].

Thus there is general agreement that information technology in vehicles will be of major economic importance in the future, and its current importance is not negligible.

11.3 WHERE THE RESEARCH IS LEADING

11.3.1 Beneficial Effects of Applying It

Chapter 1 suggested that information technology applied to road transportation would bring various benefits; these benefits have been described in subsequent chapters, and some of the more important are listed in Table 11.5.

Table 11.5
The Beneficial Effects of IT

Information Technology Applied	*Benefit*
Cellular radio and paging	• better communication • more efficient use of time and resources • avoidance of congestion
Congestion warnings by traffic information broadcasts	• reduces congestion • reduces secondary accidents • saves time
Vehicles cover less mileage because of effective route guidance	• fewer accidents • less fuel consumed • pollution reduced • time saved
Urban traffic control	• less congestion • less road-building needed • more efficient use of existing roads
"Breakerless" electronic ignition and timing control	• reduced carbon monoxide and hydrocarbon pollution
"Lean-burn" engines	• better fuel economy • less pollution
Antiskid brakes	• saves lives • reduces accidents • reduces injury
Electronic displays and instrumentation	• improved information for the driver

In the UK, there are currently 5000 deaths per annum on the roads, and 313,000 casualties; accidents cost the nation $6 billion per annum; congestion costs $16 billion per annum [Institution of Civil Engineers, 1989]. Other countries have similar figures. Inefficient route-finding and the excess travel that results costs the US $45 billion every year [King and Mast, 1987]. Table 11.5 shows that information technology can help to save lives, and improve the quality of life, by reducing the number of deaths and accidents. Information technology also helps to reduce the quantity of fuel consumed and the output of noxious emissions, thereby reducing pollution. This follows from a combination of features, including:

- route guidance and UTC,
- avoidance of congestion,
- more efficient automobile engines, and
- antilock braking.

This is not to say that information technology is a total substitute for road building or other accident prevention measures, of course. However, some systems, such as the Integrated Motorist Information System in the Long Island area of New York are ". . . implemented as an alternative to new construction, as the State does not plan to provide new capacity in this corridor. As such, it is typical of countless other urban-suburban locations in the United States with increasing traffic demand and where construction of new capacity is not seen as the acceptable solution" [Saxton, 1985]. Information technology also brings other benefits (quantifiable and unquantifiable) to the individual and the nation, such as more efficient use of time.

11.3.2 The Importance of Mobile Communication

We are in an age where communication is more important than ever before. People are therefore demanding communications in the mobile environment—enabling them to be contacted and to make telephone calls when they are in vehicles, in airplanes, or just in the street—as well as at home or in the office. Less obvious, but stressed in Chapters 5 through 9, is the need for communications to vehicles to permit efficient hazard warnings, route guidance and traffic control. Some of this communication is in digital form.

11.3.2.1 Voice Communication

By the year 2000, we will have passed from a situation where a mobile telephone was a luxury item (the "*precellular age*"—the early 1980s), through a period when it became an essential business tool for many people (the "*cellular age*"—the late 1980s), to a period (the 1990s) when it will become a consumer item available at lower cost than a video recorder or even an in-car stereo unit.

One of the factors in this change to wider availability will be the "telepoint" or "phonepoint" systems, also referred to as CT2 (i.e., the second generation of cordless telephones). As we saw in Chapter 3, CT2 phones will be usable in the home, but will be small enough and portable enough to be carried around in the pocket as well. Unlike cellular phones, they cannot be used anywhere, but only within a hundred meters of a base station. Base stations are expected to be at railway stations and other busy town center locations, motorway service areas, and other places where a mobile user might want to make a telephone call. CT2 permits only outgoing calls, because the phones, unlike cellular phones, do not regularly "register" with the system, which therefore cannot know where to direct calls.

One possible future development is the combination of a pager with a CT2 phone, to alert a subscriber to make a call. The pager would need an alphanumeric or at least a numeric display, to indicate the number to call. More generally, mobile communication technology will converge into the *Universal Mobile Telecommunications System* (UMTS) described in Chapter 3.

11.3.2.2 Digital and Data Communications

The move to digital technology is a general trend, driven by the increasing power and declining cost of microprocessors and semiconductor memory. We saw in Chapter 3 how mobile voice telephony is going digital to take advantage of this power and cost reduction—the pan-European digital cellular radio system.

RDS is also digital communication—digital data sent over the analog radio channel. And a widely available two-way communication system is fundamental to the development of the IRTE described in Chapter 9.

In addition to the use of digital techniques for voice and other communications, mobile data communications will also increase in importance. At present, data has been used, particularly in the PMR field, to make more efficient use of the available radio channels; a name and address can be transmitted much faster and more accurately as data than by verbal dictation over the air. But cellular applications are now starting to appear. For example, field service engineers have used portable computers and cellular telephones to send electronic mail and telexes, and to access company databases; cellular data has also been used for telemetry [Cole *et al.*, 1988].

11.3.3 Research Programs and Demonstrator Projects

Some of the systems described here, such as cellular radio and paging, are already in operation, and therefore familiar to at least some sectors of the population. Others, such as route guidance, are much less so (apart from in James Bond films!). The implications and ramifications of the latter are, therefore, much less clear. Their wider effect and their visibility to the public will first appear through demonstrator projects. For example, the LISB and AUTOGUIDE route guidance systems already have demonstration systems in place; there are five junctions equipped in London, with more to follow during the pilot phase, and 230 junctions equipped in Berlin.

AUTOGUIDE is now closely linked to the PROMETHEUS Eureka project, which itself has various demonstrator projects within it (see Chapter 9), including the CARMINAT project, combining the CARIN route guidance project (see Chapter 8) with the French "Atlas" project for display of gazetteer and vehicle status information.

The Mobile Information Systems Large Demonstrator project, sponsored by the UK Information Engineering Directorate, is investigating various aspects of mobile cellular voice and data communications, including GSM telephones, mobile electronic mail [Cole *et al.*, 1988], and the collection and broadcasting of traffic information [Allport, 1988a,b,c].

As part of the RACE program, a consortium of European organizations (British Telecom, GEC-Marconi, Matra, Philips, Standard Electrica and Strathclyde University) is studying mobile communications; the initial results of their deliberations into the UMTS is reported in Chapter 3. Mobile communication also figures in the European "Esprit" program and the UK government's LINK funding program.

11.3.4 Standardization Prospects

We referred in Chapter 1 to the increase in both personal and business mobility. The likelihood is that this will increase. For example, in Western Europe there is the European Economic Community, which will increase travel and trade between the nations of Western Europe; there is also the construction of the channel tunnel between southern England and northern France.

International travelers will expect their cellular radio to work wherever they are; expect their route guidance system to work just as well abroad as it does at home; and will expect to receive traffic information broadcasts in their own language. Currently, travelers are disappointed, because none of these things are possible at the present time.

This is why there has been a move to standardization of various aspects of mobile communications. Examples of this effort are the pan-European cellular radio and paging systems and the Radio Data System. The LISB and AUTOGUIDE trials are using common technology. The PROMETHEUS and DRIVE programs have large standardization components to them. Apart from user convenience, standards result in faster acceptance of new technology, and larger markets for manufacturers. Thus the prospects for standardization are good within Europe; wider standards are another matter, although CCIR Study Group VIII is already investigating worldwide standards for cellular and public aeronautical radiophones.

11.4 FUTURE DEVELOPMENTS

Here we will do some "crystal-ball-gazing," bringing together the predictions from other chapters, identifying some common themes, and introducing other possible future developments. However, coming up with accurate predictions, particularly regarding time scales, is very difficult, as exemplified by the mayor of a US midwestern town, who said, on seeing a demonstration of Alexander Graham Bell's invention, "One day, every town in America will have a telephone!" He was correct, but with hindsight we can see that his prediction was very limited.

11.4.1 Paging

Chapter 2 makes the following predictions:

- A combined pager and cordless telephone, because the latter, unlike cellular telephones, are "outgoing calls only" when they are away from their "home";
- On-air programming of pager addresses, to change the type of information received, or the group-team to which the user belongs;
- Message input by speech
  - by speech recognition if it ever becomes reliable, and
  - by compressed digital speech or vocoder, so that a spoken message can be transmitted and produced as output;
- Technical developments are making possible the acknowledgment of a paging call, but market studies are needed to see whether there is a demand for this feature;
- Direct satellite paging may be adopted in countries with low population density. (The technical feasibility of satellite paging has already been demonstrated in trials by Racal for the British National Space Centre [Casewell *et al.*, 1988a, 1988b].)

Chapter 2 also refers to the communication medium shared by paging and broadcast data (RDS), in which a paging system uses a subcarrier superimposed on normal broadcast radio channels. There is also the possibility of using alphanumeric display pagers (with group addresses) for traffic information broadcasting, either instead of (or, in addition to, if the pager has multiple addresses) for person-to-person communication. (An interesting development of this theme is described in Section 11.4.4.) We might also see display pagers built into car dashboards.

Chapter 2 stressed the importance of efficient use of radio spectrum, and has emphasized that paging is very good in this respect. Section 11.5.1 discusses this, and the related issue of frequency allocation.

11.4.2 Mobile Radio Communication

Mobile radio communication has emerged from an unfashionable technological backwater with mainly military or broadcast applications to become one of the most dynamic and exciting areas of electronic technology. This is partly a result of the ubiquitous "chip," but mainly because of the change in attitude of the radio regulatory authorities in making radio frequency spectrum available for commercial applications.

We have seen that CCIR is already at work on a worldwide cellular radio standard. So, the day may come when your cellular telephone will work anywhere in the world. However, this is unlikely to happen in practice before the 21st century, because the standard will not be agreed upon before 1994. Even the pan-European system, which will come into operation in 1991, will take several years to reach maturity.

There will be a gradual and evolutionary convergence of various forms of mobile communication into "personal communication." The UK will take the first step with Telepoint networks, which will permit the use of the CT2 telephones outside the home when in range of Telepoint base stations. A second step could come with the CT3 or *Digital European Cordless Telephone* (DECT), which might be designed to operate on cellular frequencies at 900 MHz as well as on the new CT3 frequencies between 1 and 2 GHz. This convergence will result in the UMTS discussed in Chapter 3.

One issue that personal communication raises is *personal numbering*. We frequently want to call a particular person, regardless of where he or she is. Therefore, the number should attach to the person and not to the location of a telephone at home or in the office (although, conversely, there are times when we want to call the location). Carpenter [1988] predicts personal numbering in use in the UK by 1992.

Pan-European paging will become a reality in the early 1990s, although, as Finnie [1989] indicates, there are currently some uncertainties, both technical and commercial, about the system that will be adopted.

The allocation of more radio spectrum for cellular has already been an issue in the UK, where the phenomenal usage rate created congestion on the system in London, leading to the allocation of additional frequencies—the ETACS system. The GSM system will also alleviate congestion when it comes into operation, but, if subscribers continue to increase at the rate in which they have in the past, there may well be future pressure for even more spectrum.

Cellular phones to date have largely been the prerogative of the developed Western countries and the urban environment. As costs of the handsets reduce, we may see increasing use in rural areas, where radio systems may be cheaper to install than the land lines used in the "normal" PSTN.

Cellular or other mobile telephones are unlikely ever to achieve the wristwatch size we see in science fiction or other movies. The semiconductor industry has achieved miracles of miniaturization, but some of the components used, including the antenna, loudspeaker, battery, and filter circuits, are not semiconductors, and so are unlikely to shrink much further. CT2 systems have an advantage in this respect, in that incoming calls are not received away from "home," and range is limited to 100 m from base stations. Therefore, the unit need only be switched on to make an outgoing call. Also, transmitting power is small, so less powerful and much smaller batteries can be used, as well as fewer and smaller circuit components. In any case, there is no reason to make equipment that is smaller than the distance between ear and mouth.

Cellular data communication will increase in importance, as we have discussed, although, at least initially, data applications will lag behind voice applications.

11.4.3 Satellite Mobile Communication

The major growth area for mobile satellite communication is land mobile; the Americas, Australia, Africa, and parts of Asia cover such large areas that satellite communication is the most effective for long-distance road traffic. The problem of integration with cellular radio has been addressed by Phillips and Wright [1988], whose opinion is that cellular and satellite land mobile services are complementary rather than competitive. These authors also discuss the integration of cellular and satellite terminal equipment, the use of common algorithms, and network interfaces. Phillips and Wright foresee integration of cellular and satellite services after 1995, and, as with cellular, the aeronautical system will need to provide not only passenger telephony, but also simple data communication.

CT2 subscribers will want their personal portable telephones to be compatible with the satellite systems available on aircraft and boats, so that users can make telephone calls from aircraft using their own handsets. This is another step on the way to the dream of a UMTS that can be used anywhere; although, as with CT2, to permit outgoing rather than incoming calls will always be easier.

Finally, as usage of mobile satellite systems increases, there will be the resultant pressure to make more spectrum available.

11.4.4 Traffic Information, Route Guidance, and UTC

Forecasts for traffic information, route guidance, and urban traffic control are particularly difficult to make, because the former two at least are still in their infancy.

Traffic information broadcasting is still a largely unautomated and low-priority item, relying on information gathered as a secondary task by police officers and the patrols of automobile organizations. Traffic reports are broadcast only when there is a convenient slot between normal radio programs (which also means that there is a disincentive to collect traffic information, because the means of disseminating it in a timely manner has always been lacking). With the advent of RDS, and particularly the traffic message channel (RDS-TMC), there at least will be one dedicated channel over which traffic information can immediately be broadcast. This is also likely to stimulate more and better means of collecting the information, including links to dynamic route guidance systems such as AUTOGUIDE. A novel and interesting means of extracting traffic information from data gathered for other purposes is described in Section 11.4.7.

Other possible means of disseminating traffic information include display pagers, as mentioned in Section 11.4.1. One interesting variant of this idea was set up in June 1989 in London by the UK's Automobile Association in conjunction with a paging operator, Racal Vodapage. A display pager has been connected to a large "sign-board" display by the side of one of the busy exit routes from central London. The sign-board displays messages sent from the AA's "Roadwatch" studio about traffic conditions on the main westbound routes leaving the city (i.e., the M3 and M4 motorways).

There are also dial-up services which give information about traffic in a particular area, depending on the number dialed (although studies have shown that drivers are reluctant to use such services *before* they set out on a journey). There are also signs that the type of services operated in the US by Shadow, Inc. (see Chapter 5) is beginning to appear elsewhere. A service operated jointly by the Automobile Association, Racal Vodafone (a cellular radio operator) and Radio 210 (a local radio station) was set up in late 1988 in the Thames Valley area of the UK, using commuters with cellphones as one source of information about congestion during the rush hour.

Another possible dissemination mechanism for traffic information is the "broadcast short message service" of the GSM cellular radio system. GSM, unlike other cellular systems, has a "group call" facility, in which short alphanumeric traffic messages can be broadcast to subscribers and displayed on their cell-phone.

One of the PROMETHEUS demonstrator projects is investigating the integration of RDS-TMC with mobile data systems and mobile telephones. A DRIVE project called SOCRATES (see Chapter 9, Section, 9.4) is investigating the potential of cellular radio, and particularly the GSM system, for improved traffic efficiency and safety, especially with regard to traffic information and route guidance. Other PROMETHEUS and DRIVE projects are concerned with traffic management and control. PROMETHEUS and DRIVE, in any case, are concerned with the future, as is the Intelligent Car (see Chapters 9 and 10).

These developments are primarily occurring in Europe (apart from the Japanese AMTICS system mentioned in Chapters 1 and 6). The US has concentrated more on UTC. Saxton [1985] discusses the future of traffic information and traffic management systems in the US. His opinion is that traffic information will be provided primarily by pretrip sources, variable message signs, HAR, and local radio stations. Saxton writes, ". . . a nationally-sponsored in-vehicular motorist communication system will not be implemented in the United States in the near future."

11.4.5 The Intelligent Car

Some specific developments not addressed in Chapter 10 are covered here.

Doppler Radar Speedometer

As technology advances, there will undoubtedly be developments in sensor technology (especially, but not limited to, semiconductor sensors). For example, the Plessey Company in the UK announced in 1989 a doppler radar speedometer, based on an electronic sensing head fitted underneath the car. Unlike conventional speedometers, the device measures true ground speed to an accuracy of 1%, which is much better than electromechanical devices, and is not affected by tire wear or inflation pressure, which can lead to errors in conventional speedometers that measure wheel rotation. Because the radar is not connected to the engine, transmission, or drive wheels, the device is easy to fit. Furthermore, it can be fitted to all models of car, thus reducing inventory for the manufacturer.

Apart from measurement of velocity, the doppler unit can be used in combination with conventional wheel rotation sensors to detect skidding; the radar could therefore be used as part of an antilock braking system or active suspension (see Section 10.10.1).

Additional units could also be used to detect other vehicles, and so could be a component of a collision avoidance system (i.e., PROMETHEUS). There will also be adaptation to route guidance systems, which need an accurate measure of distance covered.

Fiber Optics for Headlights

Fiber optic data buses in vehicles have already been mentioned in Chapter 10. A different application has been announced by Ford of the US at the 1989 Geneva Motor Show—a "concept car," the Ghia Via, which uses fiber-optic headlights. The conventional quartz halogen light source can be located anywhere in the vehicle. The light is guided to the "headlights" by fiber-optic light pipes. One advantage is more freedom in styling, because the "headlights" can be as small as one centimeter in height. Another advantage is in aerodynamics, because car fronts can be made smaller to reduce drag (which therefore reduces fuel consumption and noise). More flexibility in beam shaping and more efficient use of the light generated are also possible.

The initially increased cost of the lamps means that they will probably appear first in sports models, replacing the "pop-up" headlights (and presumably contributing to reliability by removing electromechanical components).

Increased Use of Semiconductors

Semiconductor chips may have the ability to diagnose their own faults as well as faults elsewhere in the car, as Chapter 10 indicates. This is useful when, according to Toyota, 10% of electronics modules replaced by mechanics are, in fact, not defective.

One development is the use of custom ICs or *Application-Specific Integrated Circuits* (ASICs) to replace several standard components, thereby reducing costs and power consumption, and increasing reliability. ASICs can also mix analog and digital functions. For example, Texas Instruments has produced ASICs for driver information systems, clocks, temperature displays, smart windshield wipers, and speech applications.

A market is also developing, particularly in the US, for replacement programmable read-only memory chips, (PROMs) and other semiconductor devices that can improve the performance of a car over the factory-provided version; one is reported to improve the time for a Porsche 911 to travel a quarter-mile from 14.85 to 14.25 s [Hampton *et al.*, 1988].

Microprocessors can do plausibility tests to check that sensor readings (or combinations of readings) are possible. In new BMW cars, the electronic systems (engine and transmission control, central body control, trip computer, dashboard control, antitheft system, and air-bag control) have on-board diagnostic capabilities. A future development will be an on-board local area network interconnecting the different electronic control and diagnostic units.

Test equipment used in the garage will become more sophisticated, and will also plug into the LAN, and expert systems for fault diagnosis will be introduced [Schurk and Fournell, 1987].

Head-up Displays

Although some authorities have cast doubt on the usefulness of head-up displays, manufacturers continue to research them. Nissan has introduced one on its Sylvia vehicle, and General Motors and other manufacturers have plans to introduce them.

11.4.6 Merging of Aerospace and Automobile Industries

A recent trend in the business world has been the merging of automobile and so-called "high-technology" aerospace companies to apply the advanced information technology skills of the latter to the somewhat lower-tech former (and perhaps to apply the mass-production skills of the former to the latter's products). For example,

- General Motors bought Hughes Aircraft;
- Daimler-Benz bought Dornier;
- British Aerospace bought BL (British Leyland);
- Volvo owns Saab, which makes planes as well as cars;
- Chrysler bought Gulfstream Aerospace.

Whether there are real benefits to either party other than for advertising copy, and whether the trend will continue by the amalgamation of other aerospace and automobile companies remains to be seen.

11.4.7 The Use of Artificial Intelligence

We saw in earlier chapters the dramatic and revolutionary effects of using computers and microprocessors for engine control, diagnostics, urban traffic control, *et cetera*. One of the "coming technologies" of the computer world is artificial intelligence (AI). It is expected to have application in various areas of mobile information systems, including engine diagnostics. AI also forms a part of the PROMETHEUS program—PRO-ART is one of the basic research themes (see Figure 9.10), and heading control by using AI is one of the demonstrator projects.

One application of AI that has been under investigation for several years is in the domain of traffic information. Chapter 5 primarily described the problems of giving information to the motorist, and the various ways in which it can be (and, in some cases, is) done. Chapter 5 emphasizes, however, that a major problem is the gathering of this information in the first place, and alludes to a means of doing so by using artificial intelligence. Allport [1988a, 1988b, 1988c] describes an AI program called a *Traffic Information Collator-Condenser* (TICC), which has access to textual information in a police database. This information is logged in by the police as a traffic incident (such as a road accident) occurs, and is used by police operators primarily to monitor the course of the incident as resources are allocated and the situation changes over a period of hours. The TICC "reads" the text, and tries to understand those parts of it which are relevant to other motorists, such as indications that a road is blocked and delays are inevitable. The TICC then prepares a message for broadcasting to motorists, warning them to avoid the area. Although still experimental, the initial results are promising, and the system appears able to gain a sufficient understanding of the incident to be useful in practice. If combined with a system such as RDS, traffic information could be automatically broadcast to motorists, without human intervention, and in a much more timely manner (see also Section 5.6.1).

Expert systems technology, a branch of AI, has also been investigated for its applicability to MIS. Bell (1988) has proposed that an expert system should be the next step in UTC. Expert systems could be used both for planning purposes and as an aid to on-line, real-time control. Birkinshaw *et al.* [1988] aim to produce an expert-system-based program for traffic signal design, especially as few written guidelines exist on this topic. Ramache and Bell [1988] are developing an expert system to advise a traffic engineer on the location of roadside objects (e.g., lighting columns) so as to minimize the risk of accidents.

Jenkins [1988] has investigated *knowledge-based systems* (KBS) within a vehicle. Higher levels of on-board intelligence will improve the capability of vehicle and crew. Jenkins particularly, presumably has military applications in mind, but he points out that other applications include:

- adaptive user interfaces,
- planning of the journey and its associated task,
- monitoring and diagnosis of the vehicle.

11.5 PROBLEMS TO BE OVERCOME

The individual chapters have highlighted problems specific to their topic. Some problems, however, recur throughout several chapters, either implicitly or explicitly. These problems include the shortage of radio frequency spectrum, and the largely uninvestigated topic of the effect on the motorist of all those high-technology products competing with the actual driving task for his attention.

11.5.1 Frequency Spectrum Allocations

Frequency spectrum is in short supply, and the more people have, the more they want and need. The company that supplies its managing director with a cellular telephone soon realizes that its sales force will be more productive if they have them as well. Hence, there is a continual increase in demand for the service, which paradoxically is fueled by the greater availability of the service, and ultimately can be met only by the provision of additional frequencies. Thus, there is increasing pressure on the frequency spectrum. However, the spectrum is obviously a very limited resource—far more so than oil or gold. To support more users, there are very few options:

- Move to higher frequencies to open new frequency bands for communication applications;
- Use the existing bands more efficiently by
  (1) using narrow-band techniques to put more communication channels into the same spectral bandwidth,
  (2) sending data messages rather than voice, with preprepared messages (for example, names and addresses sent as ASCII text will be transmitted much faster and more accurately than if dictated by speech over the air), and
  (3) ingenious frequency reuse techniques, such as cellular radio and PARS.

Unfortunately, even with these techniques, demand for frequency spectrum usually exceeds supply. Frequencies are usually allocated to specific operators, for specific applications, by national governments, within the framework agreed by international bodies such as WARC. For historical reasons, some operators appear to have a disproportionately large allocation. This situation has led to the suggestion that radio frequency spectrum be auctioned to the highest bidder, either to use for a particular application, or possibly even to use as the acquirer sees fit (and profitable!), provided that the user observes certain safeguards, such as not causing interference to others.

Such schemes have been under consideration in the US, UK, and Australia, although New Zealand appears to have progressed the furthest to date [Foster, 1989]. Whether other nations will follow suit remains to be seen.

11.5.2 Ergonomics and the Human Interface

The introduction into motor vehicles of route guidance, traffic information broadcasts, cellular radio, and diagnostic messages from the engine, in addition to distraction from the driving task, creates the danger of cognitive overload for the driver, and thus may make him or her less rather than more safe. This concern has received very little attention to date, although it is one of PROMETHEUS's thematic projects. More work is necessary on the ergonomics of vehicles and the operation of information technology systems. Dashboard design must take into account the additional display requirements; speech synthesis must be investigated to assess where it is an alternative to a display, and where it is complementary; speech recognition may allow the driver to operate equipment in a "hands-free" mode (although speech recognition is difficult enough in a quiet room, but will be more so in the noisy environment of a moving automobile).

There are also more psychological issues, such as whether these systems may make drivers more careless, relying on the on-board anticollision systems and antilock braking to keep them out of trouble. Further ergonomic and psychological studies are needed.

11.5.3 Other Problems

More general questions and issues include:

- Will all this electronics "fail soft," like conventional electromechanical systems?
- Will maintenance costs fall if there are extensive on-board diagnostic systems? Conversely, will there be enough skilled technicians or mechanics to service these new systems?
- Where is the legal liability if a vehicle radar and anticollision system fail to notice another car or a pedestrian, or fail to hold their place in a "road-train"?
- Are these electronic systems susceptible to influence or damage by *electromagnetic pulse* (EMP) radiation? (Strong electromagnetic fields, as encountered, for example, in the vicinity of a powerful radio transmitter, may induce spurious signals in electrical circuits, or even damage some of the semiconductor devices.)

Some of these issues are already being researched [Schurk and Fournell, 1987].

11.6 CONCLUDING REMARKS

Mobile information systems is a complex and wide-ranging subject—from the interior of automobile engines to the deliberations of international organizations. Each chapter in the book could easily have been one in its own right. We have tried to indicate the range and complexity of the subject at a crucial stage in its development. Some of the systems considered are already in service, and have been for some time. Others are still at the research and development stage, and the extent to which they are commercially successful will depend on the perceived need on the part of potential users, and overcoming of technical problems at an economic price.

The subject of mobile information systems is certain to continue to develop and to increase in importance, and it will affect all aspects of our lives.

REFERENCES

Allport, D. [1988a], "Interpreting incident reports," *IEE Colloquium on Applications of Expert Systems in Road Transportation,"* London, UK, 15 January 1988. IEE Colloquium Digest No. 1988/9, pp. 5/1–5/4.

Allport, D. [1988b], "The TICC: parsing interesting text," *Proc. 2nd Conference on Applied Natural Language Processing,* Association of Computational Linguistics, Austin, TX.

Allport, D. [1988c], "Understanding RTA's," *IT88 Conference,* Swansea, UK, (IEE, London) pp. 323–326.

Bell, M.C., "The fundamental issues of an expert system for urban traffic control," in *Applications of Expert Systems in Road Transportation,* IEE Colloquium Digest No. 1988/9 (IEE, London) pp. 7/1–7/6.

Bergqvist, J.T., "The colourful world of mobile communications," *Discovery* (*Nokia Telecommunications Magazine*) Vol. 16, February 1989, pp. 35–39.

Birkinshaw, D.E., H.R. Kirby, and F.O. Montgomery "Traffic signal design using an expert system shell," in *Applications of Expert Systems in Road Transportation,* IEE Colloquium Digest No. 1988/9 (IEE, London), pp. 8/1–8/3.

Carpenter, P., "From mobile to personal communications," *Third Nordic Seminar on Digital Land Mobile Radio Communications,* 12–15 September 1988, Copenhagen, Paper No. 1.7.

Carrington, J., "From mobile to personal communications," *Fourth International Conference on Land Mobile Radio,* Warwick, UK, December 1987 (IEE, London), pp. 11–12.

Casewell, I.E., I.C. Ferebee, and M. Tomlinson, (1988a) "A Uni-directional satellite paging system for land mobile users," *JIERE,* vol. 58, No. 3, May 1988 pp. 92–98.

Casewell, I.E., I.C. Ferebee, and M. Tomlinson, (1988b) "A satellite paging system for land mobile users," *Fourth International Conference on Satellite Systems for Mobile Communications and Navigation,* 17–19 October 1988, IEE Conference Publication No. 294.

Cole, R., C. Hall, M. Hassall, A. Pell, and J. Walker, "Demonstrating the mobile office," *UK IT88 Conference,* Swansea, UK, (IEE, London) pp. 597–600.

Finnie, G., "Ermes: ready to deliver?," *Telecommunications,* April 1989, pp. 61–63.

Foster, R., "Selling the airwaves," *Communications International,* May 1989 pp. 61–62.

Goodall, S., "On the road to a boom in car components," *Electronics Times,* December 1, 1988, p. 8.
Green, J "Cellular for France," *Communications International,* May 1989, p. 10.
Guppy, R., "The market for digital cellular radio in Europe—demand and distribution," *Third Nordic Seminar on Digital Land Mobile Radio Communications,* September 12–15 1988, Copenhagen, Paper No. 1.3.
Hampton, W.J., N. Gross, D.C. Wise, and O. Port, "Smart cars," *Business Week,* June 13, 1988, pp. 68–74.
Hartley, J. "Japan gears up to sensor control," *Electronics Times,* December 10, 1987, p. 22.
Hudson, A.D., "The new Private Advanced Radio Service," in *Mobile Radio Networks,* IEE colloquium Digest No. 1988/36, (IEE, London), 11 March 1988, pp. 8/1–8/8.
Institution of Civil Engineers "Congestion," Thomas Telford, London, 1989.
Jenkins, P.G. "Towards on-vehicle real-time knowledge based systems," in *Applications of Expert Systems in Road Transportation,* IEE Colloquium Digest No. 1988/9, (IEE, London), pp. 3/1–3/3.
King, G.F. and Mast, T.M. "Excess travel: Causes, extent, and consequences," *Proc. 66th Annual Meeting of the Transportation Research Board,* Washington, DC, 1987.
Manuel, G., "A pager for Europe," *Communicate,* March 1989, p. 37ff.
Norbury, J.R., "Trends in satellite systems," *Telecommunications,* September 1988, pp. 59–67.
Phillips, R.O., and D. Wright, "Complementarity between terrestrial and satellite mobile radio systems," *Third Nordic Seminar on Digital Land Mobile Radio Communications,* September 12–15, 1988, Copenhagen, Paper No. 8.2.
Ramache, A. and M.G.H. Bell, "The location of roadside objects: an expert system to assess the safety aspect," in *Applications of Expert Systems in Road Transportation,* IEE Colloquium Digest No. 1988/9 (IEE London), pp. 10/1–10/6.
Saxton, L., "Traffic systems in the United States: present and future developments," *Electronics and Traffic on Major Roads* (European Conference of Ministers of Transport), Paris, France, June 4–6, 1985, pp. 8–15.
Schurk, H.E. and H.D. Fournell, (1987) "On-Board Diagnosis of electronics; a contribution to vehicle reliability," *37th IEEE Vehicular Technology Conference,* pp. 343–358.
Stewart, R.A., "Public radio-paging in Europe—from a UK perspective," *Telecommunications,* November 1987, pp. 35–47.
Utley, A., "UK first in digital short range radio," *Electronics Times,* March 16, 1989, p. 2.
Wilson, R. [1989a], "The RACE for a mobile Europe," *Electronics Weekly,* January 18, 1989, p. 22.
Wilson, R. [1989b], "Mobile coms look to the future," *Electronics Weekly,* March 1, 1989, p. 24.

FURTHER READING

Calhoun, G., *Digital Cellular Radio,* Artech House, Norwood, MA, 1988.
IEEE Transactions on Vehicular Technology and the annual *IEEE Vehicular Technology Conference* cover some of the topics in this book,—particularly cellular radio and the "intelligent car."
The U.S. Society of Automotive Engineers publications are useful and readable.
IEEE Spectrum has authoritative and up-to-date review articles relevant to the subjects covered here (as well as on many other subjects).

Appendix A
Glossary

Term or Acronym	*Definition*
AA	The Automobile Association, a UK motoring organization
ABS	Anti-lock Braking System
ac	Alternating Current
ACSSB	Amplitude Companded Single Sideband
AEEC	Airlines Electronic Engineering Committee
AFIS	Airborne Flight Information Service
AHAR	Advanced Highway Advisory Radio
AI	Artificial Intelligence—computer programs written in languages such as LISP and Prolog which emulate "intelligent" behavior.
AM	Amplitude Modulation
AMSC	American Mobile Satellite Consortium
AMPS	Advanced Mobile Phone Service (US)
AMTICS	Advanced Mobile Traffic Information and Communications System, under development in Japan.
AOR	Atlantic Ocean Region
ARI	Autofahrer Rundfunk Information—a traffic information broadcasting system used in West Germany.
ARINC	Aeronautical Radio Incorporated
ARQ	Automatic Repeat Request—i.e., a request for retransmission of data which has been corrupted in transmission.
ASCII	American Standard Code for Information Interchange

Term or Acronym	*Definition*
ASIC	Application-Specific Integrated Circuit—an integrated circuit containing standard devices that can be interconnected to produce a particular electronic function, replacing several standard electronic components. Also called "custom chips" or "gate arrays."
AUTOGUIDE	An automatic route guidance system currently being tried in London
AVL	Automatic Vehicle Location
baud rate	Rate at which data is transmitted; may be equal to bits per second.
BER	Bit Error Rate
BPSK	Binary Phase-Shift keying
BT	British Telecom
C/I	Carrier to Interference ratio; the ratio of wanted to unwanted signal.
CARIN	CAR Information and Navigation system (Chapter 8).
CCIR	International Radio Consultative Committee
CCITT	International Telephone and Telegraph Consultative Committee
CDLC	Cellular Data Link Control—a cellular data transmission protocol based on HDLC (High Level Data Link Control).
CDMA	Code Division Multiple Access
CD ROM	Compact Disc Read-Only Memory
CEPT	The European Committee of PTTs
CO	Carbon monoxide
cochannel interference	Interference from signals on the same frequency but from an unwanted transmitter (e.g., from a cellular radio base station in a different cell).
codec	Coder-decoder
COMSAT	Communication Satellite Corporation
COSPAS/SARSAT	Committee on Space Research-Search and Rescue Satellite
CRT	Cathode Ray Tube
CVT	Continuously Variable Transmission
CT2	Second-generation Cordless Telephone; Also used in "telepoint-phonepoint" systems (Chapter 2)
dB	Decibels—a logarithmic measure of signal strength or loudness; 20dB indicates a ratio of 10, 20dB a factor

Term or Acronym	*Definition*
	of 100, 30dB a factor of 1000, relative to some other signal.
dB/K	Decibels (dB, logarithmic) per kelvin (K, thermal) units
dc	Direct Current
DECT	Digital European Cordless Telephone (or Telecommunications)
diplexer	An electronic device which prevents the signal from a transceiver (e.g., a cellular radio) from getting into the receiving part and overloading the circuits.
DQPSK	Differential 4-phase-shift keying—a modulation scheme proposed for the ERMES pan-European paging system
DRIVE	Dedicated Road Infrastructure for Vehicle safety in Europe (Chapter 9)
DSSR	Digital Short-range Radio
DTMF	Dual Tone Multi-Frequency dialing (as used in modern push-button telephones)
ECTEL	European Committee of Telecommunications and Electronics professional industries
EBU	European Broadcasting Union
ECU	European Currency Unit, approximately equal to one US dollar
EDAC	Error Detection And Correction
EGO	Exhaust Gas Oxygen (sensor)
EIRP	Effective Isotropic Radiated Power
EMP	Electromagnetic Pulse radiation—the intense electromagnetic radiation emitted by some high-power transmitters (and by nuclear explosions!) which can damage semiconductor devices
EPIRB	Emergency Position Indicating Radio Beacon
ERMES	European Radio Message System; the developing pan-European paging system (Appendix 2E)
ESA	European Space Agency
ESPA	European Selective Pager (manufacturers') Association
Esprit	European Strategic Programme for research and development in Information Technology—a European research initiative in information technology—see Appendix B.8
ESRO	European Space Research Organization

Term or Acronym	*Definition*
ETACS	Extended TACS; congestion on the TACS cellular radio system in London led to the allocation of additional radio frequencies (Chapter 3)
ETSI	European Telecommunications Standards Institute (see Appendix B)
Eureka	A European initiative in collaborative information technology projects
FAA	Federal Aviation Administration
fade	loss of radio signal—due, for example, to multipath effects (see Section 2.3.1)
FANS	Future Air Navigation Systems
FDM	Frequency Division Multiplex—sending different sets of signals along the same communications channel at different frequencies
FDMA	Frequency Division Multiple Access
FEC	Forward Error Correction
FM	Frequency Modulation
FSK	Frequency-Shift Keying—a means of sending digital information over an analog communications channel by using one frequency to represent binary "1" and another to represent binary "0"
G/T	Gain-to-noise temperature (a figure of merit for a receiver)
GHz	Gigahertz—billions of cycles per second
GMDSS	Global Maritime Distress and Safety System
GMSK	Gaussian Minimum Shift Keying
GPS	Global Positioning System—a satellite-based navigation and location system
GSC	Golay Sequential Code—see Appendix 2.C
GSM	Group Special Mobile—the CEPT sub-committee which is specifying the new Pan-European cellular radio system
HAR	Highway Advisory Radio—used in the US and Japan to warn of hazards on the road network
HC	Hydrocarbon
HF	High Frequency—also refers to the radio frequency band between 3 and 30 MHz.
HFR	High Frequency Regeneration
HFT	Hands-Free Telephone
HPA	High Power Amplifier

Term or Acronym	*Definition*
IC	Integrated Circuit
ICAO	International Civil Aviation Organization
IFRB	International Frequency Registration Board
INMARSAT	International Maritime Satellite Organization
INTELSAT	International Telecommunication Satellite Organization
intermodulation interference	Transmitters on other frequencies combining to generate unwanted signals.
IRTE	Integrated Road Transport Environment
ISDN	Integrated Services Digital Network
ISO	International Organization for Standardization—the global top-level standards-making body. Its members are the national standards bodies (e.g., the National Bureau of Standards in the US, the British Standards Institute in the UK, and DIN in West Germany).
IT	Information Technology
ITU	International Telecommunication Union—the parent body of CCITT and CCIR—see Appendix B
KBS	Knowledge-Based System—a branch of artificial intelligence
kbs	kilobits per second
kHz	kilohertz—thousands of cycles per second
L-band	A microwave frequency band
LCD	Liquid Crystal Display
LED	Light-Emitting Diode
LISB	An automatic route guidance system currently being tried in Berlin
LNA	Low Noise Amplifier
MARECS	Marine Communications Satellites
MARISAT	The first satellites dedicated to the provision of a maritime service (see Chapter 4).
MHz	megahertz—millions of cycles per second
MHS	Message Handling System (electronic mail)
microsecond	one-millionth of a second
millisecond	one-thousandth of a second
MMIC	Monolithic Microwave Integrated Circuit
modem	MOdulator-DEModulator—a device which converts digital signals to analog form for sending along a telephone wire
MSC	Mobile Switching Center—a telephone exchange in a cellular radio system

Term or Acronym	*Definition*
MSK	Minimal Shift Keying
multipath	The phenomenon in (cellular) radio propagation where radio waves reflected from obstacles interfere with the direct waves at the receiver, causing loss of signal (i.e., a "fade")
NAMTS	Nippon Automatic Mobile Telephone System
NASA	National Aeronautics and Space Administration
NEC	Nippon Electric Company
NMT	Nordic Mobile Telephone (system)
NOx	Nitrogen Oxides
OSI	Open Systems Interconnection—the seven-layer model developed by ISO which standardizes data communications
PABX	Private Automatic Branch telephone eXchange
PARS	Private Advanced Radio Service (now called DSRR)
PCM	Pulse Code Modulation
Phonepoint	See "telepoint"
PMR	Private Mobile Radio
POCSAG	Post Office Code Standardization Advisory Group—the body set up by the then British Post Office to standardize a paging code (Chapter 2)
PROM	Programmable Read-Only Memory
PROMETHEUS	Programme for a European Traffic with Highest Efficiency and Unprecedented Safety (Chapter 9)
PSDN	Public (or Packet) Switched Data Network
PSTN	Public Switched Telephone Network
PTT	Post, Telegraph and Telephone—the standard acronym for the public corporation which runs postal and telephone services in a country (although in many countries these operations are now operated by two separate organizations).
QPSK	Quadrature Phase-Shift Keying
RAC	The Royal Automobile Club, a UK motoring organization
RACE	Research and development into Advanced Communications technologies for Europe (see Appendix B)
RAM	Random-Access Memory
RDS	Radio Data System (Chapter 5)
RDSS	Radio Determination Satellite Service

Term or Acronym	*Definition*
RELP	Residual Excited Linear Predictive, an algorithm used in vocoders to code voice to data and vice versa
RES4	a CEPT working group (see Appendix 2.E)
RIC	Radio Identity Code (i.e., a pager address)
rms	Root Mean Square
ROM	Read-Only Memory
RPC1	CCIR Radio-Paging Code number 1
RTI	Road Transport Informatics
SACCH	Slow Associated Control CHannel in the GSM cellular radio system
SAT	Supervisory Audio Tone (used in cellular radio for signaling purposes).
SCPC	Single Channel per Carrier
SITA	Societe Internationale de Telecommunication Aeronautique
ST	Signaling Tone (used in cellular radio).
SWAP	System Wide Area Paging—an early Bell Canada paging system
TACS	Total Access Communication System (UK)
TAMS	Telephone Answering Message Service
TCH	Traffic CHannel—the speech channel in the GSM cellular radio system.
Telepoint	An extension of the cordless telephone concept which puts base stations in public places, allowing owners of the new generation of cordless phones to make outgoing calls when away from home. Also known as "phonepoint" (see Chapter 3).
TDMA	Time Division Multiple Access
TICC	Traffic Information Collator-Condenser (see Chapter 11)
TRRL	The UK Department of Transport's Transport and Road Research Laboratory
trunking	The allocation of a group of communication channels to a particular service. A channel is allocated to a user when he requires it, but is available to other subscribers at other times. For example, there are far fewer long-distance telephone lines ("trunks") than there are telephone subscribers; but fortunately not everyone wants to make a phone call at the same time; the telephone company just has to provide sufficient capacity for the "peak period."

Term or Acronym	*Definition*
UHF	Ultra-High Frequency; also refers to the radio frequency band between 300 MHz and 3 GHz
UMTS	Universal Mobile Telecommunication System
UPT	Universal Portable Telephone
UTC	Urban Traffic Control (Chapter 7)
VAN	Usually means Value-Added Network, but can also mean Vehicle Area Network (Chapter 9)
VANS	Value-Added Network Service
VDX	Vehicle Distributed eXecutive (Chapter 9)
VHF	Very High Frequency (also refers to the radio frequency band between 30 and 300 MHz)
VLSI	Very Large Scale Integration of semiconductor devices such as microprocessors, memory and ASICs
VMACS	Vodafone Mobile Access Communications Service (Chapter 3)
vocoder	Voice coder—a device which converts analog speech signals into digital form and vice versa—used to reduce the bandwidth needed for voice communications
WARC	World Administrative Radio Conference
XMTR	Transmitter

Appendix B
International Standardization Bodies and Their Activities

International standards are extremely important because they ensure that equipment from different manufacturers is soundly based, will not exceed a reasonable degree of interference with each other, and will work functionally together. This assurance that good compatible equipment can be obtained from several sources gives confidence to component and equipment manufacturers, system providers, and users, and greatly accelerates market take up. Some of the bodies involved in international standardization are described below. A thorough review of standards bodies and the standardization process can be found in Wallenstein [1990].

B.1. CCIR (INTERNATIONAL RADIO CONSULTATIVE COMMITTEE)

On behalf of the United Nations, the CCIR, which is a part of the International Telecommunication Union (ITU), controls the internationally agreed upon Radio Regulations and makes Recommendations on all international aspects of radio. It covers all possible main services on land, sea, air and satellite (e.g., broadcasting, mobile services, radio links, amateur services, radio location, radio astronomy).

The CCIR, in conjunction with the International Frequency Registration Board (IFRB, another constituent part of the ITU), has divided the world into three zones: Europe and Africa; Asia and Australasia; and the Americas. The whole radio spectrum is partitioned into blocks, and agreed service classifications are allocated to each block. These allocations vary somewhat between the zones (and to a lesser degree, on a national basis within zones), but are common as far as possible. The agreements made in the CCIR are observed very seriously by all countries, almost having the force of international treaties.

Before permitting transmissions at any frequency, each country must ensure that the potential interference will not exceed these agreements in neighboring countries. For instance, they must check to see if their neighbors use that frequency. This coordination is not generally a problem to large or isolated countries such as US or New Zealand, because geographical distances often reduce their radio fields to negligible proportions in neighboring countries. In Europe, the nearness of neighboring countries gives rise to continual need for coordination.

For the purpose of making Recommendations, the CCIR is divided into Study Groups. Study Group VIII is concerned with mobile services. The main CCIR Plenary Meeting is held every four years (the most recent plenum being in 1990) and each Study Group usually meets twice between Plenary Meetings to draft and amend Recommendations and Reports for ratification at the Plenary. Voting never takes place in Study Group meetings, with any agreement being on a consensus basis. Each country sends a delegation, frequently including "experts" from industry. Voting can take place in the Plenary Meeting, but is nearly always avoided so that the Recommendation is unanimous. Recommendation 584 was approved, with Brazil and Japan expressing reservations.

CCIR Recommendation 584 details standard radiopaging codes. So far, there is only one entry, the CCIR Radiopaging Code No. 1. This code, proposed by the UK, was chosen from four candidate codes. CEPT and ETSI activity (see below) may provide a second CCIR standard radiopaging code from ERMES.

Among the projects which Study Group VIII is currently studying are a world standard for public land-based cellular radiophone, and another for public aeronautical radiophone. It is unlikely that either will be completed before 1994.

B.2. CCITT (INTERNATIONAL TELEPHONE AND TELEGRAPH CONSULTATIVE COMMITTEE)

This body is the largest constituent of the ITU, and performs a function similar to that of the CCIR (and in a similar manner), but covers telecommunication by wire rather than by radio. CCITT Recommendations for transmission format and quality, system control, numbering, *et cetera,* have enabled the various telephone networks of the world to be interlinked to form the largest machine known to man. CCITT also standardizes the signals for much of the terminal equipment which can be attached to the telephone network (e.g., data terminals and facsimile machines).

B.3. CEPT (THE EUROPEAN COMMITTEE OF POSTS AND TELEGRAPHS)

CEPT includes all Western European telecommunication and postal Administrations (PTTs or Post, Telegraph, and Telephone authorities). It is modeled along CCIR lines, being organized into various study groups. There is no voting in CEPT; all agreements are arrived at by concensus.

CEPT has no official standing in CCIR. Despite this, all the CEPT countries could propose (i.e., in the 1990–94 study session) the pan-European paging code or GSM as a CCIR Recommendation.

Until recently, CEPT drafted all its Recommendations, including those for design. Recently, it has delegated its responsibility for design recommendations to ETSI (see below). New CEPT activity now is restricted to operational radio Recommendations on matters such as frequency allocation and coordination.

Originally, CEPT never included experts from industry in its meetings but lately, especially in the RES and GSM projects, industrial experts were welcomed. However, CEPT recognized that the role for such experts needed to expand, and that ETSI would be a better forum for technical design matters.

B.4. ETSI (EUROPEAN TELECOMMUNICATIONS STANDARDS INSTITUTE)

The European Telecommunication Standards Institute was formed in 1988. Its headquarters are near Nice, France. Various ETSI projects have communally funded project teams and also study groups.

Among other projects, ETSI is engaged in drafting the design Recommendations for two large mobile projects, pan-European paging (ERMES) and 900 MHz pan-European digital cellular radiotelephone, known as Groupe Special Mobile (GSM). The ERMES design specifications are available in 1990. The GSM cellular system is described in Chapter 3.

Fee-paying ETSI members include PTT administrations, registered private operating organizations, certain learned and academic bodies, manufacturers' associations, and some large individual manufacturers (fees are generally too large for smaller manufacturers).

Unlike CCIR or CEPT, in ETSI there are "indicative voting" arrangements. In Technical Committees, any member organization present can vote, and a simple majority could be sufficient; but chairmen generally seek a very decided vote in order to persuade the outvoted party to agree with the majority. In the higher Technical Assembly, voting is on national lines and at least a 70% majority is required.

B.5 ECTEL (EUROPEAN COMMITTEE OF TELECOMMUNICATIONS AND ELECTRONIC PROFESSIONAL INDUSTRIES)

This is a manufacturers' association aimed at harmonizing the views of the various appropriate European national manufacturers associations (e.g., the Electronic Engineering Association (EEA) in the UK). ECTEL has ETSI membership and participates in several standardization projects.

B.6 PAGING ORGANIZATIONS

B.6.1 ESPA (European Selective Pager (Manufacturers') Association)

ESPA is associated with ECTEL, and aims to promote all the common interests of manufacturers of radiopagers. ESPA was consulted by UK Post Office when the CCIR Radiopaging Code No. 1 was being designed, and CCIR took this into consideration. ESPA is an ETSI member.

B.6.2 Organizations of Paging Operators

Telocator Network of America and the UK Paging Operators Association are two organizations of paging operators.

B.7 ITU (THE INTERNATIONAL TELECOMMUNICATION UNION)

The ITU was formed in 1865, and became a specialized agency of the United Nations in 1947. The ITU currently comprises 164 member countries with equal status and voting rights, plus about three hundred organizations such as industrial companies and international organizations which participate in its work, particularly under the auspices of the CCIR and the CCITT, which are "subsidiary" organizations of the ITU, as is the International Frequency Registration Board. The ITU also organizes the World Administrative Radio Conferences (WARCs) and World Administrative Telephone and Telegraph Conferences (WATTCs). More details on the ITU will be found in Finnie [1988] and in Codding and Rutkowski [1982].

B.8 ESPRIT

ESPRIT is the *European Strategic Programme in Information Technology,* and was created to promote cooperation in precompetitive research and development in information technology by industrial and academic organizations within the European economic community. Its aim is to make Europe competitive in IT on world markets. Esprit is a ten-year program, in two five-year phases, begun in 1983–84. To encourage cooperation, the Commission of the European Communities provides 50% funding for approved projects, and has a budget of 750 million ECUs (about $750 million) for phase one, to be matched by an equal sum from industry. Even larger sums are expected to be allocated to the second phase.

The Esprit work program was initially divided into five major areas:

advanced microelectronics (silicon and compound semiconductors, displays and optoelectronics);

software technology (tools, methods, management, environments);

advanced information processing (artificial intelligence, architectures, signal processing, human-machine interfaces);

office systems (workstations, storage and retrieval, communication, human-machine interface); and

computer-integrated manufacturing (computer-aided design, engineering, and manufacturing).

In the second phase, Esprit II, there will be some amalgamation; the three areas of activity are microelectronics, software and applications. There is also more emphasis on industrial aspects of the program.

B.9 RACE (RESEARCH AND DEVELOPMENT INTO ADVANCED COMMUNICATIONS TECHNOLOGIES FOR EUROPE)

RACE is, like Esprit, a European precompetitive and collaborative program, but focuses on telecommunication, and in particular the "integrated broadband communication network." Various research topics have been identified as important to RACE, including:

- high-speed and complex integrated circuits,
- optoelectronics for communication,
- broadband switching,
- displays,
- communication software.

RACE's "definition phase," from July 1985 to December 1986, cost 14 million ECU, and was followed by a five-year principal phase.

REFERENCES AND FURTHER READING

Anon. (1986) "An introduction to the work of the CCIR and CCITT," *British Telecommunications Engineering,* vol. 5, October 1986, pp. 200–201.

Codding, G., and A. M. Rutkowski, *The ITU in a Changing World,* Artech House, Norwood, MA, 1982.

Finnie, G. "The changing face of the ITU," *Telecommunications,* October 1988, pp. 49–50.

Wallenstein, G., *Setting Global Telecommunication Standards,* Artech House, 1990.

Appendix C
Radio Propagation Above 100 MHz

There have been many attempts to predict radio propagation accurately for mobile services. Typically, a mass of field measurements are taken and then a "law" or model has been derived to express the statistics of the measurement results.

These models tend to take the form of a number of factors all multiplied together:

$$P_r = P_t \cdot G_t \cdot (H_t)^2 \cdot G_r \cdot (H_r)^2 \cdot D^x \cdot C_b \cdot S_d \cdots$$

where

P_r = received power
P_t = transmitted power
G_t = directional "gain" of transmitter antenna
H_t = height of transmitting antenna above surrounds
G_r = directional "gain" (or loss) of receiver antenna
H_r = height of receiver antenna above ground
D = distance between transmitter and receiver
X = a *negative* power, generally between −3.3 and −4
C_b = an urban "building clutter" loss, e.g., 20 to 30 dB
S_d = log-normal street shadowing distribution factor, e.g., 6 dB standard deviation

Other effects such as building penetration loss, and multipath (or Rayleigh) fading distribution may factor into this equation. Models frequently include diffraction losses for shadowed areas caused by geographical features [Jakes, 1974, see References Chapter 2].

Important points to note are:

(a) Within the coverage area, distance has an approximately inverse fourth power law effect on received power, but beyond the radio horizon the values suggested for factor X no longer apply, and field strength reduces even more rapidly with distance. Distance is a dominating factor in mobile radio propagation.

(b) Antenna heights have a square-law effect on received power.

(c) Transmitted power has a smaller effect than these other factors.

Note that distant high ground or a hill may often approach or rise above the radio horizon, thus giving a patch of comparatively high field strength far outside that coverge area which is regarded as "continuous." Clearly such patches interfere with any cochannel transmissions wanted in those areas. Similar remarks apply to parks and other areas which are clear of buildings. Thus, the interference areas shown in Figure 2.5 are not as solid as indicated, but tend to consist of patches of interference which are large and close together near the coverage boundary. The patches reduce in size and are farther apart from each other as the observer travels away from the coverge boundary.

Antenna Gain

An antenna can be shaped to radiate uniformly in all directions, but much of the radiation will thereby be wasted into the sky or earth. Often antennas are shaped to shift some radiated energy from the skyward and earthward directions into the horizontal plane, so that there is a horizontal gain (and concomitant vertical loss). The radiation pattern then tends toward a pancake shape. Further shaping is also possible to concentrate the energy into a beam. Note that this shaping process is entirely passive, and the "gain" is directional and achieved without any power amplification. Gain is measured in dBs, being the ratio between the highest energy density in any direction from that antenna and the energy density from a theoretical omnidirectional antenna driven by the same total power.

INDEX

The Artech House Telecommunications Library

Vinton G. Cerf, *Series Editor*

Advances in Computer Systems Security: 3 volume set, Rein Turn, ed.

Advances in Fiber Optics Communications, Henry F. Taylor, ed.

Bibliography of Telecommunications and Socio-Economic Development by Heather E. Hudson

Broadband LAN Technology by Gary Y. Kim

Codes for Error Control and Synchronization by Djimitri Wiggert

Communication Satellites in the Geostationary Orbit by Donald M. Jansky and Michel C. Jeruchim

Current Advances in LANs, MANs, and ISDN, B.G. Kim, ed.

Design and Prospects for the ISDN by G. DICENET

Digital Cellular Radio by George Calhoun

Digital Image Signal Processing by Frederic Wahl

Digital Signal Processing by Murat Kunt

E-Mail by Stephen A. Caswell

Expert Systems Applications in Integrated Network Management, E.C. Ericson, L.T. Ericson, and D. Minoli, eds.

Fiber Optics and CATV Business Strategy, by R.K. Yates, N. Mahé, and J. Masson

Handbook of Satellite Telecommunications and Broadcasting, L.Ya. Kantor, ed.

Innovations in Internetworking, Craig Partridge, ed.

Integrated Services Digital Networks by Anthony M. Rutkowski

Introduction to Satellite Communication by Bruce R. Elbert

Introduction to Telecommunication Electronics by A.Michael Noll

Introduction to Telephones and Telephone Systems by A.Michael Noll

The ITU in a Changing World by George A. Codding, Jr. and Anthony M. Rutkowski

Jitter in Digital Transmission Systems by Patrick R. Trischitta and Eve L. Varma

The Law and Regulation of International Space Communication by Harold M. White, Jr. and Rita Lauria White

Manager's Guide to CENTREX by John R. Abrahams

The Manager's Guide to the New Telecommunications Network by Lawrence D. Gasman

Mathematical Methods of Information Transmission by K. Arbenz and J.C. Martin

Measurement of Optical Fibers and Devices by G. Cancellieri and U. Ravaioli

Minimum Risk Strategy for Acquiring Communications Equipment and Services by Nathan J. Muller

Optical Fiber Transmission Systems by Sigfried Geckeler

Optimization of Digital Transmission Systems by K. Trondle and G. Soder

Principles of Secure Communication Systems by Don J. Torrieri

Principles of Signals and Systems: Deterministic Signals by B. Picinbono

Private Telecommunication Networks by Bruce Elbert

Quality Measures and the Design of Telecommunications Systems by John H. Fennick

Radiodetermination Satellite Services and Standards by Martin Rothblatt

Setting Global Telecommunication Standards: The Stakes, The Players, and The Process by Gerd Wallenstein

Signal Theory and Processing by Frederic de Coulon

Techniques in Data Communications, by Ralph Glasgal

Telecommunications: An Interdisciplinary Text, Leonard Lewin, ed.

The Telecommunications Deregulation Sourcebook, Stuart N. Brotman, ed.

Telecommunications in the U.S.: Trends and Policies, Leonard Lewin, ed.

Telecommunication Systems by Pierre-Girard Fontelliet

Television Technology: Fundamentals and Future Prospects by A. Michael Noll.

Terrestrial Digital Microwave Communications, Ferdo Invanek, ed.

Traffic Flow in Switching Systems by G. Hebuterne

Troposcatter Radio Links by G. Roda

Voice Teletraffic Systems Engineering by James R. Boucher

World Atlas of Satellites, Donald M. Jansky, ed.

World-Traded Services: The Challenge for the Eighties by Ramond J. Krommenacker